biology

SIXTH EDITION

bju press®

Greenville, South Carolina

The authors and the publisher have made every effort to ensure that the laboratory exercises in this publication are safe when conducted according to the instructions provided. We assume no responsibility for any injury or damage caused or sustained while performing activities in this book. Conventional and homeschool teachers, parents, and guardians should closely supervise students who perform the exercises in this manual.

NOTE: The fact that materials produced by other publishers may be referred to in this volume does not constitute an endorsement of the content or theological position of materials produced by such publishers. Any references and ancillary materials are listed as an aid to the student or the teacher and in an attempt to maintain the accepted academic standards of the publishing industry.

BIOLOGY Teacher Lab Manual
Sixth Edition

Writers
David M. Quigley, MEd
Christopher D. Coyle

Writer Consultants
Cheryl Batdorf, MEd
Elwood Groves
Robert Hill, EdD
Rachel Santopietro, EdD
Caitlin Ueckert

Biblical Worldview
Bryan Smith, PhD
Tyler Trometer, MDiv

Academic Integrity
Jeff Heath, EdD

Instructional Design
Rachel Santopietro, EdD
Danny S. Wright, DMin

Editor
Rick Vasso, MDiv

Cover and Book Designer
Sarah Lompe

Design Assistant
Tara Baughn

Illustrator
Sarah Lompe

Production Designer
Maribeth Hayes

DesignOps Coordinator
Lesley Ramesh

Permissions
Maria Andersen
Sharon Belknap
Sarah Gundlach
Stacy Stone
Kathleen Thompson
Carrie Zuehlke

Project Coordinators
Heather Chisholm
Chris Daniels
Gary Santiago

Postproduction Liaison
Peggy Hargis

Teacher Lab Manual Photo Credits
Key: (t) top; (b) bottom
i Eric Isselee/Shutterstock.com; **iv** Antagain/E+ via Getty Images; **v** Cook Shoots Food/Shutterstock.com; **vi**t GL Archive/Alamy Stock Photo; **vi**b Olha Rohulya/Shutterstock.com; **viii** Ground Picture/Shutterstock.com; **ix** Marco VDM/E+ via Getty Images; **x** kosmos111/iStock/Getty Images Plus via Getty Images; **xi** Solskin/DigitalVision via Getty Images

Lab Manual photo credits appear at the end of this book.

The text for this book is set in Adobe Gothic, Adobe Minion Pro, Adobe Myriad Pro, Bungee by David Jonathan Ross, FF Uberhand by Jens Kutilek, Free 3 of 9 by Matthew Welch, LiebeRuth by Ulrike Rausch, Raleway by The League of Moveable Type, Sketchnote by Mike Rohde, and Times New Roman PSMT.

The cover photo is a close-up of the feathers of a green-winged macaw.

© 2024 BJU Press
Greenville, South Carolina 29609
Fifth Edition © 2017 BJU Press
First Edition © 1985 BJU Press

Printed in the United States of America
All rights reserved

ISBN 978-1-64626-117-8

15 14 13 12 11 10 9 8 7 6 5 4 3 2 1

CONTENTS

CONTENTS

An Unexpected Discovery

"One sometimes finds what one is not looking for."

Alexander Fleming in his laboratory in London during World War II

A *Penicillium* culture

When Alexander Fleming woke up on September 28, 1928, he wasn't intending to revolutionize medicine, but, as he would later conclude, "I suppose that was exactly what I did." The funny thing is that it was a complete accident.

Alexander Fleming was a Scottish biologist with a reputation for being a brilliant scientist, though he let his laboratory get, well, a bit messy. He returned there after a family vacation and was checking on some petri dishes holding growing bacteria. One culture looked different from the others. An airborne fungus had evidently traveled to the dish and had begun to grow in that culture. It had killed the bacteria growing around it. "That's funny," Fleming remarked. The fungus was *Penicillium*, and this unexpected discovery was the beginning of a new era in medicine. Penicillin, the antibiotic derived from *Penicillium*, would save thousands of people from diseases like scarlet fever, diphtheria, and pneumonia. Fleming received the 1945 Nobel Prize for his discovery.

Though you won't be playing with dangerous bacteria this year in biology, you can develop skills to help you work like a biologist at your own level. As you work through this lab manual, occasionally you'll have to think about how to solve a problem on your own without any procedures spelled out in the lab activity. These activities will have the word "inquiring" in the subtitle. You'll learn how to obtain useful data from the tools you have to work with, things as different as smartphone apps, microscopes, and good old-fashioned glassware.

This manual is your guide as you learn to use the tools and think the way a biologist does. Be sure to check the appendixes to learn about safety rules, equipment techniques, and writing formal lab reports. Read through the procedures and follow them. Answer questions interspersed with the procedures. Measure carefully. Ask if you aren't sure what to do; check your textbook for more information. Record data carefully in the tables, drawing areas, and graphing areas provided at the end (usually) of a lab activity. And keep your mind engaged! You never know what you may discover when you do.

But what good is all of this? Why is it important to develop the skills and mindset of a biologist? If you are a Christian, you're not just a student or even a student biologist. A Christian should see science as an amazing tool to glorify God and help people by obeying God's command to wisely use His creation. We should do the work of biology within the context of a Christian worldview and use it as God intended in ways that harmonize with His Word.

So ... let's get into the laboratory!

Safety Icons

Pay close attention to these icons whenever you see them.

Animal
Animals that you are to observe or collect may inflict stings or bites.

Body Protection
Chemicals, stains, or other materials could damage your skin or clothing. Wear a laboratory apron or laboratory coat and gloves as directed by your teacher.

Chemical Fumes
Chemical fumes may present a danger. Use a chemical fume hood or make sure that the area is well ventilated.

Electricity
An electrical device (hot plate, lamp, microscope) will be used. Use the device with care and watch for any frayed cords.

Extreme Temperature
Extremely hot or cold temperatures may cause skin damage. Use proper tools to handle laboratory equipment.

Eye Protection
There is a possible danger to the eyes from chemicals or other materials. Wear safety goggles.

Fire Hazard
A heat source or open flame is to be used. Be careful to avoid skin burns and the ignition of combustible materials.

Pathogen
Organisms encountered in the investigation could cause human disease.

Plant
Plants that you are to observe or collect may have sharp thorns or spines or may cause contact dermatitis (inflammation of the skin).

Poison
A substance in the investigation could be poisonous if ingested.

Sharp Object
When using equipment in this lab activity, be careful to avoid cuts from sharp instruments or broken glassware.

To the Teacher

Whether in a laboratory or in the field, your biology students will learn practices and techniques that aren't found in any other laboratory course. Your students will observe and collect data to grasp the big ideas of biology specifically and of science generally.

The Laboratory Experience and a Biblical Worldview

Of course, the biology laboratory experience is important for a student's science education. But does it have any other significance? What exactly is Christian about laboratory work?

BIOLOGY AS MODELING

As your students read the introductions to many of the lab activities in *BIOLOGY Lab Manual 6th Edition*, they may begin to notice something unusual: there is a lot of history in the book. In this lab manual you and your students will see that biology is all about modeling. Some of the most exciting and revolutionary ideas in biology began in the laboratory. As each scientist struggled to make sense of a phenomenon, he or she was working with models—either existing or newly emerging ones.

The proper way to evaluate a scientist's thinking and accomplishments is to comprehend the worldview that framed the scientist's models. In fact, the theories of biology that we use today make sense only when we understand them against the backdrop of the rich history of biology; a look at the study of life shows that we use it to make sense of God's world. Science is not a progression toward greater truth; it is a quest for more workable models. God's Word alone is the source of absolute truth.

Your students will learn to make their own models to help them grasp the big ideas of biology. They'll use existing models along with qualitative and quantitative observations and graphs to form hypotheses and make predictions that they can test. At their own level, they are behaving as biologists.

Biology developed over time, both out of innate curiosity and pure practicality. Your students will see that biology is a powerful tool to exercise good and wise dominion over God's world. Many things that we take for granted these days, such as a readily available food supply and good health, are products of an applied understanding of biology that helps people.

Dominion challenges us to use a hands-on approach to learn about God's creation, and the laboratory is a place where this can begin as students see science come alive. Ideas that seem abstract in class become concrete when reinforced by a lab activity. The laboratory experience also provides hands-on skills. As students perform each lab activity, they will develop proficiency in using laboratory instruments and making measurements. They'll learn to focus and refine their observational skills as they collect data and build models. A particularly important component of biology is fieldwork. Plants and animals don't live in laboratories—they live in the great outdoors, and understanding biology often requires getting out to where the plants and animals are.

Students will encounter many practical instances where biology helps us fight disease, produce an abundant food supply, and manage Earth's precious resources. This practical approach to biology will show students why this class and the laboratory experiences they have are important tools to help them become better servants of God.

Features of BIOLOGY Lab Manual 6th Edition

The lab manual contains forty-nine lab activities. Each one corresponds to a relevant chapter in the Student Edition to help you give your students a consistent laboratory experience throughout the course.

DEEP THINKING IN THE LABORATORY

The lab activities follow a linear approach through each topic; that is, instead of being isolated at the beginning and end of the activity, questions are mixed in with the procedure steps. Questions follow a logical progression, with their answers often suggesting the next procedure step. Arranging the material in this way encourages students to think deeply and see the question-answer process as integral to the learning experience rather than something to get out of the way. Students also learn to think like a scientist in the way that they observe and record their data.

Many questions specify that students explain their answers. This requirement allows them to think through the reasons behind their answers. Encourage thorough explanations in complete sentences. Questions that require longer responses may be done out of class as homework. These kinds of questions develop writing skills in the science content area.

TECHNOLOGY IN THE LABORATORY

Students will occasionally graph the data that they've collected during a lab activity. The graphs they create will be analyzed through questions that follow. You may choose to have students use the graphing areas provided or graph in a spreadsheet program. The latter will allow students to gain experience with the technology that many scientists use.

The Lab Manual also includes several activities that lend themselves to data collection using probeware. Probeware allows students to explore almost any area of science in new and innovative ways.

To the Teacher

INQUIRY IN THE LABORATORY

Several lab activities encourage students to think about how to solve a problem on their own, without any specific procedures spelled out in the activity. These lab activities will have the word "inquiring" in the subtitle. These will have some guiding questions to jump-start students' thinking before they formulate a plan and carry it out. Like the other lab activities, these will be completed in student groups when possible. The teacher guides for these lab activities will help you guide students through the inquiry process, teaching them about the process of science.

SAFETY IN THE LABORATORY

Biology Lab Manual 6th Edition is written with your students' safety in mind. Make sure that you point out the safety icons at the beginning of each lab activity and go over the laboratory safety rules in Appendix A at the beginning of the school year. Your students may also find Appendix D helpful as they learn to perform laboratory techniques safely.

While the Lab Manual encourages a culture of safety, you are ultimately responsible for your students' safety; please do not leave students unattended while working in the laboratory. Learn more about your legal responsibilities as a lab instructor as well as about hazard information systems for chemicals and local guidelines about chemical storage and waste disposal. Make sure that you label containers in ways that promote safe handling and disposal. Assemble and maintain an up-to-date, easily accessible SDS (Safety Data Sheet) notebook containing a data sheet for every chemical in your laboratory. Equip your laboratory with basic safety equipment.

For more detailed guidelines about chemical labeling and safety, visit the Occupational Safety and Health Administration (OSHA) website. You may also want to perform a general online search on high-school laboratory chemical safety, HMIS labeling, the Globally Harmonized System of Safety, and the NFPA fire diamond.

TEACHER NOTES

Each lab activity comes with detailed teacher notes in the margins. These address a wide variety of topics ranging from equipment and material substitution suggestions to answers for those hard questions that students tend to ask. Before performing any lab activity for the first time, you should read it through completely, looking at the relevant teacher notes as you go along. Teacher notes include icons that identify their purpose. The key below explains each one.

Biblical Worldview Shaping
This note introduces a discussion or idea that reinforces a Christian worldview of biology.

Color Vision
This icon alerts you to an activity that requires students to distinguish between different colors. Students with color vision impairment may require accommodations.

Demonstration
Some lab activities (or parts of them) may be better for you to present as demonstrations to give students an idea of what they will observe. Demonstrations don't take the place of hands-on laboratory work, but some lab activities may be used this way if your time or equipment is limited.

Difficult Concept
This icon alerts you to concepts that students may find difficult to grasp. Information and suggestions will be given on how to clarify or reinforce such material.

Equipment
Notes flagged with this icon provide extra information about the equipment and materials needed for the lab activity. Acceptable substitutions, local sources, or preparation instructions are included.

Helpful Tip
Helpful tips provide you with hints and ideas for running the activity that will help students get better results and have a smoother experience.

Recall
Recall icons show that prior knowledge of content will help a lab activity run more smoothly. They alert you to the need to review a concept or section of the Student Edition before proceeding further.

Resource
This note directs you to a helpful resource that provides additional insights into the topic or suggests ways to take the topic further.

Safety Alert
When you see this icon, be sure to read the note since it contains crucial information that will help you complete the lab activity safely or avoid a potential safety problem.

To the Teacher

ASSESSMENTS

There are lots of questions in the Lab Manual, and you have all the answers! Answers to questions asked within a lab activity are provided in magenta overprint on the reduced student page. Tables and graphing areas at the end of each activity give students a place to display data that can be analyzed later. Sample data is usually provided to give you a sense of what collected and analyzed data should look like. We don't recommend grading students on the accuracy of their collected data since it will vary with the equipment and technique used. However, students should be held accountable for how they present and analyze their data.

EQUIPMENT LISTS

In addition to lab activity notes on equipment sources and preparation, there are also five equipment lists in the back of this Teacher Lab Manual. These lists are compiled from the equipment lists at the beginning of each activity. You will also find these lists in a spreadsheet, available online as a digital resource.

How to Use This Book

PLANNING AHEAD

Don't feel like you need to do all forty-nine lab activities in this book! Most states require that about 20%-30% of total class time in upper-level science courses be dedicated to lab activities, hands-on experiences, and demonstrations. Some activities can be done in one class period (about fifty minutes), but others will require several lab periods (or outside class time) to complete. Set a goal to do a minimum of one lab activity for each chapter. Choose activities that fit your time and equipment resources. The BIOLOGY Teacher Edition 6th Edition includes a lesson plan overview to suggest how to fit the lab activities into your schedule and coordinate them with different sections in the Student Edition.

See the Teacher Edition notes in many of the chapters for guidelines as to when you should do a lab activity; some activities may be done at any time within the current textbook chapter, while others are better performed before or after you've reached a certain place in the chapter.

OBTAINING EQUIPMENT AND MATERIALS

Another aspect of good preparation is ensuring that you have the necessary equipment and materials to perform each lab activity. We suggest that you spend some time planning ahead in this area before the school year begins. Confirm that you have the required equipment or that the materials you need are available locally (excepting materials that must be ordered online). Since it's generally more economical to place a few large orders than many small ones, working out the year's equipment and materials at the start of the year is recommended.

If you are teaching in a smaller school or are homeschooling, don't let the equipment guidelines discourage you—remember that there are many possible alternatives and substitutions for most of the equipment. Often, everyday items or locally obtainable materials may work just as well. For this reason, teacher notes flagged with the equipment icon often include suggestions for suitable alternatives.

Obtaining chemicals often poses a challenge, particularly for homeschoolers and smaller Christian schools. In many cases, however, chemicals can be obtained locally from common sources, such as drug stores, home improvement stores, and grocery stores. A few chemicals have no local source and must be ordered from a science equipment supplier. To make this task easier, we have created a master list of chemicals used for BJU Press secondary science courses. This list, *Chemicals Database*, is available as a digital resource. Chemicals are listed alphabetically along with alternative names and both local and online sources. When appropriate, helpful notes about the chemical may also be included.

Finally, while most of the materials specified in this manual are reasonably safe, a few do require special care and handling. Never order a chemical unless you are comfortable with its safety guidelines. This includes having a suitable storage location that can be locked and that maintains the proper environment. If you are unsure regarding safety, seek proper guidance first. And if you just don't feel comfortable with a lab activity's requirements, choose another lab activity in its place.

Let's Get into the Laboratory!

You are the model for how students should think, behave, and feel about the laboratory experience. Keep your wonder and love for science alive because passion is contagious to students! View laboratory experiences as an opportunity to give your students a front row seat to observe God's orderly and beautiful creation.

1A LAB

A Method to This Madness

Scientific Inquiry

Laura wants to use scientific inquiry to test whether practice can improve reaction times. To be sure that her results are valid, she decides to conduct a controlled experiment. A controlled experiment typically has two groups that are identical except for a single factor, called the *experimental*, or *independent*, *variable*. The group exposed to the independent variable is called the *experimental group*. The second group, the *control group*, is not exposed to the independent variable. During the experiment, the researcher measures a factor in both groups. This factor is called the *dependent variable*—it results from, or is dependent on, the independent variable. The experiment is designed to determine whether there will be a measurable difference in the dependent variable between the two groups.

Six students at Laura's school agree to be part of her experiment. To consistently measure their reaction times, she decides to use a falling meter stick. The test subject will have to catch the falling meter stick after seeing it begin to fall. The distance that it falls before being stopped will indirectly measure reaction time. She formulates a hypothesis: The reaction time of students who have practiced catching the meter stick will be less than their reaction time before having practiced.

Each of the six students places a thumb and forefinger on the edge of a ring stand and keeps eyes closed. Laura positions the meter stick in the center of the ring stand so that the end is level with the student's fingers. At Laura's instruction, the student opens his or her eyes and Laura drops the meter stick. Her friend records how far the meter stick drops before the student catches it. Laura repeats these procedures with each student and averages the results.

Laura repeats the experiment five more times and plots the six averages on a bar graph to see whether student reaction times have improved with practice. The independent variable is the amount of practice, and the dependent variable is the reaction time.

How do scientists find answers to their questions?

Equipment

meter stick

PROCEDURE

Now it's your turn. You will conduct a different experiment with a different experimental variable. You will test whether a person can catch a meter stick faster if given a heads-up that it is about to be dropped. Your classmates will form both the experimental and control groups.

QUESTIONS

- How can you use scientific inquiry to answer questions?

- What is a controlled experiment?

- How is a controlled experiment used to answer scientific questions?

- Define controlled experiment.
- Describe the process that scientists use to answer questions.

Using Meter Sticks

You may use meter sticks made out of wood, plastic, or metal for this lab activity without affecting the results in significant ways. But be sure that all lab groups are using meter sticks made of the same material.

Understanding Scientific Inquiry

The purpose of this experiment is to help students understand and carry out scientific inquiry. Students will need to be able to understand and use scientific terms, such as *variable*, *hypothesis*, *problem*, and *experiment*, which they have learned in Section 1.2 of the Student Edition.

The experiment described in the introduction and the one that students perform in this lab activity use the same people as both the experimental group and the control group. This is not possible in most experiments. Studies involving humans typically use large numbers of people to try to average out individual factors that might affect results. This produces better results, but the goal of this activity is less about getting highly accurate results and more about helping students become familiar with doing scientific inquiry.

Experimental and Control Groups

Keen-eyed students may note that this lab activity describes two groups in an experiment but subsequently uses only one group. Point out that in this instance Laura's control group will subsequently function as her experimental group once they have had time to practice. What's truly important are the variables—the control variable is without practice, while having practice is the experimental variable.

Forming Lab Groups

If the class has an odd number of students, one group can have an additional member who can act as data recorder.

Question 2 Answer
Answers will vary but should be similar to one of the following:

1. The catchers will catch the meter sticks more quickly if they are alerted when it is about to be dropped.

2. The catchers' knowing that the meter sticks are about to be dropped will not affect their catching speed.

3. The catchers will catch the meter sticks more quickly if they are not alerted when it is about to be dropped.

Question 3 Answer
Answers will vary. *Examples:* (1) The catchers are all either sitting or standing. (2) The meter sticks are identical. (3) The catchers all hold their thumb and fingers the same distance apart. (4) The droppers all hold the meter sticks at the same height.

Recording Data in Table 1

A dropper-catcher pair may work together to fill in Table 1. Distances should be recorded to the nearest 0.1 cm.

Analysis Steps D and E

The ability to create and interpret graphs is very important in science and other fields as well. This is the first of many graphing exercises that students will be asked to complete throughout this course. Use this opportunity to familiarize them with your expectations regarding graphs (e.g., neatness, completeness). Have students consult Appendix E (Graphing Techniques).

If you are short on time, students can skip Step E without jeopardizing the rest of the activity. If they don't collect data from the other lab groups in the class, they will not need to use Area B for graphing. To save time, you could draw Table 2 on the board and have students record their results on the board. Once all the students have done that, each student can take a quick photo of Table 2 to complete the analysis.

1. Scientists often start out with a research problem—the question they are trying to answer. What is your research problem?

   Is the reaction time required to catch a falling meter stick, as measured by how far a dropped meter stick falls, decreased when the individual knows when the meter stick is about to be dropped?

2. What is your hypothesis?

   See margin for answer.

Your teacher will be the "caller" and will divide the class into teams consisting of a "dropper" and a "catcher." Each team should record its own results. Remember, each lab group is part of two groups: the experimental group and the control group.

3. What are three variables that you need to keep constant?

   See margin for answer.

4. What is the independent variable?

   whether the catcher is aware of when the meter stick will be dropped

5. What is the dependent variable?

   how far the meter sticks drop before the catchers catch them

After each pair of droppers and catchers is provided with a meter stick, the droppers will hold it up for the catchers to catch, keeping constant the variables identified in Question 3.

A *Experimental Group:* The caller calls out a sequence such as "1, 2, 3, drop," or "Ready, set, go." This will alert droppers to drop their meter sticks in unison. Record the distance that the meter stick fell before it was caught in the *Trial 1* row of Table 1.

B *Control Group:* The experiment will now be repeated, but the caller will stand where not visible to the catchers and will signal silently to the droppers to drop their meter sticks. Record the distance the meter stick fell in the *Trial 1* row of Table 1.

C Repeat Steps A and B four times, recording the results for Trials 2–5 in Table 1.

ANALYSIS

D For each team, average your experimental group scores and your control group scores and record the averages in Table 1. Make a bar graph of the data in Graphing Area A.

E Copy the averages from Table 1 to the first row of Table 2. Obtain the experimental group averages and the control group averages from your classmates and record them in Table 2. Average these averages and record the results in Table 2. Create a bar graph of the data in Graphing Area B.

6. What conclusions can be drawn from the data?

   Answers will vary. *Example:* There was a slight improvement in reaction time when the droppers knew that the meter stick was about to be dropped.

7. Does the data support the hypothesis? Explain.

   Answers will vary.

PERSONAL OBSERVATIONS

8. Did your repeated trials yield similar results? If not, suggest some sources of error that might have affected your results.

   Answers will vary. If the students practiced good technique, the repeated trials should yield similar results. If not, some possible sources of error could include the dropper holding the meter stick too high or too low or making a mistake in recording the data.

Scientific Argumentation (Question 7)

Argumentation is a skill that scientists use all the time and students should learn. Arguments take the form of a claim and evidence. Students' answers to Question 7 should state a definite claim regarding their hypotheses and then support their claim with evidence, which should consist of references to specific data points or analysis items from their experiment.

9. What could you do to make the repeated trials of your experiment closer in value?

Answers will vary. Students may give answers such as having the droppers practice dropping the sticks to get them to fall straight.

10. Can you think of any additional changes that need to be made?

Answers will vary. *Examples:* Include more trials, providing better support for the hypothesis. Improve the lighting in the room. Screen out test subjects who spend extensive time using handheld games as they may have already improved their eye-hand coordination to the point where further changes are negligible.

11. On the basis of your experience, what other experiments dealing with response time would you like to try?

Answers will vary.

GOING FURTHER

You read a news article about a scientific study into whether Kentucky bluegrass (*Poa pratensis*) or timothy (*Phleum pratense*) is a better grass for raising horses. A herd of thirty thoroughbreds was pastured in a field of Kentucky bluegrass in Tennessee, and a herd of thirty mustangs was pastured in a field of timothy in Arizona. At the end of the yearlong study the average weight gain of the thoroughbreds was greater than the average weight gain of the mustangs. The study concluded that Kentucky bluegrass is a better fodder for horses.

12. This study has some problems, primarily with keeping the two groups the same except for the experimental variable. Redesign this experiment in a way that corrects these problems.

Answers will vary. At a minimum, students should mention that the experiment should use the same breed of horses and that the horses should be kept in fields near each other rather than in very different parts of the country. The study description mentioned very little about variables such as water, shelter, or exercise, and some students may mention these as well.

Analyzing a Scientific Article

If you have the time, your students can learn about the scientific process by analyzing a real scientific study. The *Science Daily* link is available as a digital resource and can serve as a database of readable summaries of recent scientific studies.

Evaluating Contrary Thinking (Question 12)

Any issues in a published scientific paper will not be as obvious as the example in Question 12, so you should tell your students to be especially on the lookout for how the researcher's worldview affected his observations and interpretation in the article. Have them ask themselves the question, "What is wrong with the ideas in this article when compared with biblical teaching?"

Sample Data

The data shown in these tables and in the graphs on page 5 is typical. Students' results may be different.

TABLE 1 Group Data

	Experimental Group Distance (cm)	Control Group Distance (cm)
Trial 1	22.9	20.3
Trial 2	17.8	22.9
Trial 3	21.6	21.6
Trial 4	17.8	29.2
Trial 5	19.1	35.6
Average	19.8	25.9

TABLE 2 Class Data (Averages)

	Experimental Group Distance (cm)	Control Group Distance (cm)
Your Averages	19.8	25.9
Classmate's Averages	13.3	32.0
Classmate's Averages	14.9	18.4
Classmate's Averages	12.3	20.8
Classmate's Averages		
Average	15.1	24.3

GRAPHING AREA A

Distance Dropped in the Experimental and Control Groups

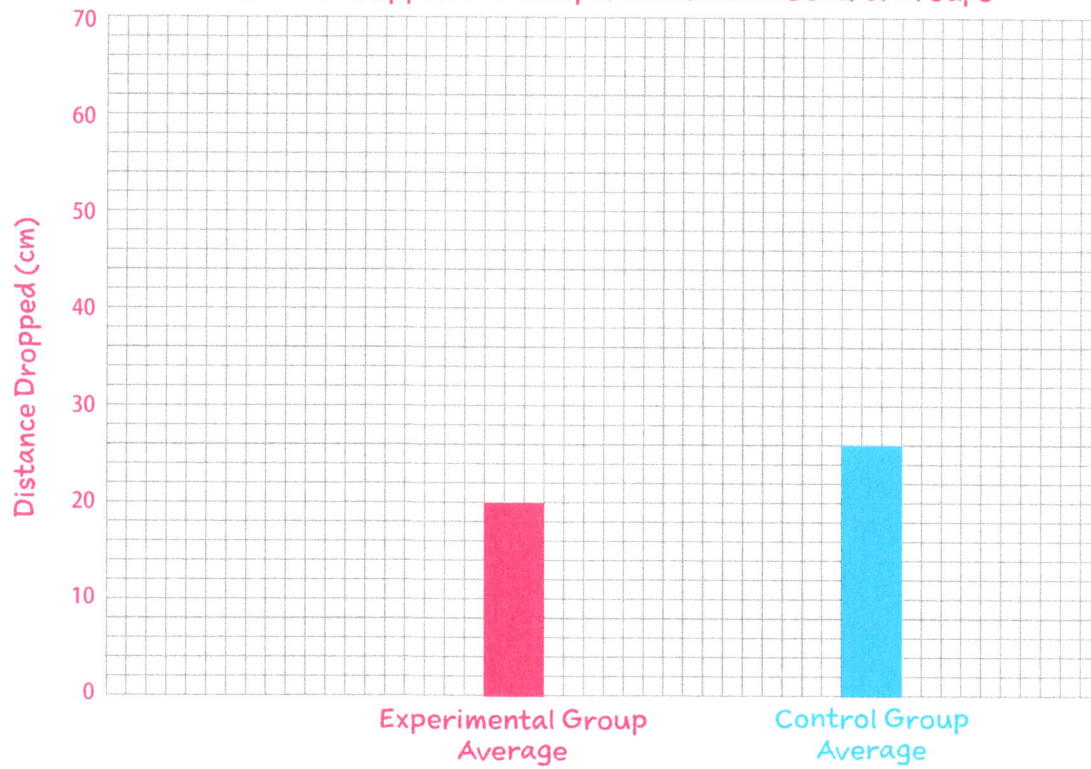

Distance Dropped (cm) (y-axis: 0, 10, 20, 30, 40, 50, 60, 70)

Experimental Group Average

Control Group Average

GRAPHING AREA B

Distance Dropped in the Experimental and Control Groups

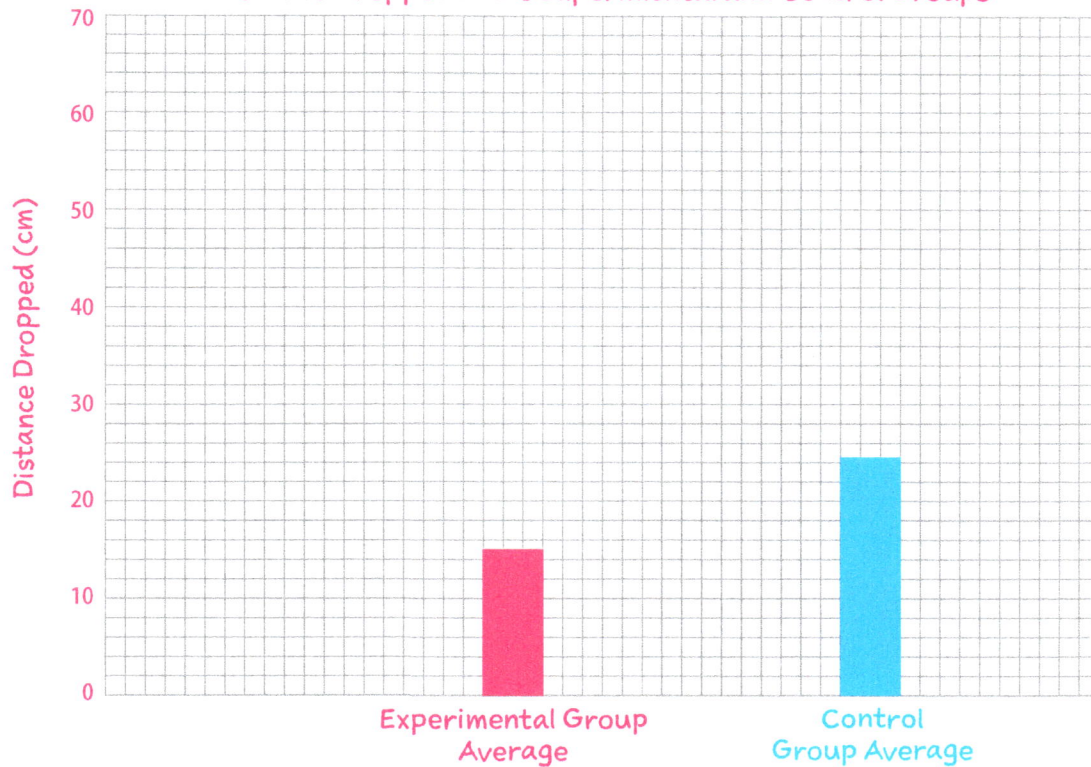

Distance Dropped (cm) (y-axis: 0, 10, 20, 30, 40, 50, 60, 70)

Experimental Group Average

Control Group Average

1B LAB

More Than Meets the Eye

The Microscope

In 1676 a Dutch cloth merchant named Antonie van Leeuwenhoek was investigating what gives pepper its hot taste. His hobby was making lenses, and with them he had already discovered many things too tiny to see with the unaided eye. But on this particular day he saw creatures much smaller than he or anyone else had ever seen before: bacteria.

Van Leeuwenhoek made and used simple (single-lens) microscopes because the low-quality compound microscopes of his day distorted images. His simple microscopes could some-times render clear images at magnifications four times greater than existing compound microscopes. The secret to van Leeuwenhoek's success was his skill at making lenses, using a technique that he kept secret. But by the middle of the 1800s lens designers had greatly improved the compound microscope, allowing microbiologists to see things with greater clarity. Today almost all light microscopes are compound microscopes; simple microscopes are usually called *hand lenses* or *magnifying glasses*. A compound microscope is a very important tool in almost any field of biology.

In this lab activity you will learn how to use a microscope to study tiny life-forms, much as van Leeuwenhoek did almost 350 years ago.

Antonie van Leeuwenhoek

How can we use a microscope to learn about cells and microscopic organisms?

QUESTIONS

- What are the parts of a microscope?
- What is the proper care for a microscope?
- How is a microscope used?

Equipment

microscope
lens paper
tissue

preserved slide of desmids or diatoms
immersion oil

LAB 1B OBJECTIVES

- Label the parts of a microscope.
- Describe how to take care of a microscope.
- Draw an image of a microorganism from a microscope slide.

Microscopes

This lab activity is written to correspond with the microscope shown on page 8, which has a fixed body, movable stage, four-objective nosepiece, and binocular eyepiece. Not all student microscopes have all these features; if necessary, adjust the procedure for use with your type of scope(s).

Ideas for Prelab Discussion

This is the first of several lab activities that require students to draw what they see in the microscope, so now is a great time to let them know your expectations for their drawings. Please reference the section on microscope drawings in Appendix D.

Immersion oil is necessary only if you will have students use an oil-immersion lens during the activity; this is typically the 100× objective on a student microscope. You may choose to skip that portion of the activity if time is limited.

Many students may find it easier to draw from an image rather than from the micro-scope itself. They can take a photo of the image with their smartphones or with a built-in camera on more expensive micro-scopes. This may be especially helpful for those who have trouble drawing to per-spective. If you choose not to emphasize drawing, consider having students paste pictures onto the drawing areas provided.

Lab Activity Rubric

You may choose to develop a lab activity ru-bric to use for the remainder of this course. If so, share it with students so that they can see how their work will be assessed.

Using Science Tools to Serve Others

Finding bad bacteria is just one of many important uses of the microscope. It's the first step in responding to many diseases and ailments. If you have time, discuss with your students how the biblical belief that people are valuable encourages them to take advantage of every means possible to improve the health of God's image-bearers.

PROCEDURE

The Structure of Your Microscope

As you study the infographic below, find each part on your microscope and check the box that goes with that part.

Adjustment Knobs ☐
Most microscopes have two types of adjustment knobs, one of each on each side of the microscope. The larger coarse adjustment knob allows you to rapidly change the distance between the specimen and the objective. The smaller fine adjustment knob is usually found underneath or centered on the coarse adjustment knob. It allows you to change the distance between the specimen and the objective slightly, producing a sharper focus.

Arm ☐
This "backbone" of the microscope supports the body tube.

Body Tube ☐
For a compound microscope to function properly, the two lenses must remain a certain distance apart. This long, narrow tube maintains that distance. In microscopes with an inclined body tube, mirrors are used to bend the path of the image.

Eyepiece ☐
This holds the top lens of the two lenses in a compound microscope. It is sometimes called the *ocular lens.*

Nosepiece ☐
This rotating disc at the bottom of the body tube holds the objectives, the metal cylinders that contain the bottom lenses of a compound microscope. Most microscopes have several different objectives that you can interchange by rotating the nosepiece. Each objective has a different power of magnification.

Stage ☐
This platform positioned directly below the objectives and above the light source supports the specimen.

Diaphragm and Substage Condenser ☐
These are located between the stage and light source. The diaphragm regulates the amount of light that passes through the specimen. The substage condenser provides a clearer image by bending and concentrating light before it reaches the specimen.

Stage Clips ☐
These fasteners on top of the stage hold the slide containing the specimen firmly in place.

Base ☐
This large rectangular or horseshoe-shaped structure supports the microscope and keeps it steady.

Light Source ☐
This is typically some type of light bulb, and in many microscopes you can adjust the amount of light being produced.

1. Which part of the microscope does the specimen rest on?

   the stage

2. Which two structures hold lenses?

   See margin for answer.

Caring for a Microscope

It is important to carry the microscope properly. Excessive jarring or bumping may knock the lenses out of adjustment.

A Carry the microscope with two hands—one under the base and the other on the arm.

B Keep the microscope close to your body in an upright position so that the ocular lens does not slip out of the body tube.

C Place the microscope gently on the table away from the edge. Remove the dust cover, if present, and plug in the power cord. Leave the light source turned off for now. Your microscope should have been stored with the stage in its lowest position and its nosepiece rotated to the lowest-power objective (typically 4×). Check to see that this is the case and make adjustments if necessary.

You will have a better experience using the microscope if you prepare it properly. Your microscope may need to be cleaned before you begin to use it.

D Use lens paper to clean lens surfaces and the light source.

3. Why should you not use a tissue to clean a lens surface?

   Answers will vary. A rough material like a

   tissue might scratch the lens.

E Wipe the lens in one direction across the diameter of the lens. Do not use a circular motion—doing so might grind dust into the lens, scratching it.

F Consult your teacher if any material remains on your objectives. Never use your fingernail or another object to chip away hardened material.

G Never attempt to take your microscope apart.

H Check that there is no slide on the stage.

Using Your Microscope

To have a clear view of your specimen, follow these procedures exactly. Keep in mind that it will probably take some practice for you to become proficient in using a microscope, so don't get discouraged if your first attempt at viewing a specimen doesn't give you the results you expect. With practice, using a microscope will become almost second nature, and you will be able to get a clear image quickly.

I Make sure that the nosepiece is rotated back to the scanning objective and that the stage is still lowered.

J When using your microscope, it helps to know the powers available. Power is the number of times larger the image of the object appears. You will find the powers written on the eyepiece and the objectives. As the power increases, the size of the image increases, but the field of view shrinks. Record the power of each objective in the appropriate rows of Table 1. Then compute, as directed in Table 1, the total magnification obtained using each objective and record the results.

K Obtain a preserved slide. Make sure that the slide is dry and free of cracks or other damage.

L Place the slide on the stage with the cover slip up, directly over the opening in the stage, and secure it with the stage clips. Turn on the light source.

M Position the slide so that the specimen is centered over the opening of the stage.

Now that you have the slide in place, you can focus your microscope.

N Double-check that the scanning objective (4×) is directly below the body tube. You should be able to feel that it is clicked into place.

O Looking at your microscope from the side, turn the coarse adjustment knob until the objective is just above the slide. (Some microscopes have safety devices that will prevent the objective from striking the slide and damaging the lens.)

P Look into the ocular lens and slowly turn the coarse adjustment knob until the specimen comes into focus. Make sure that the distance between the objective and the stage is increasing. You do not want the objective to strike the slide, potentially cracking or even breaking the slide.

Question 2 Answer
the eyepiece, or ocular lens, and the objectives

✓ Focusing on Focusing

Students often become frustrated while trying to focus their microscopes. Have them read the steps on these pages and then do each step while you watch. Sometimes this will help them self-diagnose their problems and encourage independent laboratory work.

If students experience discomfort by squinting one eye to observe through a microscope, you may want to encourage them to observe specimens with both eyes open to avoid eye strain. Students with glasses may find it best to take off their glasses while observing. If a lab partner wants to observe, he will need to readjust the microscope to suit his own eyes.

✓ Specimen Drawings

Drawing areas for specimen drawings have been provided. Students should consult the section of Appendix D on making biological drawings to help them complete this procedure. Make sure that they have each made a legitimate effort to correctly draw the outline of a desmid or diatom. It is helpful if students add titles to drawings.

🧪 Using Oil-Immersion

Oil immersion objectives are necessary for advanced microscopic work, but not all microscopes have them. If your classroom microscopes do not, then skip this section.

Q Once the specimen has been brought into focus, slowly turn the fine adjustment knob to get a sharper image.

#QUICK NOTE

If no image comes into view, consider the following scenarios.

- You may have turned the coarse adjustment knob too fast and passed the point of focus. Start at Step O and try again more slowly.

- You may not have a specimen in your field of view. Check to make sure that there is something on the slide directly in the center of the hole in the stage. Then start at Step O and try again.

- You may have too much light. If you have a microscope with an adjustable light source, reduce the amount of light. If not, ask your teacher to adjust the diaphragm.

If you still cannot focus your image, ask your teacher for help.

4. If you move the slide to the right, how does its position change in your field of view?

It moves to the left.

5. Turn the fine adjustment knob slowly. Describe what happens to the image that you are viewing. Why do you think this occurs?

Answers will vary. Different regions go in and out of focus. The specimen has depth, and the lenses can focus only on a fraction of the depth at a time.

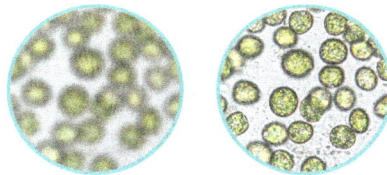

A slightly out-of-focus image (left) and a properly focused image (right)

R Draw the outlines of three specimens in Drawing Area A using the instructions provided in Appendix D; ignore the internal structures. Make sure that you are drawing typical specimens, not just odd globs that may be on your slide. Confirm the correctness of your drawings with your teacher.

Using High Power on Your Microscope

Now that you've learned to focus your microscope, you can switch to the next higher power.

S Center the desmid or diatom that you want to examine in the microscope field.

6. Why is it essential to position the specimen in the center of the field?

The field of view shrinks when the power increases, so if the specimen is not in the center of the microscope field, it will not be visible when the microscope is changed to a higher power.

T Focus your microscope using the scanning objective (4×) and then rotate the nosepiece to the low power objective (10×). If necessary, focus the image using the fine adjustment knob.

U Rotate the nosepiece again to the high-dry (40×) objective. As you do so, watch the stage from the side to ensure that the slide is not touched as the objective clicks into place.

V Prepare a specimen drawing of a section of your specimen in Drawing Area B, including the internal structures. If the specimen in your field of view is one that you drew in Step R above, you may add the details to your outline.

Using the Oil-Immersion Power on Your Microscope

The oil-immersion (100×) objective allows you to view structures too small to see clearly with a high-dry objective. However, an oil-immersion objective is also more difficult to use, so you may need to practice using it several times before you become proficient. Follow these instructions exactly!

W Once you have focused the image on high-dry power, turn the nosepiece halfway toward the oil-immersion objective.

X Place one small drop of immersion oil on the center of the cover slip.

Y Turn the nosepiece so that the oil-immersion objective touches the drop of oil. Be sure to watch from the side to ensure that the objective doesn't hit the slide.

Z Using the fine adjustment knob, change the focus very slowly until you obtain the proper focus.

AA If the objective is raised so high that the oil separates from the objective, you have passed the point of focus. Repeat Steps Y and Z.

BB Observe a desmid or diatom using oil-immersion power and draw a portion of the specimen in Drawing Area C. Include the internal structures.

If you still cannot focus the image, you may have too much light. Adjust the light source or ask your teacher to adjust the diaphragm.

CC When you finish, clean the microscope and the slide carefully. Remove excess oil from the slide with a dry tissue. Clean the objective carefully with lens paper. (Again, never use anything but lens paper on a microscope lens.) Clean the slide with a wet tissue, being careful not to wet the label. Dry the slide thoroughly before returning it.

7. Scientists use scientific instruments to make better observations about the world. Write a paragraph on the ways that the microscope extends our ability to see things in our world.

Answers will vary. Students should mention that microscopes allow us to see cells and organisms too small for us to see with our unaided eyes. They might also mention that microscopes are used in medicine to identify bacteria that are causing an infection and to help doctors perform very precise surgery.

To ensure that your microscope lasts a long time, you should always store it properly. Carefully return the microscope to its storage place. Cover the microscope with a dust cover if one is available.

TABLE 1 Magnification of a Microscope

	Ocular Lens		Objective		Total Magnification
Scanning Objective	10×	×	4×	=	40×
Low-Power Objective	10×	×	10×	=	100×
High-Dry Objective	10×	×	40×	=	400×
Oil-Immersion Objective*	10×	×	100×	=	1000×

*Some microscopes do not have an oil-immersion objective. If yours does not, skip this line of the table.

Post-Lab Check

1. **What two parts of a microscope magnify an image?** *the objectives and the eyepiece*

2. **How should a microscope be carried?** *with both hands, one on the arm and the other beneath the base*

3. **In what order should the focus knobs on a microscope be used?** *coarse focus first, then fine focus*

4. **Why should the body tube always be raised before placing a new specimen on the stage to be examined?** *to prevent cracking or breaking the slide when the microscope is focused*

5. **After placing a specimen on the stage, no image is seen when looking through the microscope. Suggest a reason for this and describe how to address it.** *Answers will vary. Possible reasons include the specimen being out of focus (repeat the focusing procedure), the specimen not being positioned correctly over the hole in the stage (reposition slide), and the light source being too bright (adjust the light source or diaphragm).*

DRAWING AREA A

DRAWING AREA B

_____

DRAWING AREA C

2 LAB

Lost in the Woods

Designing a Water Treatment System

You are alone and lost in the wilderness. You were with a group of people backpacking in the woods, but you fell behind and then took a wrong turn on one of the trails. A thunderstorm and a few blisters later, you discover that you are lost. You know that you need to stay in one spot to improve your chance of being found. You have plenty of food, a tent, and a light source. Temperatures are mild, so you don't have to worry about staying warm. Your biggest problem—drinking water. Thankfully, you've camped near a stream, but is the water safe to drink?

The water in natural areas often lacks the chemicals and pollutants that water in urban areas has. Trees and vegetation in watersheds act as natural filters for contaminants. But water from a stream can contain bits of gravel, sand, and soil. It can also contain bacteria and other critters that could make you sick.

The water you drink every day has likely been treated at a water treatment plant to remove contaminants that you would find if you scooped up water straight from its source, such as a local lake or river. People around the world need clean, fresh, disease-free drinking water. Let's explore how we can use changes in matter to purify water for drinking.

How can chemical and physical changes be used to purify water?

Equipment

laser pointer

universal indicator paper

petri dish and lid

nutrient agar

sterilized cotton ball

untreated water from a local water source

PROCEDURE

You are tasked with designing a water treatment system to filter a 250 mL sample of water. Your system must produce drinkable water as indicated by a series of water quality tests. You will be constrained by the types and quantities of materials that you can work with. As is always the case in real life, you will also be constrained by the amount of time that you have to design and build your system. Your teacher will specify a set of design parameters for you: the materials that you can use to construct your device and the time limitations for you to design and build it.

QUESTIONS

- What is dangerous about drinking unpurified water?
- How can you remove from water material that you can't even see?
- What tests can be done to verify the quality of water?

- Explain the biological and theological importance of providing treated water.
- Design a water treatment system that produces safe drinking water.
- Test the water treatment system through a series of water quality tests.

Timing

Lab 2 can be done at any time during your teaching of the material in Chapter 2.

A Teacher Guide for this activity can be found on page 18*a*.

Laser Pointers

Laser pointers are widely available and inexpensive. A flashlight may be substituted for the laser pointer.

Extension Question

The city of Greenville, South Carolina, is known for exceptional water quality. The city processes about 625 million liters of water every day. How long would it take your treatment system to treat all the water for the residents of Greenville, assuming that your treatment system would be able to hold up?

To answer the question, students should multiply 625 million liters by 4 (the number of 250 mL samples in 1 L) and then by 15 (the number of minutes given as an example in Question 3 to purify each sample). On the basis of a fifteen minute purification time, it would take 37.5 billion minutes, or roughly 71,347 years, to filter 625 million liters.

Planning the Design

A Start by researching how water treatment plants treat water from natural sources to make it drinkable. This may give you some ideas for your design.

B On the basis of what you have learned, design a way to treat water using supplies that could be carried in a backpack or found in the woods. Draw a diagram of your design on a separate sheet of paper. Indicate that this is a *preliminary design*.

C Share your research findings with your group. Reach consensus on a design that your group will build and submit for testing.

D Have someone in the group draw a full-size diagram of your group's agreed-upon design on a separate sheet of paper. You will use this during construction and submit a copy of it with your lab report. Indicate that it is the *initial production design*.

1. How will your mini water treatment facility eliminate particulates in water?

   Answer will vary. Students may mention flocculation and filtration.

2. How will your device kill any bacteria in water?

   Answer will vary. Students may mention boiling the water, treating it with a disinfecting solution, or irradiating it with sunlight.

Constructing the Treatment System

E Now build it! Pull together the materials you need and construct your water treatment device. Take a picture of your finished treatment system and paste it into Photo Area A.

Testing the Design

F Obtain a 250 mL water sample to be treated from your teacher. Time how long it takes the water to run through your device.

3. How long did it take?

   Answers will vary. It may take 15 minutes or so.

Now you need to prove that you have drinkable water—without drinking it! You will need to prove that there are no contaminants in your water.

G Test your water for particulates by shining a laser pointer through the water.

4. What did you observe about the clarity of your water?

   Answers will vary. Pure water will not produce a visible beam. Particulates will cause a beam to be visible.

H Test your water for harmful chemicals by using indicator paper to test the pH.

5. Report the pH of the water that you tested.

   See margin for answer.

Question 5 Answer
Answers will vary. Water should be at a pH of about 7, though it may be slightly acidic or basic.

I Swab the agar in a petri dish with a sterilized cotton ball that has been dipped in your water sample to test your water for bacterial contaminants. Cover it with the lid and leave it undisturbed for two days.

6. Report the results of the bacterial test of your water.

   Answers will vary. If bacteria are present, they will produce visible colonies on the nutrient agar.

GOING FURTHER

7. Considering your treatment system's performance, what would you say are its strengths and weaknesses?

   Strengths: Students may comment on the portability of their devices or the ease of constructing them.

   Weakness: Students may mention their devices' inability to handle large quantities of water quickly.

8. What changes could you make to your treatment system to improve its performance?

   Answers will vary depending on test results.

J If you have the opportunity, redesign, rebuild, and retest your design. Submit a drawing of this as the *modified production design*.

9. How could you use the way that nature filters water in the environment in your own water treatment device?

   Using a mix of foliage, sand, and gravel would help filter out particulates in the water.

10. The Greenville Water company owns 100% of the watershed that supplies Greenville, South Carolina, with water. In light of what you have learned about how trees and vegetation naturally filter water in a watershed, why do you think it is important to protect these areas from development?

    Protecting a watershed from development allows the soil, trees, and vegetation to continue to filter runoff water. This can have an extensive effect on the surrounding land, even if these areas are not protected.

11. How is establishing protected areas an example of good and wise dominion through conservation?

See margin for answer.

_____

_____

_____

_____

PHOTO AREA A

2 LAB *Teacher Guide*

Lost in the Woods

Lab 2 is a STEM lab activity. Students are presented with the problem of designing a water treatment system that can purify one cup of water for drinking. This lab activity is a good introduction to changes in matter used for a very practical purpose: water treatment. Most of the changes in matter used to treat water are physical changes, but a few may be chemical. A chemical change to water would lead to something other than water and may not be beneficial to drink, but chemical changes could be done to substances within water to eliminate them. Additionally, chemical change is significant to bacteria since it kills them. You can reinforce these concepts by asking students to consider whether the process and design they arrive at to treat their water sample uses physical or chemical changes.

Collaboration is an important part of the engineering process. Students will need to work together to design a workable solution. You may choose to have students do part of this activity as homework. For example, students may do their research as homework, use class time to collaborate on a design, build their devices at home, and finally test their designs in class.

For this activity students will model what actual engineers must regularly do; that is, they will design something that can efficiently perform a prescribed task after being made only from available materials and within a given set of constraints. The effectiveness of their design at completing the task will be assessed through testing.

STEM activities are a great opportunity to help students recognize where the STEM areas (science, technology, engineering, and mathematics) intersect with living a God-honoring life. We are confronted with problems all the time, and the STEM process is one that man has developed to solve many of them. We are showing our love for others when we seek to help them solve problems. All STEM problems include constraints. We are called to be good stewards of what God has given us. We should always seek a solution that manages resources in the best way possible. Most students think that STEM projects are a lot of fun, and they are. But they also help us develop skills that we can use throughout life to help others and glorify God.

If your students are new to STEM, you may need to give them additional time and guidance to complete the activity.

MATERIALS

Suggested materials for this project are listed below. You may add or substitute other materials at your discretion according to availability, safety concerns, or modifications that you may want to make to the process. Students may suggest alternative materials as well; make sure that they check these with you first before using them.

Suggested Materials for the Treatment Systems

- plastic cups or bottles (for water samples and device bodies)
- alum, disinfectant solution, dilute bleach, or iodine solution (for chemical treatment of water)
- sand, aquarium gravel, paint filter, cheesecloth, sterile cotton (for physical filtration)
- plastic wrap (for physically containing filtering media)
- heat source (for boiling water)

How can chemical and physical changes be used to purify water?

?

2 LAB *Teacher Guide*

Materials for Testing the Treatment Systems

- flashlight or laser pointer (groups may share a few of these)
- universal indicator paper
- petri dish and lid (two per group)
- nutrient agar
- cotton ball, sterilized

Students will of course need water to filter. Ideally, you should obtain this water from a local source such as a pond or stream. It is helpful for the water to have visible contaminants such as suspended particulates or algae. The absence of these later in the filtered sample will serve as a visual indicator of a system's effectiveness. Remind students that their filtered water may *look* safe to drink, but may still contain contaminants, including pathogens.

It would be best if you as the teacher prepare the agar plates for Step I so that there is no contamination. Provide a sterile agar plate for students as a control for the experiment in addition to the sterile plate that they will use to test their treated water. Make sure that you have bleach or another disinfecting solution on hand for killing cultured bacteria. Exercise caution in culturing bacteria since some may have harmful effects on people. If you are limited in equipment, you may choose to eliminate Step F.

PROCEDURE

Planning the Design

Real-life engineers must work within certain design constraints, so you will have to specify the design parameters. There are many ways to treat water. Shown on the left is one suggestion for a filtration system that incorporates successively finer substrates for filtering out particulates. You may research other materials and methods of testing for emergency water filters. Make sure that students don't lose sight of the fact that their treatment system needs to *retain* the filtered sample. After all, the objective is to achieve water suitable for drinking, and the water cannot be drunk if it is lost during the filtering process!

You will need to specify the desired characteristics of the filtered water: having no particulate matter, being within a safe pH range, and being free of bacteria. If desired, you may obtain water testing kits that can check for other characteristics, such as the presence of mineral ions. Most surface water is not free of such ions. Natural minerals such as calcium and magnesium can leach into groundwater from the soil. Students' treatment systems will not normally remove these ions.

If you test for ions, you may discuss with your students whether removing them would be desirable (many serve as necessary electrolytes for human health). If you choose to test for such minerals, you may want to test the untreated source water to confirm the presence of these minerals in the source water. You may also want to constrain your students by giving them a limitation on the amount of material that they may use. One of the easiest ways to do this is to require that the entire device be housed in a vessel of a particular size, say a two liter bottle or a gallon milk jug. Time is always a constraint, so think about how much time you want to give students to design and build their systems.

dirty water

gravel

coarse sand

fine sand

cheesecloth

clean water

Constructing the Treatment System

You may allow students to build their treatment systems at home or during class at your discretion. Having them build their treatment systems in class will make it easier for you to set and monitor a time constraint.

Testing the Design

Testing the designs and swabbing the agar plates should not take an entire class period. Depending on how much class time you have available and how much has been allotted for construction, you may either have students build and test their treatment systems in the same lab session, or have students do the testing portion at the beginning of the next class period.

Testing for bacteria *after* treatment does not of course establish whether the untreated sample was contaminated in the first place. You may decide to swab an additional agar plate with a sample of untreated water to confirm the presence or absence of bacteria. Testing for pathogenic microorganisms must be done by trained technicians and is thus beyond the scope of this activity.

Refining and Retesting the Design

A key aspect of the engineering design process is refining and retesting a design. If time permits, have students analyze how and why their filters performed as they did. Then suggest that they come up with one or more modifications that they believe would improve their designs. Students can then redesign, rebuild, and retest their devices. If students update their designs, they should complete new drawings labeled *Final Production Design*.

Lab 2 Lost in the Woods—A Grading Rubric

Excellent—Student's design demonstrates thorough understanding of the task. Work may include minor flaws in design or execution that do not affect completion of the task.

Good—Student's design demonstrates good understanding of the task. Work may have some flaws.

Fair—Student's design demonstrates some understanding of the task. Work may have major flaws.

Needs Work—Student's design demonstrates little or no understanding of the task.

Point values are left blank to allow teachers to assign values within their grading system.

	Excellent (_____ pts)	Good (_____ pts)	Fair (_____ pts)	Needs Work (_____ pts)
Design				
Design	Detailed and properly labeled design drawings are provided.	Design drawings are provided. A few details or labels may be missing.	Partial design drawings are provided. Many details and labels may be missing. One or more diagrams may be missing entirely.	No design drawings are provided.
Product				
Execution of Design	Treatment system design demonstrates evidence that much thought and planning have gone into the design. Suggested improvements are reasonable, practical, and based on sound scientific principles.	Treatment system design demonstrates evidence that a good amount of thought and planning has gone into the design. Suggested improvements are mostly reasonable, practical, and based on sound scientific principles.	Treatment system design demonstrates evidence that some thought and planning have gone into the design. Suggested improvements are mostly reasonable but may be impractical and/or not based on sound scientific principles.	Treatment system design demonstrates evidence that little or no thought has gone into the design.
Workmanship	Treatment system demonstrates excellent use of materials and quality construction.	Treatment system demonstrates a good use of materials and construction.	Treatment system demonstrates a fair use of materials and construction. Treatment system may have a shabby appearance or exhibit flaws in construction that weaken the treatment system.	Treatment system exhibits poor use of materials and/or shoddy construction. Flaws in construction seriously weaken the treatment system or otherwise impede its ability to meet the performance standard.
Performance	Treatment system exceeds all performance expectations.	Treatment system meets all performance expectations.	Treatment system meets some performance expectations.	Treatment system meets few or no performance expectations.
Quality				
Timeliness	completed on time			late

3A LAB

Tag!

Mark-and-Recapture Sampling and Population Size

Fisheries biologists often need an idea of how many fish are in a pond or lake. They may need to determine, for example, whether a certain lake needs to be restocked with fish before fishing season opens. The most certain way to do this would be to catch and count all the fish in the lake, but this would be nearly impossible. So instead of fruitlessly trying to catch all the fish in the lake, biologists use a technique called *mark and recapture* to estimate the population size in a body of water. You will use the same technique to estimate the size of a population of crickets.

How can scientists estimate a population size without actually counting all the individuals in the population?

Equipment

terrarium or large bucket   dip net or cup
live crickets   correction pen

PROCEDURE

A From your terrarium or large bucket, randomly capture some crickets in a dip net or cup.

B Using the correction pen, mark each cricket that you've captured on the back of the thorax above the wings. See sketch on the right.

C Count the number of crickets as you mark them.

D Record the number of crickets that you marked in the first row of Table 1.

E Return the crickets to the terrarium and allow them to reintegrate into the population.

F After five minutes, randomly capture a second sample of crickets from the terrarium.

G Record the total number of crickets in your second sample in the second row of Table 1.

H Record the number of marked crickets in your second sample in the third row of Table 1.

QUESTIONS

- What is mark and recapture?
- How can mark and recapture be used to estimate population size?
- Is mark and recapture an accurate sampling method?

mark here

Tag! 19

LAB 3A OBJECTIVES

- Explain how mark and recapture can be used to estimate population size.
- Collect data by mark and recapture to answer a scientific question.
- Describe the limitations of the mark-and-recapture method of sampling.

Scheduling

Since populations are defined in Section 3.1 of the Student Edition, this lab activity is best scheduled after covering the material in that section.

Setup

Before the lab activity begins, prepare one or more containers of crickets. A small aquarium or medium-to-large bucket will work well. If you have a small class, you could use just one container of crickets and have your class work on the activity together. For a larger class you might find it helpful to divide students into two to four large lab groups, each group with its own bucket.

Counting Crickets

For students to calculate the percent error of their estimates, they will need to know how many crickets are in the terrarium. Live crickets can be hard to count, but there are several ways to make this easier.

You could simply give students the number on the package the crickets came in. While this is almost certainly an estimate, it should be close enough to allow students to calculate the percent error of their estimates.

Another option is to buy a cricket counter. These scoop-like devices allow you to trap a large number of crickets and dispense them one at a time, making counting them much easier. Some of these devices can also be used to gather the initial crickets to be marked.

Finally, you could also count the crickets by hand, but this is not recommended as it could be a long, frustrating process. If you do decide to use this method, enlist students' help. The exact number of crickets you choose to place in the terrarium is not important, but there should be enough of them that students cannot visually count them in short order.

✓ Analysis

Once the lab groups have gathered their raw data, you might have them complete the Analysis section individually or in pairs. They can compare their answers with other members of their group.

✓ Sample Calculations (Question 1)

We have left room for student calculations in the Lab Manual. The solution shown on this student page and those shown throughout this Teacher Lab Manual are based on the data collected while developing the lab activities. You may have a particular way for students to show their work. The worked-out solution shown here follows one standard method. Student answers will differ, but their setup should be similar.

ANALYSIS

One way to analyze data from mark and recapture is to use a formula called the *Lincoln-Petersen method*, named for two biologists who studied large populations of fish and birds. The Lincoln-Petersen method begins with the proportion

$$\frac{n}{N} = \frac{k}{K},$$

where n = the number of organisms that were originally marked, N = total population size, k = the number of marked organisms that are recaptured, and K = the total number of organisms that are captured. In other words, this proportion says that the ratio of marked organisms to the total population (n/N) is the same as the ratio of marked recaptured organisms to the total number of captured organisms (k/K).

When solved for N, the proportion takes the form below.

$$N = \frac{Kn}{k}$$

1. Apply the Lincoln-Petersen method to your data in Table 1. What is your estimate of the total number of crickets? Show your work below.

   Answers will vary.

   What we know: $n = 9$, $K = 10$, $k = 3$
   Unknown: N
   Write the formula and solve for the unknown.
   $$N = \frac{Kn}{k}$$
   Plug in the known values and evaluate.
   $$N = \frac{(10)(9)}{3}$$
   $$= \frac{90}{3}$$
   $$= 30$$

While the Lincoln-Petersen method tends to give accurate estimates for larger populations, it is less accurate for small populations such as the crickets in the terrarium. Douglas G. Chapman found that the method works much better if we add 1 to each factor.

First, the Lincoln-Petersen formula is set up as a proportion.

$$N = \frac{Kn}{k}$$

$$Nk = \left(\frac{Kn}{k}\right)k$$

$$Nk\left(\frac{1}{kn}\right) = Kn\left(\frac{1}{kn}\right)$$

$$\frac{N}{n} = \frac{K}{k}$$

$$\frac{n}{N} = \frac{k}{K}$$

Then, 1 is added to each factor.

$$\frac{n + 1}{N + 1} = \frac{k + 1}{K + 1}$$

Finally, the Chapman modification is solved for N.

$$N = \frac{(K + 1)(n + 1)}{(k + 1)} - 1$$

2. Apply the Chapman modification to your data in Table 1. What is your estimate of the total number of crickets? Show your work below.

Answers will vary.

What we know: $n = 9$, $K = 10$, $k = 3$
Unknown: N
Write the formula and solve for the unknown.

$$N = \frac{(K + 1)(n + 1)}{(k + 1)} - 1$$

Plug in the known values and evaluate.

$$N = \frac{(10 + 1)(9 + 1)}{3 + 1} - 1$$
$$= \frac{110}{4} - 1$$
$$= 28 - 1$$
$$= 27$$

3. For the Lincoln-Petersen method to be accurate, certain assumptions must be true. First, the population must be constant. What are some events that would make this assumption false?

Answers will vary. Students should mention that some animals may be born or die or may enter or leave the area.

4. Is it likely that this assumption is true in this experiment? Explain.

No crickets could enter the terrarium, and no crickets could escape—we hope! And since only five minutes passed between the time when the crickets were marked and the time of the second sample, it's unlikely that any crickets hatched or died during that brief period.

5. A second assumption is that the animals captured and marked are just as likely to be captured the second time as the animals that have not been marked. Suggest a way of marking crickets that would violate this assumption.

Answers will vary. An example would be cutting a segment of one of the cricket's back legs, hindering its movement and possibly making it more likely to be recaptured.

Since you have a captive population of crickets, you can calculate how close your estimates are to the actual population. A common way of determining the accuracy of an estimate is to calculate percent error using the formula below.

$$\% \ error = \left(\frac{|value_{estimated} - value_{actual}|}{value_{actual}} \right) 100\%$$

6. Calculate the percent error of your Lincoln-Petersen method estimate compared to the actual number of crickets provided by your teacher.

What we know: $value_{estimated} = 30$, $value_{actual} = 37$
Unknown: % error
Write the formula and solve for the unknown.

$$\% \ error = \left(\frac{|value_{estimated} - value_{actual}|}{value_{actual}} \right) 100\%$$

Plug in the known values and evaluate.

$$\% \ error = \left(\frac{|30 - 37|}{37} \right) 100\%$$

$$= \left(\frac{|-7|}{37} \right) 100\%$$

$$= \left(\frac{7}{37} \right) 100\%$$

$$= 18.9\%$$

7. Calculate the percent error of your Chapman modification estimate compared to the actual number of crickets.

What we know: $value_{estimated} = 27$, $value_{actual} = 37$
Unknown: % error
Write the formula and solve for the unknown.

$$\% \ error = \left(\frac{|value_{estimated} - value_{actual}|}{value_{actual}} \right) 100\%$$

Plug in the known values and evaluate.

$$\% \ error = \left(\frac{|27 - 37|}{37} \right) 100\%$$

$$= \left(\frac{|-10|}{37} \right) 100\%$$

$$= \left(\frac{10}{37} \right) 100\%$$

$$= 27.0\%$$

8. Which method was more accurate?

See margin for answer.

Question 8 Answer
Students should indicate that the method that yielded the answer with the lower percent error was more accurate.

GOING FURTHER

Now that you've finished using mark and recapture to estimate the number of crickets in your laboratory's terrarium, let's apply what you have learned to help your state fish and wildlife agency estimate the white-tailed deer population in a particular county.

The agency will use this fall's deer hunting season—when hunting of both bucks and does is allowed—as the recapture event by comparing the numbers of marked and unmarked deer killed. Propose a way to mark deer, keeping these requirements in mind.

» You must mark the deer in a way that will not make them more likely or less likely to be killed by hunters this fall.

» You must mark the deer in a way that will not spoil the meat. Hunters will not appreciate killing a deer only to find that they cannot eat the venison because the state agency adopted a poorly conceived sampling plan.

» You must mark the deer in such a way that the mark will not be lost or wear off in the time between now and the end of hunting season. (For simplicity's sake we will assume that the deer will not be shedding antlers or coats between now and then.)

» Finally, you must mark the deer in such a way that hunters can easily determine that the deer has been marked.

9. What is your plan?

Check to see that students' plans fit the criteria laid out above. Expect to see a wide range of ideas. Application of the principles taught in this lab activity are more important than real-world feasibility (e.g., cost).

✓ **Going Further**

Students will probably need the entire lab activity time to finish the procedure part of the activity, so consider assigning the Going Further section as homework.

Exercising Dominion

Exercising wise dominion over Earth, an integral part of the Creation Mandate, requires a thorough understanding of what exactly it is that we are trying to exercise dominion over. In the example of game management, wise use of a resource requires knowing how much of the resource is available. The method illustrated in Lab 3A is just one of many techniques that game managers use to estimate the numbers of certain kinds of animals whose exact numbers are, due to various reasons, difficult to know exactly. A Christian fish or game manager shouldn't be doing his job just to discern how much game is available for sportsmen; he should also honor his Creator by taking care to conserve a valuable part of the creation that God lovingly prepared as a habitation for us. Doing so shows thankfulness to God for His gift in addition to manifesting a godly concern for the generations to follow that will also make use of Earth's resources.

Question 12 Answer

While it is true that humans and their well-being are more important than animals, it does not follow that there is no place for good stewardship of animals. Inevitably, the stewardship of our natural resources will have a positive or negative effect on people as well. For example, mismanagement of animal populations can lead to a loss of food sources, significant negative changes to the environment, or an increase of pests.

10. The agency put your plan into action and marked 2000 deer at the beginning of the hunting season. Out of 6578 deer killed during this year's deer season, 459 of them were marked. Using the Lincoln-Petersen method, estimate the deer population in this county at the end of the hunting season.

What we know: $n = 2000$, $K = 6578$, $k = 459$
Unknown: N
Write the formula and solve for the unknown.

$$N = \frac{Kn}{k}$$

Plug in the known values and evaluate.

$$N = \frac{(6578)(2000)}{459}$$
$$= \frac{13\,156\,000}{459}$$
$$= 28\,662$$

11. It would be difficult to ensure that no animals were added to or lost from the population during the length of this experiment. List two ways that individual deer could have joined or left the population.

Possibilities include deer crossing the county line, animals dying from natural causes, animals being killed by predators, and animals being born (since the time of year when the marking was done is not specified).

12. Someone tells you, "Counting fish and deer is a waste of time. We should be spending our money on more important things, like people who are sick or homeless." How would you respond?

See margin for answer.

TABLE 1

	ESTIMATE
n	9
K	10
k	3

3B LAB

Must You Be So Competitive?

Inquiring into Growth Rate

If you were to walk through an older forest, you would probably notice that there were few young trees. Plants need sunlight to live and grow, and relatively little sunlight makes its way through the canopy formed by mature trees. Young trees tend to grow slowly in the forest's shady depths.

Sunlight is just one of the resources in an ecosystem, and the availability of resources governs how fast a population can grow. When conditions are ideal, the number of organisms in a population can soar. Conversely, if even one factor is limited, a population's growth may stagnate or even decline. In this lab activity you will examine how differences in an environmental factor affect plant growth. Duckweed is a small plant that usually reproduces asexually. This makes it an excellent test subject for population growth since you won't have to wait for it to flower or bother with pollinating it.

How does a plant's environment affect its growth rate?

QUESTION
- How do differences in environmental factors affect a population's growth rate?

Equipment
duckweed

PROCEDURE

Planning/Writing Scientific Questions

A Together with your team, decide on an environmental factor that you will test in relation to duckweed growth. Possible factors include but are not limited to the following.

  » duration, intensity, or color of light exposure

  » amount or type of fertilizer

  » moving water versus still water

B Brainstorm with your lab group about how you can test the selected environmental factor while keeping other factors constant.

C Write specific questions related to the environmental factor that you could answer by collecting data.

LAB 3B OBJECTIVES

- Design and conduct an experiment to evaluate the effect of a selected factor on the growth rates of plants.
- Evaluate the experimental design on the basis of collected data.

✓ Scheduling

Though related to the material in Section 3.2 of the Student Edition, this is a multi-week activity, so it will need to be started well in advance of covering the material in that section. After the initial setup has been completed, subsequent gathering and recording of data should not require much class time. Plan to do the setup roughly four to five weeks in advance. The Analysis section may be assigned as homework.

A Teacher Guide for this activity can be found on page 26a.

🧪 Equipment Notes

Duckweed can be gathered from a local pond, but doing so can also bring unwanted organisms into the laboratory. It would be best to obtain your duckweed from a science equipment supplier.

✓ Communicating Results

This activity provides an excellent opportunity to practice writing formal lab reports. See Appendix F.

Designing Scientific Investigations

D Write procedures to collect data that will allow you to answer the questions that you wrote in Step C.

E Have your teacher approve your procedures.

Conducting Scientific Investigations

F Collect data according to the procedures that you have written, then answer the questions that you wrote.

Developing Models

G Use your data to develop models of how changes in the environmental factor affect the growth of duckweed.

H Propose at least one way that you would change your experimental procedures to obtain better results.

Scientific Argumentation

I Explain what you know about the relationship between changes in the environmental factor that you have chosen and its effect on duckweed growth. Support your claims with evidence from the data that you collected.

3B LAB *Teacher Guide*
Must You Be So Competitive?

An inquiry lab activity allows students a degree of freedom, but the teacher must guide the process to achieve the desired educational outcomes. As teachers start using inquiry lab activities, they often struggle with this process of guiding students. The best method is to think through the goals that you want students to achieve and then use questions to guide them to reach those goals. You want to balance students' freedom to investigate the question in the manner they choose with the need to meet your educational goals.

Inquiry lab activities provide you with a degree of flexibility in deciding how the activity should be carried out. Students will need to collaborate to design a workable test; you may choose to have students do this within their group, with each group deciding on its own question and test, or you can have a class discussion resulting in a single question and test. If the latter, allow students to generate the ideas. Prompt them with questions to guide the discussion in directions beneficial to the activity. You may choose to have students do part of this activity as homework. For example, students may do their research as homework, use class time to collaborate on a test design, conduct their tests at home, and finally present their findings in class.

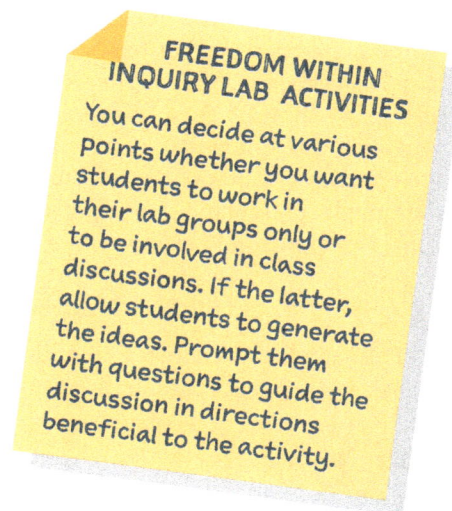

> **FREEDOM WITHIN INQUIRY LAB ACTIVITIES**
>
> You can decide at various points whether you want students to work in their lab groups only or to be involved in class discussions. If the latter, allow students to generate the ideas. Prompt them with questions to guide the discussion in directions beneficial to the activity.

PROCEDURE

Planning/Writing Scientific Questions

Each lab group selects an environmental factor to investigate (e.g., light, nutrients). To simplify this activity, you may choose to preselect the factor to be tested. The groups will then brainstorm ways in which they can test the growth rate of duckweed in response to variations in the selected factor. They may think of the following:

- duration of light exposure
- intensity of light exposure
- color of light exposure
- incidence of light exposure (the angle of incoming rays)
- type of fertilizer added to the system
- amount of fertilizer added to the system
- moving water versus still water
- temperature of the water

Other tests are, of course, possible.

Each lab group should write a series of questions related to the environmental factor that they have chosen, which they could answer by collecting data.

Designing Scientific Investigations

Once the groups have written their questions, they need to develop specific procedures to answer them. This may be a new task for some students, so you may want to do part of this task as a class discussion or review their procedures prior to their starting to collect data.

Help students think through how they will alter only one variable at a time. For example, if students are testing the effect of light duration, they must be able to regulate daily cycles of light and dark without affecting other conditions.

Conducting Scientific Investigations

Once the groups have good procedures, allow them to collect data.

Developing Models

Students should analyze their data and identify any trends. Encourage them to work with other researchers who may be examining the same environmental factor so that they can pool data.

An excellent way to analyze data for an experiment of this type is to create a graph.

Students should always assess their experimental procedures to determine whether they could have done something different to obtain better results.

3B LAB *Teacher Guide*

Scientific Argumentation

Students should document what they can conclude about the relationship between the environmental factor they have chosen to test and its effects on population growth. Writing a formal lab report would be an excellent way to do this, but you may opt to have students present their findings in another format, such as an oral report or multimedia presentation. As always, you should expect students to support their claims with evidence from their data.

Sample Procedure

This sample procedure tests the relationship between growth and nutrient concentration.

Equipment

graduated cylinder, 100 mL

fluorescent grow light

lighting timer

plastic cups, 16 oz (16)

disposable pipette

pond water

distilled water

Miracle Grow liquid plant food

duckweed (*Lemna aquatica*)

A Label four groups of four cups each as Groups 1–4.

B Fill each Group 1 cup with 350 mL of pond water. This is the control group.

C Fill each of the remaining cups with 350 mL of distilled water. To each Group 2 cup add 1 drop of liquid plant food. Add 3 drops of liquid plant food to each Group 3 cup and 5 drops to each Group 4 cup.

D Add three duckweed plants to each of the cups and place them under the grow light.

E Follow the instructions on the lighting timer and set it for 10 hours of exposure daily. After the setup is complete, cups may be periodically topped off with distilled water. (This will add more water but not nutrients.)

F After one week, count the number of duckweed plants in each cup and record the data.

G Continue monitoring the duckweed growth for four additional weeks.

This activity gives wide latitude to students to choose for themselves what environmental factor to test. What might they observe depending on the factor that they choose to examine?

- *Light Color.* Like other green plants, duckweed makes its own food through photosynthesis. Chlorophyll—the pigment that makes photosynthesis possible—absorbs light energy mainly in the blue and red portions of the electromagnetic spectrum. Students who choose to test the effect of light color on duckweed growth should see greater growth in blue or red light than in other colors.

- *Light Intensity and Duration.* As might be expected, students should see an increase in duckweed growth with higher light intensities and durations, though the increase in growth rate may decline at higher values.

- *Temperature.* The optimal temperature range for duckweed is between 20 °C and 30 °C. Less growth can be expected in temperatures outside this range, especially as temperature drops below 17 °C or rises beyond 35 °C.

- *pH.* Duckweed can tolerate pH values between 3.0 and 10.0, but optimal growth is seen between values of 5.0 and 7.0.

- *Nutrients.* Duckweed does not require a lot of nutrients for optimized growth—a mere ¼ tsp of Miracle Grow 24-8-16 liquid plant food in 5 gal of water will suffice. To help prevent students from providing a growth medium that is too "hot," you may choose to mix up some stock nutrient solution in advance. Generally, students can expect to see a direct relationship between nutrient concentration and growth rate. But despite its prevalence in ponds, duckweed does not tolerate elevated nutrient levels well, and students may observe less growth in high nutrient concentrations compared to low or medium concentrations.

No matter what factor is being tested, growth of any duckweed group should taper off as nutrients are depleted. Students were not asked to address carrying capacity or limiting factors, but if one or more groups produce evidence that the growth of their duckweed populations tapered off after an initial increase, you may want to discuss with the class why that happened.

4A LAB

Forest or Farm?

A Mathematical Model of Biodiversity

Imagine a rainforest. Towering trees draped in a tangle of vines, dotted with exotic orchids and bromeliads. Brilliantly colored macaws call raucously from the canopy while jewellike hummingbirds flit from flower to flower in search of nectar. Tapirs softly tread well-worn paths through the forest, keeping a wary eye out for the occasional jaguar. Caimans drift silently in the rivers, scattering schools of shimmering fish. And everywhere there is an astonishing variety of insects, from ants to beetles to butterflies.

Now imagine a cornfield—acres and acres of corn planted in neat rows. Thanks to herbicides and insecticides, the field is free of weeds and relatively free of insects. It's mainly corn—and not much else! That's the way the farmer wants it.

One of these two ecosystems has many different kinds of organisms living together in the same area. The other ecosystem consists almost entirely of only one kind of organism. Scientists say that the rainforest has a high biodiversity, while the biodiversity of the cornfield is low. Biodiversity is a measurement of the variety of life in a particular ecosystem. It's pretty easy to tell just by looking that the biodiversity of a cornfield is low, but what about other kinds of ecosystems? What about a forest, a meadow, or even a vacant lot near you? Assessing the biodiversity of an ecosystem gives biologists a valuable piece of information that can be used to help manage the system wisely. In this lab activity you will measure the biodiversity of an ecosystem by collecting and analyzing data with the help of a special formula from the biologist's mathematical toolbox.

How do scientists decide whether an ecosystem is biologically diverse?

Equipment
field tape
field notebook
camera

PROCEDURE

A Select a site for sampling, such as a forest, field, or other ecosystem.

1. Give a brief description of the site you've chosen to sample, including its location and the date of the sample.

   Answers will vary. *Example:* We sampled the abandoned field on the east side of Anytown High School on September 15, 2024.

QUESTIONS
- How do scientists measure biodiversity?
- How do scientists obtain representative samples from very large systems?

LAB 4A OBJECTIVES

- Carry out a field transect.
- Analyze the diversity of an ecosystem using Simpson's Diversity Index.
- Interpret the meaning of the Simpson's Diversity Index value for a particular ecosystem.

Introduction to Lab 4A

Several of the lab activities in *Biology* Lab Manual 6th Edition, including this activity, were written to introduce students to actual methods used by scientists in the field. For more information on transects, see page vii in *Biology* Student Edition 6th Edition.

Equipment Notes

A field tape is the best equipment option for an exercise like this since it is generally longer than a tape measure and will lie loose. A rope marked in suitable increments is an inexpensive option. The increments used will depend on the size of the sampling area and the number of samples needed.

You may choose to have students use a field notebook to collect data if they have been instructed on how to do so. Otherwise, they may collect their field data in Table 2. Instruct them to copy additional rows for Table 2, if necessary, onto a separate sheet of paper and attach it to their lab activity handout.

In addition to verbal descriptions of sites, photographs are useful for providing additional information about a site. You may choose to direct students to provide such photos by placing them on a separate sheet and attaching it to their lab activity handout.

✔ Sampling Considerations

Careful thought will need to be given to choosing a sampling site. Sampling a lawn would definitely be easy, and Simpson's Diversity Index can certainly describe one, but a lawn might not be the most interesting habitat for students to sample. Try to find a habitat that appears to have at least a moderate variety of organisms.

If your students are sampling a large area or using a long transect line, you might want to decide ahead of time how large the sample size needs to be. Many biological sampling protocols sample as little as 5%–10% of the area or population being surveyed. If you have time, it might be profitable to have students assess the same site using different sample sizes to see whether any significant difference is produced between the values of Simpson's Diversity Index for each sample.

⚠ Avoid the Itch!

You might want to make sure that the area you choose to sample is free of plants known to cause contact dermatitis, such as stinging nettles or poison oak.

Question 6 Answer
A 100% sample will take longer to do, especially if the transect touches a large number of organisms. A 100% sample might also produce an overly large data set, making analysis unnecessarily difficult or tedious.

Question 7 Answer
A smaller sample size may not include every kind of organism along the transect and may not produce a data set that is representative of the transect as a whole.

2. Why do you think it is important, especially for field studies, to include the date of your sample?

    In many areas, biodiversity is seasonally variable.

    _____

3. Do you think the biodiversity of your sampling site is high, low, or somewhere in between? Write your answer in the form of a hypothesis.

    Answers will vary. *Example:* The biodiversity of this forest is

    high.

B Lay your field tape in as straight a line as possible across your sampling area. This technique is known as *conducting a transect*—a method for collecting a subset of data from a much larger potential sample.

C Decide on the number of observations that you will make along your transect (each organism you tally is one observation). If your sampling area is small, you might choose to tally every organism that your transect touches. This is sometimes referred to as 100% sampling.

4. What is the advantage of a 100% sample?

    A 100% sample will include every kind of organism touched by

    the transect and will give a definitive number of each kind of

    organism.

5. What is a possible disadvantage of collecting a 100% sample?

    A 100% sample will take longer to do, especially if the transect

    touches a large number of organisms. A 100% sample might also

    produce an overly large data set, making analysis unnecessarily

    difficult or tedious.

D If the sampling area is large, an alternative to a 100% sample is to conduct a smaller sample. For example, if your transect is 25 m long, you might decide to make an observation once every meter or half-meter along the length of the line.

6. What is the advantage of a smaller sample size (less than 100%)?

    See margin for answer.

    _____

    _____

7. What is a possible disadvantage of a smaller sample size?

    See margin for answer.

    _____

    _____

    _____

    _____

E Starting at 0 m on your transect line, begin listing and tallying the kinds of organisms that your transect line touches. Count only those organisms that the line actually touches. Use Table 2 to list each new kind of organism that you encounter along your transect. Tally the number of each kind that you encounter. If you need additional rows, copy Table 2 onto a separate sheet of paper and attach it to your lab report.

ANALYSIS

The mathematical tool referred to in the introduction is a measure of biodiversity known as *Simpson's Index*. The formula for the index is

$$D = \frac{\sum n(n-1)}{N(N-1)},$$

where D is the index value, the Greek letter sigma (Σ) means "sum of," n is the number of individuals of any one species in the sample, and N is the total number of organisms in the sample. The result is a decimal number whose value is between 0 and 1.

F Calculate $n-1$ and the product $n(n-1)$ for each observed species and record in Table 2.

G Add up all the values for $n(n-1)$ calculated in Step F. This is $\Sigma n(n-1)$. Record this value in Table 2.

H Add up all the values for n; this sum is N. Record the value for N in Table 2.

I Multiply the total number of observed organisms (N) times that number minus one ($N-1$). This is $N(N-1)$. Record this value in Table 2.

J Calculate the value of Simpson's Index for your sample and record this value in Table 2.

Answers will vary. *For the given sample data:*

What we know: $\Sigma n(n-1) = 14$, $N(N-1) = 90$
Unknown: D
Write the formula and solve for the unknown.

$$D = \frac{\sum n(n-1)}{N(N-1)}$$

Plug in the known values and evaluate.

$$D = \frac{14}{90}$$
$$= 0.16$$

K Simpson's Diversity Index, the actual number that you need to assess the diversity of your sample, is slightly different than Simpson's Index. To obtain the value of Simpson's Diversity Index for your sample, subtract the value that you calculated in Step J from 1 (i.e., $1 - D$) and record this value in Table 2.

✔ Identifying Organisms

Students do not need to be able to identify each species of organism encountered, but they do need to keep track of each different kind of organism in their sample. For example, a student may not recognize a certain plant as a dandelion; they may describe it in their data as a "yellow flower." Subsequently, though, they need to be able to distinguish that particular yellow flower and tally it separately from any other kind of yellow flower that they encounter.

⁉ Managing the Math

This lab activity requires students to do some mathematical calculations that should be within the abilities of most tenth-grade students. But if students are having trouble with the math, you may decide to direct them to an online Simpson's Diversity Index calculator. It's also interesting to compare the results obtained from such a calculator with results calculated by hand. You should find the two results to be in very close agreement.

Advanced students may notice that it is not absolutely necessary to calculate the value for Simpson's Diversity Index since the Simpson's Index by itself can be used to evaluate diversity. But the index will be counterintuitive to many students since a lower Simpson's Index value corresponds to higher biodiversity (an inverse relationship).

It is difficult to overstate the importance of mathematics as a tool in biology. Some students may harbor a misconception that biology is a "soft" science compared to, for example, physics or chemistry. This perception may be reinforced by the relative lack of discussion of math given in the Student Edition. However, many biological concepts are difficult, if not impossible, to adequately model without the aid of some very advanced math. Students with an interest in biology should be encouraged to take as much math as possible during their high-school years.

CONCLUSIONS

So now you have a measure of the biodiversity of your sample, but what does the measure actually mean? To answer that question, let's think about the cornfield again. Imagine a cornfield with exactly one hundred corn plants and one dandelion in it and nothing else, the almost-perfect cornfield—hardly any weeds and no pests. That's definitely a lack of diversity! Plugging the values for the entire cornfield into the formula for Simpson's Index yields a Simpson's Index value of 0.98. See Table 1 and the calculation below.

TABLE 1 Cornfield Data

Species	Number (n)	$n - 1$	$n(n - 1)$
corn	100	99	9900
dandelion	1	0	0
	$N = 101$		$\Sigma\, n(n - 1) = 9900$
	$N(N - 1) = 10\,100$		

$$D = \frac{9900}{10\,100}$$

$$= 0.98$$

Subtracting 0.98 from 1 gives us a Simpson's Diversity Index value of 0.02.

8. What does the Simpson's Diversity Index value for the imaginary cornfield tell you about the relationship between biodiversity and the values produced by the index calculation?

   Since the value for the cornfield is very low, a low value for

   Simpson's Diversity Index indicates a low biodiversity.

9. What does the Simpson's Diversity Index value that you calculated for your sample tell you about the biodiversity of the ecosystem that you sampled?

   Answers will vary. A low value indicates a low biodiversity, while

   increasingly higher values indicate greater biodiversity.

10. Does the Simpson's Diversity Index value that you calculated for your sample support your hypothesis from Question 3? Explain.

    Answers will vary. Students' answers should be consistent with

    the analysis of their data.

11. Discuss some reasons why your Simpson's Diversity Index value may not be reflective of the actual biodiversity in the ecosystem that you sampled.

Answers will vary. Students may suggest that the transect necessarily counts only sessile life in the sample and that precision might be affected by the sample size, especially for transects that do not count 100% of the organisms touched.

12. How do mathematical tools such as Simpson's Diversity Index help biologists exercise wise dominion over God's creation?

Without mathematical tools such as Simpson's Diversity Index, biologists do not have the means to assess whether a given habitat is truly diverse. They will also be unable to determine whether management policies actually improve biodiversity.

✝ Biodiversity and Dominion

As stated in a previous lab activity, it is difficult to exercise wise dominion over the earth without a basic understanding of what is being overseen. Biodiversity is a topic frequently seen in popular science literature. Scientists—and Christians—need tools to evaluate just how diverse an ecosystem is or is not. Believers shouldn't shy away from tools such as Simpson's Diversity Index—the tools themselves are worldview neutral. It is up to us to use the information gleaned from such tools in a manner that honors God and serves people.

Sample Data

The data shown in Table 2 is typical. Students' results may be different.

TABLE 2 Transect Data

SPECIES	NUMBER (n)	$n - 1$	$n(n - 1)$
pine	1	0	0
sweet gum	3	2	6
sedge	3	2	6
pokeweed	1	0	0
dandelion	2	1	2
	$N = \underline{10}$		$\Sigma n(n-1) = \underline{14}$
	$N(N-1) = \underline{90}$		
	$D = \underline{0.16}$		
	$1 - D \approx \underline{0.84}$		

4B LAB

Hale Hardwoods or Sickly Cedars?

Monitoring Forest Health

Think of the last time you had the flu. You probably had some obvious and visible symptoms, such as a fever, runny nose, and chills. How can a biologist tell whether an ecosystem is sick and in need of some attention? The signs that an ecosystem is in trouble are not always obvious to an untrained observer. A forest, for example, that looks perfectly healthy to you may, in fact, be very "sick." But just as your family doctor can evaluate your fever and sniffles, scientists have ways to evaluate the health of ecosystems. They can determine whether an ecosystem is in prime condition and, if necessary, prescribe an appropriate plan of treatment to get their "patient" on the road to recovery.

When you go to the doctor's office, you don't get a prescription for medicine as soon as you set foot in the door. The doctor or an assistant first collects some basic information about you, such as your age, height, weight, temperature, pulse, and blood pressure. Biologists do the same thing when assessing an ecosystem—they first need some basic information. In this lab activity you'll try your hand at collecting an ecosystem's vital signs.

How can forest professionals give a forest a checkup?

Equipment
field tape
flagging tape
calculator
rangefinder

QUESTIONS
- How can scientists assess the health of a forest?
- What can an ecologist learn about a forest using what he knows about the size and number of trees in a sample?
- Why is it important to manage forest resources?

PROCEDURE

Establishing a Plot

Forest ecologists, dendrologists (scientists who study trees), and foresters (scientists who manage forest resources) often need to collect forest data. Forests, of course, have many trees in them, sometimes numbering into the millions. Assessing every tree would be impossible! What scientists need is a *sample* of the trees in the forest. Scientists call this a *plot*.

For this activity you will need a forest—or at least a place with some trees! A park or wood lot will do. The plot is a circular area representing a fraction of one acre. The size of that fraction is determined by the radius of the plot circle.

- Demarcate a forest plot.
- Measure tree circumference.
- Derive characteristic data related to forest health.
- Infer forest characteristics from indirect measurements.

✅ Introduction to Lab 4B

In actual field science scientists must often make conclusions about systems or parts of systems that are difficult or impossible to measure directly. In this exercise students will have an opportunity to make similar conclusions on the basis of measurements of just a few features of a forest ecosystem.

Note that this lab activity is written for US customary units rather than metric units. American foresters still typically work in US customary units.

✅ Prelab Discussion

Review Chapter 3, if necessary. Remind students that the various parts of an ecosystem are interrelated and that processes or changes in one aspect of the system can have profound effects on other parts of the system.

✅ No Forest? No Problem!

This lab activity guides students through the collection of data in a forest. Of course, not all students will have access to a forest to assess. Keep in mind that all manner of ecosystems are assessed in ways similar to those presented in this activity, and with some modification parts of the activity can be adapted to other kinds of ecosystems.

Alternatively, since this lab activity requires no measurements of tree *height*, you can create an artificial "forest" made of various cylindrical or spherical objects, such as waste baskets, coffee cans, or playground balls, where each object represents the trunk of a tree. Measure the diameter of each "tree" to obtain its diameter at breast height (DBH). Include enough faux trees so that students can demarcate a plot that includes some trees and excludes others. Include some small "trees," such as tennis balls or small cans, so that students will have to measure and exclude some trees that don't meet the DBH requirement.

Plot Notes

There are many other ways to visually mark out a plot other than those outlined in this activity. It can be helpful to temporarily mark (e.g., with flagging) which trees are to be included in students' plots.

You can save time by demarcating plots beforehand.

Students could, of course, use circles with radii not listed in the chart, but then they would have to calculate what fraction of an acre that circle is. Best to stick with the listed sizes!

Rangefinders

The presence of trees, of course, makes it difficult for partners to walk a circle while holding both ends of the tape. An optical or laser rangefinder would be helpful in this instance. If you have access to one, be sure to instruct students in its proper use, especially if using a laser rangefinder, which has a beam that could potentially cause eye damage.

DBH Shortcut

A shortcut for Step D is to have students measure 4.5 ft above ground on themselves and then use that as a reference when assessing trees.

A Working with a partner, choose a plot large enough to include at least a few trees but not so large that you won't have enough time to assess all the trees within your plot. Use the chart below to help determine how large a plot to establish. Record your plot size in Table 1.

TABLE 1

PLOT SIZE (ACRES)	RADIUS (FT)
1/5	52.7
1/10	37.2
1/20	26.3
1/30	21.5
1/40	18.6
1/50	16.7

B Once you have determined how large a circle to make, use a field tape to mark the boundary of your plot. Have your partner hold one end of the field tape at the center of the circle while you measure out the appropriate radius. Mark the end of the radius with a piece of flagging tape.

C Repeat Step B enough times to sufficiently establish the boundary of your circle. When you are done, you should have an easily seen border between the trees inside the plot and those outside.

A 1/5 acre plot

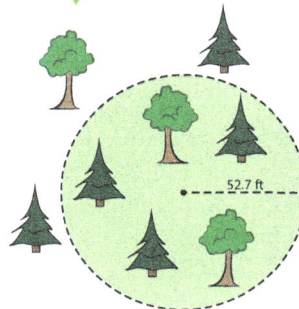

Diameter at Breast Height (DBH)

It may sound silly, but we first have to define the word *tree*. Biologically speaking, every sapling in the forest is a tree, but young trees lack many of the features of mature trees. They don't have crowns, trunks, or very much woody tissue. Foresters define a tree as a woody plant with a trunk at least 3 in. in diameter at breast height, or DBH, where breast height is defined as 4.5 ft above ground.

D Use your field tape to measure 4.5 ft above ground on the trunk of one of the trees in your plot. Wrap the field tape around the trunk at that level. The point at which the 0 in. mark on the tape aligns with the wrapped tape indicates the circumference of the tree.

E Use a calculator to divide the measurement from Step D by 3.14. Since the circumference of the tree is equal to π times the diameter ($C = \pi d$), the result is the tree's DBH.

F If the DBH is 3 in. or greater, then that tree must be included in your sample. Record its DBH in Table 1.

G Repeat Steps D–F for each tree in your plot. Do not record data for any tree whose DBH is less than 3 in. Add more rows if needed.

Stem Count

One indication of forest health is *stem count*, or the number of trees (stems) per acre. To find the stem count represented by your sample, multiply the number of trees (as defined by DBH) in the sample by the number of plots it would take to make one acre. For example, if you have five trees in a 1/20 acre plot, then multiply 5 by 20 (5 trees per plot × 20 plots to make 1 acre), giving 100 stems per acre.

H Calculate the stem count per acre for your forest on the basis of your plot data and record this value in Table 1.

1. Why do you think stems per acre is an important indicator of forest health?

   See margin for answer.
   _____
   _____
   _____
   _____

2. Do you think that different species of trees have the same ideal number of stems per acre? Explain.

   Different species have different ideal numbers of stems per acre, depending on the size of mature trees and their needs for sunlight, water, and nutrients.
   _____

Basal Area

Once you have determined that a tree in your plot is, in fact, a tree by a forester's definition, you next need to determine its basal area—a measurement of how much ground surface area the trunk of the tree covers. A shortcut for finding the basal area in square feet of a tree whose diameter is measured in inches is to take the square of the number of inches in its diameter and multiply the result by 0.005 454. For example, the calculation for a tree with a 6 in. diameter would be

$$(6 \text{ in.})^2(0.005\ 454\ \text{ft}^2/\text{in.}^2) = 0.2\ \text{ft}^2.$$

Thus, a tree with a diameter of 6 in. has a basal area of $0.2\ \text{ft}^2$.

I Calculate the basal area for each tree in your sample and record the results in Table 1.

3. What do you think the basal area for an individual tree can tell a dendrologist about the age and maturity of that tree?

   Since the diameter of a tree is directly related to the age of the tree, higher basal values correspond to older, more mature trees.

Basal Area per Acre

Using the basal areas of the trees in your plot, you now need to determine how much basal area there is per acre in the part of the forest that you are assessing.

J To calculate your forest's basal area per acre, start by multiplying the basal area for each tree in your plot by the value of the denominator of your plot size. For example, the basal areas for trees in a 1/20 acre plot would each be multiplied by 20. Do this for each tree in your plot and record the results in Table 1.

K Add up all the results from Step J to get the calculated value for the total basal area per acre for your forest and record this total in Table 1.

4. What sorts of information about forest health can a forest ecologist learn from basal area per acre? (*Hint*: Think about what might be indicated by a small basal area versus a large one.)

   See margin for answer.
   _____
   _____
   _____
   _____
   _____
   _____
   _____
   _____
   _____
   _____
   _____

5. Why would a forester need to know *both* the stem count per acre and basal area per acre for a particular forest?

   Stem count does not indicate how much area is covered by trees; there might be many stems that cover only a small area (e.g., young trees). Conversely, basal area does not indicate how many trees are present; a high basal area and low stem count would indicate the presence of a few very mature trees.

Question 1 Answer

Stems per acre is an indicator of the density of the tree population in a given area. The higher the density, the greater the competition for resources. High tree density also means that trees are closer together, putting them at greater risk of damage or death due to insect pests, disease, and fire.

⁉️ Making Inferences

Inferring is a higher-order thought process that might be difficult for some students to grasp. Since there are many inferences that can be made from a measurement of basal area per acre, don't be afraid to "expend" a couple of them in the cause of coaching students how to get from measured quantity *A* to inferred conclusion *B*. After one or two examples they will probably come up with suggestions of their own.

Question 4 Answer

Basal area per acre is an indirect measure of competition for resources: the higher the basal area, the greater the demand for resources. Higher basal areas indicate a greater volume of wood per acre and thus may be an indication of greater forest age and maturity. Higher basal areas correlate to greater shade coverage and thus to lower levels of shade-tolerant understory trees, shrubs, and groundcovers. Higher basal areas may indicate overly dense forests that might benefit from the selective removal of some trees. By indicating the level of forest maturity, basal area can also indirectly measure the amount of habitat available to species that prefer stands of a particular age.

⁉️ Managing the Math

If students are curious about how the formula for basal area is derived, show them the following derivation. We start with the area of a circle.

$$A = \pi r^2$$

Since we are measuring diameter in inches, we would have to convert square inches to square feet for basal area. Since

$$1\ \text{ft}^2 = 144\ \text{in.}^2,$$

our formula becomes

$$A = \pi r^2/144.$$

Since we are using diameter and not radius, and the radius is half the diameter,

$$A = \frac{\pi(d/2)^2}{144}$$
$$= \frac{\pi d^2}{(4)(144)}$$
$$= \frac{d^2}{183.346} \text{ or } 0.005\ 454d^2.$$

To satisfy them that the given conversion factor works, show them that the cross-sectional area of a 6 in. diameter tree ($\pi \times 3^2$) divided by 144 produces the same result of $0.2\ \text{ft}^2$.

How Much Sampling Is Enough?

A useful point of discussion would be to talk about sampling protocols. If one sample is insufficient to produce reliable data and sampling an entire system is impossible, the obvious question then is, how many samples is enough? It is sufficient for students at this age to simply be aware that there are such questions when it comes to sampling and that they will learn about issues like sample size and reliability in a course on statistics.

Question 8 Answer
Answers will vary. Some examples of possible data include percent canopy coverage, tree height, tree volume, crown height, crown width, tree spacing, stand age composition, stand species composition, soil type, soil moisture, soil drainage, slope measurement, climate data, fire history, presence/absence of diseases and pests. Other answers are possible.

CONCLUSIONS

6. On the basis of the stem count and basal area per acre that you calculated for your forest, what conclusions can you make about the age and maturity of the trees in your forest?

   Answers will vary. Students' conclusions should be consistent with their data in Table 1.

7. What aspects of this lab activity might indicate that your conclusions in Question 6 are unlikely to be very reliable? What could you do to improve the reliability of your conclusions?

   This activity extrapolates data from one plot, an extremely small sample size and one that may not be very representative of the forest as a whole. More reliable conclusions could be made by increasing the number of plots.

8. What other kinds of data could a forest ecologist collect? Describe what that data might say about the health of a forest.

   See margin for answer.

GOING FURTHER

Ponderosa pine is a large conifer (cone-bearing tree) native to the American West. Ponderosa pine seeds grow best on bare soil and in full sunlight. The life cycle of these trees is very closely bound to the periodic wildfires that regularly occur in their habitats. Fires clear the ground of competing trees and shrubs, return nutrients to the soil, and cause ponderosa pine seed cones to open and release their seeds.

9. Would stands of ponderosa pine tend to be more evenly aged (being trees of similar age) or unevenly aged (being trees of widely varying age)? Explain.

   Since ponderosa pines require full sun to grow best and heat to release seeds, the trees in a particular stand would tend to be the same age, corresponding to the last wildfire event.

10. A particular stand of ponderosa pines is found to have a low stem count and high basal density. What can you deduce about the amount of shade in this stand? Would this be a good place for ponderosa seedlings to grow?

    The trees in this stand are most likely tall, mature trees that create a lot of shade. This would not be an ideal situation for ponderosa seedlings, which need full sunlight for best growth.

11. Other than timber products, such as lumber, what other vital goods and services do forests provide?

Students may include fuel (e.g., firewood), food (e.g., berries, pine nuts, mushrooms, game), clean water, habitat for wildlife, grazing, mining, recreation, providing oxygen, and storing carbon.

12. How does understanding forest health fit into a Christian worldview?

A proper understanding of forest health that is based on quantifiable measures helps forestry professionals maintain forest resources and maximize their benefits for people both now and in generations to come. This correlates with God's command that people manage the earth. Wise use of resources demonstrates that we are committed to ensuring that those resources are available to share with others.

A stand of ponderosa pines

Forestry Careers

Students may not think of forestry when queried about jobs in science, but there are many professions that require a degree in or strong knowledge of forestry science. Students interested in forestry can search online for forestry careers to find information on the jobs that are out there and the kind of education required for them.

Forestry and Wise Dominion

Forestry is another example of the need for a basic understanding of resources in order to exercise wise dominion over them.

Sample Data

The data shown in Table 1 is typical. Students' results may be different.

TABLE 1

PLOT SIZE (ACRES):			1/5
Tree	DBH (in.)	Tree Basal Area (ft²)	Basal Area per Acre (ft²/acre)
1	19.7	2.12	10.60
2	9.2	0.46	2.30
3	6.4	0.22	1.10
4	15.3	1.28	6.40
5	22.6	2.79	13.95
6	18.5	1.87	9.35
7	6.7	0.24	1.20
8	15.0	1.23	6.15
9	28.0	4.28	21.40
10	14.0	1.07	5.35
11	14.3	1.12	5.60
12	10.2	0.57	2.85
Stem Count/Acre	60	Total Basal Area/Acre	86.25

5A LAB

Dwell on the Cell

Basic Cytology

Imagine that you are a biology student in the year 1600. The microscope has not yet been invented, and no one, including the greatest scientific minds of the day, has a clue about the existence of an entirely unseen microscopic world all around them—and on them, even in them! Like the children in *The Chronicles of Narnia* who stepped through the wardrobe, the first microscopists to peer through their primitive, homemade lenses would encounter a world that defied imagination.

In Lab 1A you learned how to use a microscope. In this lab activity you will learn an additional microscopy technique: the making of a wet mount—a temporary microscope slide in which the specimen is mounted in water or some other fluid. After you have read about how to prepare a wet mount, you will practice the technique by doing what Robert Hooke did to observe a slice of cork. You will then observe some typical plant and animal cells, in the form of onion epidermal cells and human cheek cells.

What do the parts in plant and animal cells look like?

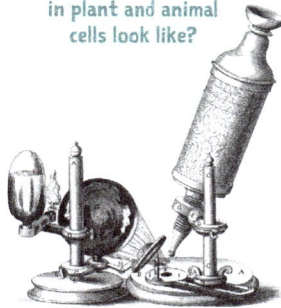

Equipment

microscope	pipette
dissection kit	onion
cork stopper	methylene blue
hexagonal metal nut, large	paper towels
razor blade, single-edged	flat toothpick
glass slides (3)	laboratory apron
cover slips (3)	goggles

QUESTIONS

- What did Robert Hooke see when he looked at cork through a microscope?
- What does a typical plant cell look like?
- What does a typical animal cell look like?

PROCEDURE

Observing Cork Cells

Over three hundred years ago Robert Hooke discovered that certain plant tissues are made up of what he called *cells*. To get a proper perspective of cytology, you will repeat his experiment. To see cork cells well, you must use a very thin slice of cork, only one to two cells thick. If suitable slices of cork are provided for you, skip Steps A and B.

A Take a small cork stopper (piece of cork) and insert it into a hexagonal metal nut. Twist it carefully so that the flat surface of the cork does not become crooked inside the nut. As the cork reaches the other side, continue to turn it until it barely protrudes beyond the nut.

B Run a single-edged razor blade along the surface of the nut, carefully cutting into the cork. You do not need to get an entire cross section of the cork, but the section you use must be very thin.

Dwell on the Cell   39

LAB 5A OBJECTIVES

- Prepare wet mounts.
- Create labeled sketches of cell structures seen under magnification.
- Compare plant and animal cells.

✔ Introduction to Lab 5A

Lab 5A introduces students to some basic microscopy skills: tissue preparation, making wet mounts, and staining. These are skills that are commonly used in advanced biology courses and in a wide variety of career fields.

✔ Prelab Discussion

It will be useful to give students some suggestions beforehand on how to make a scientific sketch and to discuss the level of detail you expect in their sketches. Students can reference Appendix D for tips on making scientific drawings. You may also choose to demonstrate the process of creating a wet mount before students attempt the procedure.

✔ A Slice of Life

For students who may be interested, the study of tissues using microscopy is known as *histology*. In a laboratory the cutting of very thin specimen slices for histological purposes is done with a special tool called a *microtome*. Some students may be familiar with computed tomography (CT) scan technology, which uses x-rays to allow the viewing of thin "slices" of a patient's anatomy.

Prepared Slides

All three of the tissues examined in this lab activity are commonly available as prepared slides from science equipment suppliers, if necessary. But avoid the temptation to take this shortcut, as making wet mounts is a common technique as well as a basic laboratory skill.

C Prepare a wet mount of cork by following the instructions in Appendix D. *Note:* If your cover slip teeters on the cork, your slice of cork is too thick. Try making a thinner slice.

D Observe the wet mount of cork on low power. Your slice is probably thinnest along one of the edges, so you might want to start exploring there first.

1. When placing wet mounts on the microscope stage, the slide must remain parallel to the floor. Why is this necessary?

   The fluid medium will drain off the slide if the slide is not level.

2. Can you see any internal cellular structures in the cork cells? Explain.

   No. Only the cell walls remain.

3. You are, of course, observing dead cork. What cellular structure are you observing?

   cell walls

Observing Onion Epidermal Cells

E Obtain the scale of a small onion. (A scale is one of the onion's layers.) The epidermis of the onion is a thin, translucent skin on the inside surface of the scale. Take the scale and break it. At the edges of the broken scale you should be able to see a portion of the epidermis.

F Peel a translucent layer from the scale using forceps. (A translucent layer allows light through but is not transparent.) The epidermis is a very thin sheet of cells, so do not crush or wrinkle it. If you do, the cells can be damaged and air bubbles will get trapped between the layers, making them hard to observe.

G Prepare a wet mount using a small piece of onion epidermis no larger than the drop of water on your slide. Place the onion epidermis so that it lies flat. If it begins to fold or curl, use probes to straighten it. Put a second drop of water on it and then put the cover slip on. The cover slip should adhere tightly. If it appears to be floating, you can draw some water off with a paper towel. If there are large air bubbles, then there is not enough water, and you can add small amounts with a pipette at the edge of the cover slip. You may need your lab partner's help.

H Observe the onion epidermis cells under low power.

4. What is the general shape of one onion epidermal cell?

   Answers will vary. *Example:* rectangular

5. What is the general shape of a group of onion epidermal cells?

   Answers will vary. *Example:* looks like a brick wall

6. What do the cork and onion cells have in common?

   Answers will vary. Characteristics include cell shapes, cell walls, and regularity of cell shapes.

7. Of the following terms, indicate the ones that apply to the onion epidermis: unicellular, multicellular, tissue, organ system.

   multicellular, tissue

I Carefully remove the slide from the stage and place it next to the microscope. Stain your onion epidermal cells by placing one drop of methylene blue on the slide at the very edge of the cover slip, in contact with the water under the cover slip. At the opposite side, touch a paper towel to the water under the edge of the cover slip, allowing the paper to absorb the water.

⚠️ **Stains**

Methylene blue stains can be removed from clothing or furnishings but with difficulty. Advise students of the importance of wearing protective clothing and handling the stain carefully.

The stain will be drawn under the cover slip. (If the stain runs over the outside edges of the cover slip, you probably used too much water when you made your wet mount. Use paper towels to absorb the excess water and try again.) When the stain has contacted the onion epidermis, blot away any excess fluids on the slide or cover slip.

J Allow the stain to remain on the slide three to five minutes before observing the specimen. This permits the stain to enter the cells.

K Observe the stained onion epidermis on low and on high power. If you don't see any changes, you may need to add another drop of stain or wait a little longer for the stain to have an effect.

8. What can you see that differs from your observation of an unstained onion epidermis?

   Students should be able to see structures inside the now-colored cell walls.

9. Look among the cells until you find a dark, circular structure inside one of them. What is it?

   a cell nucleus

10. Frequently, darker spots can be seen within this dark structure. What are they?

    nucleoli

L In Drawing Area A make a drawing of one onion cell with a few adjoining cells to show how the cells fit together. Use the power that you feel is best, but do not use the oil immersion lens. Draw the internal structures for the main cell only. Label only the structures that you see in your specimen. Be sure to indicate which power you used to prepare your drawing.

Observing Human Cheek Epithelial Cells

M Prepare a wet mount of your cheek cells or those of your lab partner. Collect some mucous epithelial cells by rubbing the blunt end of a flat toothpick back and forth inside your cheek. **Collect cells from your own mouth only!** To get the greatest concentration of cells, do not twirl the toothpick around; use only one side. Remove the toothpick carefully, collecting as little saliva as possible.

N Put one drop of methylene blue on the center of a microscope slide. Immediately tap the edge of the toothpick with the cells several times in the stain. After this is done, carefully add the cover slip.

11. Why do you need to be careful when you place the cover slip on top?

    Answers will vary. *Examples:* You want to avoid getting stain on your fingers or allowing your cells to get flushed off the slide.

O View the cheek cells under low power. Look for isolated cells, not clumps.

12. How can you distinguish the epithelial cells from the other debris that appears on the slide?

    Epithelial cells have a distinct shape, and each has a nucleus.

13. These cells are called *mucous epithelial cells*. Mucus (noun form of *mucous*) is the slimy substance produced by membranes to moisten and protect them. What does the word *mucous* tell you about the functions of these cells?

    Answers will vary. Mucous epithelial cells are part of the tissue that manufactures mucus. They help to lubricate and protect the mouth.

P Look for a suitable cheek epithelial cell to draw, center it in your microscope's field of view, focus on high power, and then draw the cell in Drawing Area B. Label all the parts that you see.

No Kebabs!

The safety issue here is, naturally, that you don't want a student inadvertently skewering his partner's cheeks!

14. What are some of the similarities between the onion epidermal cells and the cheek epithelial cells?

Answers will vary. Both cell types may form tissues. They both

cover other tissues and have nuclei.

15. What are some of their differences?

Answers will vary. Onion cells are larger and have cell walls.

Cheek cells are not as regular in shape.

CONCLUSION

16. Why don't cheek cells all have the same shape on the slide?

They are not all lying flat. Some are folded over and squashed

since they do not have a cell wall to maintain their shape.

17. If you were to stain both types of cells, which would stain more quickly? Why?

The cheek cells would stain more readily because they have no

cell walls to slow the movement of the stain.

18. What is the purpose of the plant cell wall?

The plant cell wall protects and physically supports the plant.

GOING FURTHER

19. Animal cells do not have cell walls since animals have other means of physical support, such as a skeleton. Discuss the problems that animals would have if their cells did have cell walls.

Animals would have to maintain fluid levels to maintain the

pressure in the cells; otherwise, the animals would "wilt." Animals

would lose the elastic properties of cells that are possible without

a cell wall. Elasticity helps animals move and respond to their

environment.

20. How does God's design of the cell for both plants and animals show His care for creation?

God provides for the needs of His creatures with a cell design

that supports how they live, get nourishment, and reproduce.

21. David thanked God in Psalm 139:14 for being fearfully and wonderfully made. On the basis of your own understanding of cell structure, write a sentence or two expressing thanks to God.

Answers will vary. Our understanding of the intricacy and function of cells should cause us to be thankful as well since we are even more fearfully and wonderfully made than the psalmist could have known in his time.

DRAWING AREA A

DRAWING AREA B

The Creation Gives Evidence of Its Creator

The Bible is not a book of science, but it is to be expected that God the Creator would describe His creation accurately in passages such as Psalm 139:14. David would certainly have been awed had he been able to observe the marvels of the miniature world of cells. The intricacies of cellular structure and the hierarchical order of cells working together in tissues do indeed provide strong evidence of the infinite care and forethought of a loving, wise, and powerful Creator.

5B LAB

The Pressure Is On

Investigating Osmosis

As we have learned, osmosis is a form of passive transport that helps cells regulate their internal environment. In this activity you will investigate how a difference in one factor—solute concentration—can affect osmosis. Afterward, you will extend what you have observed to other factors that might affect osmosis.

How do different solute concentrations affect osmosis?

Equipment

osmometer	paper towels
squeeze bottle	distilled water
sucrose solutions	laboratory apron
wax pencil	goggles
ruler	

PROCEDURE

Solution A has twice the concentration of sucrose as Solution B. Your teacher will task you with testing one of the two solutions.

A Gently shake the osmometer bulb to remove all water.

B Use a squeeze bottle to completely fill the bulb with the assigned solution, then set the bulb on a wet paper towel.

C Uncap the squeeze bottle and insert the bottom 5 cm of the osmometer's tube into the solution. Place your thumb over the top end of the tube to keep the solution in the tube as you transfer it to the osmometer bulb.

D While you keep your thumb over the osmometer tube, place the tube into the neck of the osmometer bulb. Tap the bottom of the membrane to remove any air trapped in the bulb. Make sure that there is no air trapped in either the bulb or the tube.

E Use a paper towel to wipe the joint between the bulb and tube. If it leaks, ask your teacher for help.

F Rinse the outside of the bulb with distilled water.

G Fill the osmometer's beaker with distilled water to a depth of 5 cm.

H Suspend the osmometer bulb in the beaker so that the membrane is covered but the water does not come near the joint of the bulb and tube (your teacher will demonstrate the method to use). Use the wax pencil to mark the level of solution in the tube.

I After ten minutes have passed, measure how far the fluid has risen in the tube. Remember to measure from the fill mark, not from the bottom of the tube or bulb. Record your measurement in Table 1.

J Repeat Step I four times.

QUESTIONS

- Is the rate of osmosis affected by solute concentration?
- Does osmosis ever reach an equilibrium?
- Can I predict an effect on osmosis on the basis of solute particle size?

LAB 5B OBJECTIVES

- Explain the effect of solute concentration on osmosis.
- Identify whether an osmotic system has reached equilibrium.
- Make predictions about how other factors may affect osmosis.

✔ Timing

As you can see, the time needed for this lab activity is significant. The differences between the two sucrose solutions should be apparent as early as twenty minutes into the procedure but will become more pronounced with additional time. You can save time by demonstrating the procedure to students ahead of lab day and having them practice the setup with plain water.

Alternatively, this activity can be done as a class demonstration and the sample data provided to students for analysis.

🧪 Equipment Notes

The osmometer described in this activity includes a tube and bulb, collectively referred to as a *thistle tube*, and a 250 mL beaker. Thistle tubes and membranes can be purchased from science equipment suppliers.

The solutions are approximately 1 molal (Solution A) and 0.5 molal (Solution B). They can be made using the formulas below.

Solution A: 25.0 g of sucrose (table sugar) in 75.0 mL of distilled water

Solution B: 15.0 g of sucrose in 85.0 mL of distilled water

✔ Joining the Tube to the Bulb (Step D)

One way to help reduce the likelihood of air being trapped in the osmometer when its parts are joined is to release your thumb from the tube right as the tube is inserted into the bulb. The liquid flowing out of the tube and momentarily overflowing the bulb will prevent any air from entering through the joint.

ANALYSIS

K Pool your data together with the results from the other lab groups in your class. Record the pooled data in Table 2. Remember to tabulate the results for Solution A and Solution B separately!

L Average the class data for Solution A and enter the result in Table 2. Repeat for the Solution B data.

CONCLUSION

1. On the basis of your pooled class results, what can you conclude about the permeability of the membrane to sucrose?

   The membrane is not readily permeable to sucrose.

2. Is there a difference in the height gain between Solutions A and B? If so, what could account for the difference?

   Answers will vary. Since Solution A is twice as concentrated as Solution B, there is more osmotic pressure for Solution A, which should rise higher in the tube as a result.

GOING FURTHER

3. If you were to let the experiment run long enough, could all the water in the beaker enter the osmometer? Defend your answer using data from the procedure.

   No. Equilibrium would occur when gravity and other forces balanced the osmotic pressure. This is demonstrated with Solution B: the column reaches a maximum height and thereafter remains unchanged. Molecules would still move across the membrane, but there would be no net movement.

4. If you were to slowly heat the water in the beaker, what might happen to the rate of osmosis? Explain.

   It would probably accelerate because of the increased activity of the molecules. Equilibrium would also be reached more quickly.

5. If you were to set up the osmometer using salt instead of sucrose, would you expect similar results? Why or why not?

   Answers will vary. Because salt dissolves into ions that are much smaller than sucrose molecules, they might be able to pass through the membrane. If the ions passed through the membrane, there would not be osmotic pressure to move the solution up the tube.

6. Sucrose is a *disaccharide*, a sugar made of two different, smaller molecules, glucose and fructose. Invertase is an enzyme that decomposes sucrose into glucose and fructose. Predict how your test results might be different if invertase were added to your sucrose solutions.

Answers will vary. If glucose and fructose are small enough to pass through the osmometer membrane (they are), then some solute would diffuse into the beaker, lowering the osmotic pressure of the system. At some point during the procedure, the level of sugar in the beaker would become detectable.

TABLE 1 Test Results

	COLUMN HEIGHT (mm)					
Solution (A or B)	Start	10 min	20 min	30 min	40 min	50 min
A	0 mm	2	4	6	7	8

TABLE 2 Class Results

	COLUMN HEIGHT (mm)					
Solution (A or B)	Start	10 min	20 min	30 min	40 min	50 min
A	0 mm	2	4	6	7	8
A	0 mm	1	3	6	7	8
A	0 mm	3	5	6	7	8
B	0 mm	2	3	4	5	5
B	0 mm	2	2	3	5	5
B	0 mm	1	3	4	5	5

	AVERAGED COLUMN HEIGHT (mm)					
Solution	Start	10 min	20 min	30 min	40 min	50 min
A	0 mm	2	4	6	7	8
B	0 mm	2	3	4	5	5

✔️ **Sample Data**

The data shown in Tables 1 and 2 is typical. Students' results may be different.

6A LAB

No Swimming Today

Oxygen and Metabolism

It's a hot, sunny summer day, and you are looking forward to jumping into the pool and feeling the cold water close over your head. You get to the pool and notice that there is no one around. A sign says, "Pool closed today—dangerous bacteria levels."

Public pools use tests to make sure that people are not swimming in water that could make them sick. Pool water must be treated to ensure that algae and bacteria don't grow in it. Bacteria can feed on organic matter in a swimming pool, and even if chlorine kills the bacteria, the water could still be polluted.

Scientists and wastewater technicians test the water we drink to make sure that it doesn't contain organic materials. When a large amount of organic material enters a body of water, the bacteria that normally decompose dead plants and other organic materials increase. This is a problem because the bacteria use oxygen to break down these organic materials, causing the oxygen levels in the water to plummet. The problem is especially bad at night when plants are not producing oxygen through photosynthesis.

When working properly, wastewater treatment facilities remove most of the organic matter before the treated water is released into a body of water. However, technicians constantly monitor the body of water to make certain that treated water is clean. One of the ways they do this is to use a test to measure *biochemical oxygen demand* (BOD)—the oxygen needed for bacteria to break down organic matter in a water sample at 20 °C over a period of time. It's a way to measure how much organic matter is in a water sample. A crystal-clear stream has a BOD of 1 mg/L of water over a five-day period. Untreated sewage in the United States has an average BOD of 200 mg/L for the same time period.

You will be measuring the BOD of a water source by comparing the dissolved oxygen content of the water right after it was collected with the oxygen content of the water five days after it was collected. This waiting period allows bacteria in the water to use the oxygen to break down any organic matter.

Equipment

LaMotte 5860-01 test kit
two water samples
nitrile gloves

laboratory apron
goggles

How do scientists use oxygen and metabolism to measure organic pollution?

QUESTIONS

- What is biochemical oxygen demand (BOD)?
- What is the link between oxygen and metabolism?
- How do we test for BOD?
- How do we use BOD to estimate the amount of organic pollution in a water sample?

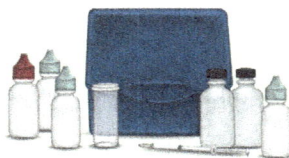

- Define *biochemical oxygen demand* (BOD).
- Relate oxygen demand to cellular metabolism.
- Measure BOD using a dissolved oxygen test kit.
- Estimate the amount of organic pollution in a water sample using BOD.

Ideas for Prelab Discussion

Be sure that students clearly understand that metabolism is related to oxygen requirement. The amount of oxygen that an organism needs is related to how big it is and the temperature at which it operates.

For a video that shows how people measure biological oxygen demand, see the *BOD Demonstration* link. The video, produced by Carolina Biological Supply, demonstrates how to use their dissolved oxygen test kit. This video is available as a digital resource.

Equipment Notes

You will have to purchase a dissolved oxygen test kit that uses the Winkler method. These can be quite pricey, but a decent one can be purchased for $50–$60. You may have to purchase additional equipment, which may raise your cost to $140. (This additional equipment will most likely be a one-time purchase.) The suggested kit is the LaMotte Dissolved Oxygen Water Test Kit (5860-01), available online and from several science equipment suppliers. The LaMotte kit contains all the equipment necessary to do this lab activity except for the water samples. Note that this kit contains only one sample bottle; you should purchase an additional bottle for each sample that you intend to take.

This lab activity was field-tested with the La-Motte 5860-01 kit. If you use a different kit, follow the manufacturer's instructions for obtaining, fixing, and titrating your samples.

To measure biochemical oxygen demand, you will need to store water samples in the dark for five days at room temperature. If you don't have a dark place to store them, wrap the bottles in aluminum foil. Note that once a sample is fixed, your students must test for dissolved oxygen within eight hours.

Water Quality Testing Program

For more information on a field-based water quality testing program, see the *Water Quality Testing* link. This link is available as a digital resource.

Bacteria in Water

Much of the bacteria in water comes from the intestines of endothermic animals such as birds and mammals. This is called *coliform bacteria*.

Reviewing Chemical Reactions in Biology

To help students answer Questions 1–5, review the material on chemical changes on pages 34–37 in the Student Edition.

Proportionality (Question 4)

You may need to clarify the ideas of direct and indirect proportionality using an example such as speed, time, and distance or a similar analogy. In $d = st$, distance is directly proportional to speed and time, but speed and time are inversely proportional to one another.

Variations on the BOD Experiment

Students will be using the Winkler method to measure dissolved oxygen (DO) and BOD. If a creek or pond is nearby, you may opt to have students collect two samples from it and test the first one in the field. On the fifth day they should test the other sample to measure how DO changes over the five-day period. The difference between the DO on the first and fifth days is the sample's BOD. Part of testing for dissolved oxygen will involve fixing the sample. Students should not fix all the samples at once; they should fix samples just before doing the dissolved oxygen titration.

If you choose to have students collect their own samples, bring markers and field test kits. They will also need fresh water for washing their hands, plastic containers for collecting water, and aluminum foil for covering the water sample that will be kept in the dark. Students will need to fix the first sample before doing the dissolved oxygen test.

1. Why do bacteria need oxygen to metabolize waste products?

   Oxygen is one of the reactants needed for the chemical processes that break down organic waste products.

2. Why does the temperature for BOD need to be specified?

   Temperature affects how fast a chemical reaction, such as metabolism, progresses.

3. Why does the length of time for BOD need to be specified?

   Bacteria can process more wastes over a longer period of time. Therefore, a specified amount of times makes one BOD measurement comparable to another.

4. Do you think the mathematical relationship between BOD and the amount of organic pollution is directly or indirectly proportional? Explain your choice.

   BOD is directly proportional to organic pollution. As pollution goes up, more bacteria can feed on organic matter, causing their need for oxygen to rise.

5. Suggest how the oxygen that an organism requires for metabolism is related to its size, such as the difference in the metabolism of an elephant versus that of bacteria. Explain your reasoning.

   Larger animals have more cells, so they require more oxygen to break down food.

PROCEDURE

Fixing the Sample

People who check water sources for safe bacterial levels usually use a BOD kit similar to the one shown on page 49. The first step will be to fix the water sample. Fixing means to treat the sample with chemicals that will lock in the oxygen that it already contains so that it doesn't change over time.

A Obtain a water sample by allowing water to flow into a dissolved oxygen (DO) bottle. Recap the bottle while it is submerged, then take the bottle out of the water. Label this sample as *Day 1*. Repeat this step to obtain a second sample and label it as *Day 5*. Store the Day 5 sample in a darkened area at room temperature until ready to test.

6. Where did your water samples come from? When did you obtain it?

   Answers will vary.

If you want to spend less time on this experiment, collect water samples and fix the first sample yourself. Keep in mind that students will need to make dissolved oxygen measurements on fixed samples within eight hours.

You may choose to have students measure dissolved oxygen each day for five days, or they could simply fix and measure the dissolved oxygen after the fifth day.

7. Describe your observations of the Day 1 water sample, including its cloudiness, or *turbidity*.

Answers will vary.

8. On the basis of your observations, predict whether your water sample will have a BOD more similar to a clear stream (1 ppm) or sewage (200 ppm).

Answers will vary.

9. Why is it important that your DO bottle not contain any air bubbles?

Air space could allow oxygen to move between the sample and the air.

B Uncap the bottle. Put 8 drops of manganous sulfate solution and 8 drops of alkaline potassium iodide azide into the water sample. Since the bottle is already full, some water will probably spill over the side of the bottle.

C Recap the bottle and mix its contents by gently inverting the bottle several times. Let the water sit for several minutes.

10. Describe your observations of the water in the DO bottle.

Students should observe a color change and the formation of a precipitate.

D Uncap the bottle and add 8 drops of sulfuric acid. Recap and gently invert the bottle several times until the cloudiness, or precipitate, begins to disappear.

11. What did you observe about the water sample in the bottle?

Students should have noticed that the solution begins to turn gold.

Notice that the second chemical you added to the water sample was alkaline potassium iodide azide. As you have added chemicals, free iodine has been produced in solution. This is similar to a dark brown solution known as *tincture of iodine* that is sometimes used to treat wounds. The word *alkaline* means that this compound acts as a base.

12. Why is it helpful to react an acid with this basic solution?

Adding the acid to the basic water sample in the right amount will neutralize the base.

Measuring Dissolved Oxygen

Now that the water sample is fixed, the dissolved oxygen in it is stable and can be measured. You will use a process called *titration* to find out how much dissolved oxygen is in your sample. Titration is a way to chemically react something of an unknown concentration with something of a known concentration. We can measure the volume of the solution of known concentration we use to determine the concentration of the unknown substance. You will use a titration syringe to measure this volume.

✔️ Factors That Affect DO Measurements

As you consider the sources for your water samples, be aware that there are several environmental factors that can influence dissolved oxygen content. Agitated, colder waters with lower mineral content and more plant life will have a higher dissolved oxygen content. Saltier water with more organic wastes will have a lower dissolved oxygen content.

If you don't want to leave the classroom for this activity, collect water samples ahead of time, fixing the first water sample yourself, or do the test on an aquarium in your classroom.

🧤 Proper Stoppering

It is important that the DO bottles are filled and stoppered properly. Trapped air bubbles will adversely affect the DO measurement. Be sure to demonstrate the proper technique to your students.

⚠️ Protect Skin

Caution students that alkaline potassium iodide azide can irritate or burn skin. They should wear gloves when working with this chemical and avoid contact with their skin.

✔️ Proper Dropping Technique (Step B)

Make sure that students hold the chemical bottles perfectly vertical. The bottles' nozzles are calibrated to dispense the proper amount of chemical when held vertically; holding the bottles at an angle will cause the solution to adhere to the sides of the nozzles.

Reviewing Chemical Equations

You may need to review the parts of chemical equations to help students answer Question 13.

You are trying to find the concentration of oxygen in your solution. The sodium thiosulfate ($Na_2S_2O_3$) in the test kit is a solution whose concentration is known. This chemical will react with the iodine in the water sample according to the reaction below.

$$2Na_2S_2O_3\ (aq) + I_2\ (aq) \longrightarrow Na_2S_4O_6\ (aq) + 2NaI\ (aq)$$

13. What does the (*aq*) in this reaction mean?

aqueous solution; These substances are dissolved in water.

Scientists usually use an indicator that makes the solution change color so that we can know when the reaction is done. You will use a starch solution to determine that the reaction above is done.

Note: The titration syringe provided in some BOD kits is marked in parts per million (ppm). Parts per million is a ratio, not a unit of volume; in this instance, each 0.1 mL of sodium thiosulfate used during titration corresponds to 1 ppm of dissolved oxygen in the water sample. Instead of measuring the amount of titration solution used and then using that information to calculate the amount of dissolved oxygen in the sample, the dissolved oxygen level is read directly from the titration syringe. This method may vary; follow the directions provided with the kit you are using.

E Pour some of your fixed water sample into the titration bottle to the 20 mL mark.

F Using your titration syringe, draw up sodium thiosulfate to the 0 ppm mark. (The syringe is marked in reverse because you will be concerned with the amount dispensed, not the amount drawn up.)

G Deliver sodium thiosulfate into the titration tube drop by drop by putting the syringe in the hole in the lid of the titration tube. You may not need to use all the sodium thiosulfate solution, so it's important to add it one drop at a time. Swirl the titration tube between drops. Keep adding sodium thiosulfate until the solution color is about the same shade of yellow as a dandelion. Placing a piece of white paper behind the titration tube will help you see the color change better.

H Now carefully remove the titration syringe and gently set it upright in its storage tube. Uncap the titration tube and place the cap on your bench top. Add 8 drops of starch solution to the titration tube.

14. What did you observe when you added the starch solution to the fixed water sample?

The solution turned a dark blue-purple color.

Starch indicator changes color in the presence of free iodine. When all the free iodine is used up, the solution will quickly change to a clear color.

I Put the cap and syringe back on your titration tube and add sodium thiosulfate drop by drop, swirling the tube gently after each drop, until the solution flashes clear. If you use up all the sodium thiosulfate in the syringe, add more, keeping track of how much you've used. Once the solution flashes clear, record the corresponding amount of dissolved oxygen in parts per million (ppm) for Day 1 of Table 1. (See note prior to Step E.)

15. You will now let the water sample sit for five days and repeat the DO test. Predict whether dissolved oxygen in the second water sample will increase or decrease in storage. Explain your reasoning.

Answers will vary. Dissolved oxygen content will decrease over time because bacteria will consume oxygen as they metabolize organic matter in the water sample.

J On Day 5 follow Steps E through I to fix and measure the dissolved oxygen for the Day 5 sample. Record the amount of dissolved oxygen for Day 5 in Table 1.

ANALYSIS

16. Comment on the accuracy of your prediction in Question 15.

Answers will vary. You may want to follow up with students who had flawed reasoning behind their predictions.

K Subtract the DO value for Day 5 from the value for Day 1. This is your BOD value. Record the BOD in Table 1.

17. What is the BOD for your water sample?

Students' values may range between 1 and 10 ppm.

18. Using the information below, describe the nature of your water samples.

» 1–2 ppm—very clean, little organic matter

» 3–5 ppm—moderately clean, some organic matter

» 6–9 ppm—poor water quality, abundant organic matter

» ≥10 ppm—very unhealthy water, large amounts of organic matter

Most water samples from natural places should be moderately clean.

19. On the basis of the information about the water source that you recorded in Question 6, explain why your water had this cleanliness rating.

Students should comment on the vegetation, wildlife, pollution, and even temperature at the source of their water sample.

20. How can we protect the environment to keep water sources around us as clean as possible?

Minimizing pollution and wastewater runoff can help keep water clean. Also protecting areas of vegetation can help naturally filter runoff water.

Applying the Creation Mandate to Clean Water

If we understand the Creation Mandate to apply to all aspects of creation, then, of course, water is another resource that must be wisely managed to ensure clean, safe water both for today and for future generations. Utilizing an understanding of clean water management to provide safe drinking water in underdeveloped parts of the world also demonstrates God's compassion to others.

Sample Data

The data shown in Table 1 is typical. Students' results may be different.

21. Explain from a biblical worldview why we should keep water clean.

Keeping water clean helps people and wisely uses the water resources that God has provided for us.

GOING FURTHER

22. Design an experiment to test different factors that affect BOD.

Answers will vary but should include information on collecting water samples from a variety of areas with varying temperature, pH, vegetation, salinity, mineral content, and water agitation.

23. How could this information be useful to a county that is considering building a public park?

The county would need this information to maintain healthy waters, especially if the park contained areas with streams, lakes, or ponds. This could involve protecting certain areas of vegetation, planning areas where trash and wastewater are disposed of, and carefully planning the location of restrooms, septic tanks, and swimming areas.

TABLE 1

COLLECTION DAY	DISSOLVED OXYGEN (ppm)	BOD (ppm)
Day 1	7.1	3.8
Day 5	3.3	

6B LAB

Hidden Code

Extracting DNA from Cells

A long-lost uncle has left you $1,000,000 in his will. All you need to do is to prove that you are his relative. But how are you going to do that?

You've been learning about how every cell stores information in the form of DNA. Scientists can extract DNA from any cell, including the cells in your body. In Chapter 9 you'll learn more about DNA fingerprinting, or DNA profiling. This is a process of forensics that involves extracting DNA from two individuals, cutting the DNA up into fragments, and comparing them to show that two people are genetically related. Because of the code hidden in your cells, you would be able to prove that you are an heir!

In this lab activity you will be extracting DNA from plant cells to see what DNA looks like and what scientists can learn from it. It's easier to extract and see DNA from the cells of some plants because their cells contain multiple copies of their DNA. Botanists can use the DNA in plants to figure out how to breed plants that have certain properties that make them more useful for people.

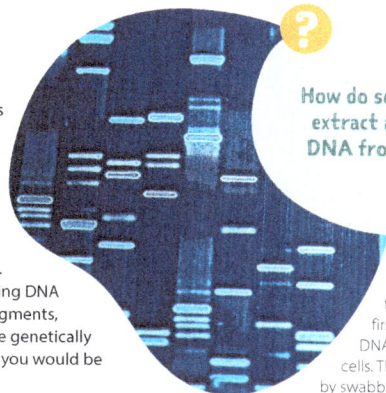

How do scientists extract and use DNA from cells?

When forensic scientists do DNA fingerprinting, the first step is to extract DNA from a person's cells. They get these cells by swabbing the inside of the person's cheek. The DNA is then cut up into fragments that can be observed and compared.

Equipment

blender
microscope
beakers, 150 mL (2)
graduated cylinder, 100 mL
stirring rod
microscope slide and cover slip
beaker, 50 mL

lentils, dried and uncooked
liquid dish detergent
isopropyl alcohol (rubbing alcohol)
methylene blue solution
salt solution, 6%
phenol red indicator solution

nitrile gloves
laboratory apron
goggles

QUESTIONS
- What substances contain DNA?
- How do scientists get DNA out of a substance?
- What does DNA look like?
- How do scientists use DNA to learn about living things?

1. Where will the DNA that you are extracting come from? Where will it not come from?

   DNA will be extracted from the plant cells. It should not come from any other chemical used to treat the cells unless there is bacterial contamination.

2. Why do the cells in plants have DNA?

   See margin for answer.

- Extract DNA from plant cells.
- Observe DNA strands separated from the nucleus.

Prelab Discussion

If you would like to connect DNA extraction with forensics and biotechnology, you could do a DNA extraction of cells in saliva. The process and equipment are identical to that which is used in this lab activity, except that students will need to swish with a 10% salt solution that is then spit into the beaker and treated. Add blue food coloring to better see the extracted DNA.

Equipment Notes

You may substitute several pieces of equipment in this lab activity. You may use a food processor, a mortar and pestle, or both to grind up lentils if needed. Instead of a beaker, you can use clear plastic cups. Substitute wooden splints or toothpicks if glass stirring rods are not available. Instead of lentils, you could use raw wheat germ, split peas, or crushed strawberries or banana, all of which will also produce an abundance of DNA. Instead of meat tenderizer, you may use pineapple juice or contact lens solution to break down the proteins.

Make sure that all glassware used for this activity is thoroughly cleaned to avoid bacterial contamination.

Adding the Alcohol and Extracting the DNA

You may want to demonstrate the process of adding the alcohol into the tilted beaker to prevent the two layers from mixing. You may also want to show a video of the extraction process to help students know what they should be looking for. Do an internet search for a DNA extraction video.

Question 2 Answer

The cells in plants have DNA that helps the plant carry on life-giving processes like making proteins and processing nutrients. Students may also mention that DNA is involved in plant reproduction.

Virtual DNA Extraction Lab Activity

To view a virtual DNA extraction, see the *Virtual DNA Extraction Lab Activity* link. This link is available as a digital resource.

Ever So Slowly (Step D)

It is critical that the alcohol be added slowly. Pour carefully down the side of the beaker. If the alcohol mixes too much with the water, students will not be able to separate out the DNA. The alcohol should form a separate layer on top.

If Students Don't See DNA

If students don't observe DNA in either beaker, it may be because they have an initial mixture of lentils that is too watery. Also, they may not have mixed the soap sufficiently or allowed enough time for the chemicals to do their work.

Comparing DNA

Students can compare their results with other groups or take pictures to show to others.

DNA in the First Beaker

If during Step E your students find evidence of DNA in the first beaker, they should complete the steps to verify the presence of DNA in that beaker as well. This evidence would indicate that there was contaminating DNA in the water or on the beakers used.

3. Why must the plants that you are using be uncooked? (*Hint:* Think about the effect of heat on large molecules.)

Cooking the lentils can degrade the large molecules they contain, such as proteins and DNA.

PROCEDURE

A Add 60 mL of water to one of the 150 mL beakers.

B Using the graduated cylinder, measure 15 mL of lentils. Pour the lentils into a second 150 mL beaker. Add 45 mL of water to this beaker. Pour the mixture into a blender and blend on high for about 15 seconds until the lentils form a kind of sludge. Deposit this back into the beaker.

4. Why do you need to add water to the lentils?

Water is added to help with the blending and to dissolve any water-soluble molecules in the lentils.

5. Why do you need to blend the lentils?

Blending the lentils increases the surface area in contact with chemicals that will be used to extract DNA.

6. What do you observe about the contents of both beakers?

The first beaker should contain just water. In the second beaker the lentils form a brown, watery sludge.

7. You will be adding chemicals to both beakers. Think about the way that an experiment is conducted. What is the function of the first beaker?

The first beaker is a control that confirms that none of the other chemicals added to the lentils contain bacterial contaminants that include DNA.

C Now add 8 mL of liquid dish detergent to both beakers. Gently swirl both beakers in a way that minimizes the amount of foam that forms on top.

8. Suggest why we are treating the lentils with soap, which will attract proteins in the cell membrane.

The purpose of the soap is to lyse or break the cell so that the DNA can be extracted from the cell's nucleus.

9. After adding the soap, where is the DNA?

DNA has been washed into the water.

D While tilting the first beaker, add 45 mL of isopropyl alcohol to it, pouring the alcohol gently along the side of the beaker so that the alcohol forms a separate layer from the water. Do the same for the second beaker. Let both beakers sit for two to three minutes.

10. What do you observe about the contents of both beakers?

In the first beaker students should observe two layers, one of alcohol and one of cloudy water. No DNA should be present. In the second beaker they should see bubbles of white, cloudy, stringy material rising through the alcohol.

If you see bubbles rising up through the alcohol in the second beaker, observe them carefully. They may be drawing white strings of thousands of DNA molecules along with them as they rise. You will physically remove one of these DNA samples from the beaker.

E Insert a stirring rod all the way to the bottom of the second beaker. Slowly turn the rod one direction, winding any white strands that you observe around the rod. Gently remove the rod to observe the DNA strands. Clean the stirring rod, then repeat this procedure with the first beaker to see whether you can observe any DNA.

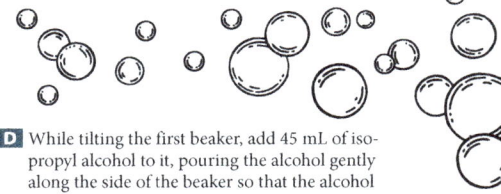

11. Why did two layers form in the beakers?

See margin for answer.

_____

_____

F Take some DNA out of solution and put it on a microscope slide to prepare a wet mount. Add 2 drops of methylene blue and wait a few minutes for it to dye the DNA. Cover it with a cover slip.

G Using the microscope, view the DNA first at low power, then at a higher power.

12. What do you observe about DNA under a microscope?

DNA looks like a narrow strand.

_____

_____

H Now measure 15 mL of 6% salt solution into a 50 mL beaker. Wind more DNA around your stirring rod and place it in this solution, turning the rod to dislodge the DNA.

13. What do you observe when you place the DNA molecules in the salt solution?

If enough DNA molecules are present, they

will clump together.

_____

Phenol red is an indicator that works on solutions that have a pH between 6.8 and 8.2. It will turn yellow in the presence of an acid, orange at a neutral pH, and pink in the presence of higher pH.

6.0   7.0   8.0

I Add 5 drops of phenol red indicator to the salt solution containing the DNA that you have dislodged.

_____

CONCLUSION

Let's consider all the information that you have gathered to learn a little more about what DNA is like as a chemical.

14. Is DNA an acid, base, or neutral? How can you tell?

See margin for answer.

_____

_____

15. What does DNA physically look like?

It is a white, strand-like substance.

_____

_____

16. The first beaker should contain no cells and therefore no DNA. If you were able to extract DNA from this beaker, what would you conclude?

Extracting DNA from a beaker with no lentils

would indicate that there was some kind of

bacterial contamination of the beaker or of

the chemicals used.

17. The DNA that you observed is clumped together, but it will readily dissolve when it is stirred in water. Is DNA a mostly polar or nonpolar molecule? Explain.

DNA is a mostly polar molecule since it

readily dissolves in a polar solvent.

_____

_____

18. You couldn't see a single strand of DNA, even though DNA is a large molecule over 2 m long. Why can you see DNA in your lentil solution?

The DNA is clumped together to form

strands, similar to the way that thread is

visible when it is wound around a spool.

_____

Question 11 Answer
Alcohol is less dense than water or the lentil mixture. (Water has a density of 1.0 g/mL, while rubbing alcohol has a density of 0.79 g/mL.)

Question 14 Answer
DNA should turn yellow, indicating that it is an acid. Students may also mention that the Student Edition indicates that DNA has a hydrogen to donate and that the name indicates that it is a nucleic acid.

Working with Indicators (Step I)

If you have a student with colorblindness, make sure that you pair him with another student who can help identify the colors of the indicators in this activity.

Reviewing pH

Remind students that acids have a pH less than 7 and bases have a pH greater than 7 (see Chapter 2 in the Student Edition). A pH of 7 indicates a neutral solution.

19. When geneticists do DNA fingerprinting or analyze the genome of a plant that they are trying to breed, the first step is to extract the DNA. The second step is to cut the DNA up into fragments. Considering what you know about DNA, why do you think they do this?

Because the molecule is so long, DNA that is extracted must also be fragmented so that it is more easily studied and analyzed.

GOING FURTHER

20. How could a geneticist move a section of DNA that coded for something useful from one organism into another organism of the same species?

The geneticist would have to extract the DNA from the first organism, cut the DNA that included the useful segment, and then insert it into the DNA of the other organism.

One DNA molecule is over 2 m long!

21. The white goop that you extracted from the lentils includes both RNA and DNA because the process you used extracts all nucleic acids. Design an experiment that you could use to answer this question: What percentage of the lentils' mass is the mass of the nucleic acids?

Answers will vary. Students should mention weighing the lentils before treatment, extracting DNA and RNA through the process mentioned, isolating the DNA and RNA through filtration, and weighing the filtered nucleic acids. The mass of each of the nucleic acids should be divided by the total mass.

7A LAB

Whatever Floats Your Leaf

Rates of Photosynthesis

Ah, Thanksgiving dinner! A turkey baked to golden perfection, creamy mashed potatoes, cranberry sauce—and Grandma's famous green bean casserole! You probably don't think too much about it, but photosynthesis makes all these delicious entrées possible. Without photosynthesis, not only would there be no Thanksgiving dinner—there'd be no dinners at all! And that would make for a sad sort of world, not to mention one in which God's creatures could not live.

Photosynthesis is a series of enzyme-catalyzed reactions that take place in autotrophic organisms such as green plants. Plants use this process to convert energy from the sun to stored energy in sugars that they can then use for cell growth. The leaves of a plant are the main photosynthetic factories. They contain chloroplasts with chlorophyll, which absorbs light energy. The overall formula shows the raw materials that plants need to produce their food.

$$6H_2O + 6CO_2 + \text{light energy} \longrightarrow C_6H_{12}O_6 + 6O_2$$

This one process provides all the energy needed for growth and survival for producers, like potato and green bean plants, and for consumers, like turkeys. Even secondary consumers, like people, ultimately get their energy from photosynthesis. In this lab activity you will use parts of leaves to demonstrate the photosynthetic process.

What factors can affect the rate of photosynthesis in plant leaves?

Equipment

hole punch	fresh spinach or ivy leaves	light source
oral syringe, 10 mL or larger	sodium bicarbonate solution, 2.5%	stopwatch
forceps	clear plastic cups	colored pencils

PROCEDURE

A Technique to Measure Photosynthesis

QUESTIONS
- How can we measure the rate of photosynthesis?
- What factors might affect the rate of photosynthesis?

PREPARING THE LEAF DISKS

Parts of leaves normally float in solution since they are filled with oxygen and carbon dioxide, but they will sink when infiltrated with a sodium bicarbonate solution. As seen in the photosynthesis equation, leaves that undergo photosynthesis produce oxygen that is released into the leaf spaces, making a leaf part capable of floating. Respiration, which consumes oxygen, is also taking place in the leaves. The measurement of leaf parts rising is an indirect way of quantifying the net rate of photosynthesis.

A Use the hole punch to cut about fifteen disks out of a spinach or ivy leaf. Avoid punching out the leaf veins.

- Observe the results of photosynthesis in leaf disks.
- Form a hypothesis about factors affecting photosynthesis.
- Test a hypothesis about photosynthesis.

Equipment Notes

The materials for this activity are inexpensive and easily sourced. Oral syringes can be obtained from a local drug store. Sodium bicarbonate is sold as baking soda. Each group will need a minimum of two to three plastic cups, depending on the range of values that they test in their experiment.

Care needs to be given to your choice of light source. Incandescent bulbs will give off considerable amounts of heat, which can affect the experiment, so fluorescent bulbs are preferred. Standard fluorescent bulbs do not normally encourage plant growth, but a grow light or full spectrum fluorescent bulb should have the desired effect.

While any green leaves can be used for this experiment, some leaves have waxy cuticles or hairs that will make them more buoyant. Spinach leaves work well and are easily obtained at the grocery store, but they need to be fresh. Ivy leaves are also an excellent choice and can be picked the day of the experiment. You will need two to three leaves per trial cup. Keeping the leaves in a dark environment for a few hours prior to the experiment will ensure that they are not actively producing oxygen until students start the procedure.

An approximately 2.5% sodium bicarbonate solution can be made using one slightly rounded tablespoon (15 g) of baking soda in 600 mL of water. You will need about 100 mL of solution for each trial cup.

Leaf Prep

Choose darker leaves; these contain more chlorophyll and will photosynthesize faster than lighter leaves will.

If students continue to have difficulty sinking the leaf disks, try adding a drop of dilute liquid detergent.

Students can better load their leaf disks into the bottom of the syringe by tapping the side of the syringe gently to knock the disks to the bottom of the syringe or by using forceps to push the disks down, being careful not to damage them.

Video Help

Searching online for leaf disk labs will yield links to videos that demonstrate the disk-sinking technique.

Save Time with a Demo

The first portion of the lab activity is essentially a "dry run" for students to make sure that they can do the technique properly. To save time, you can demonstrate both the sinking technique and the measuring technique simultaneously by letting a prepared sample of disks photosynthesize while you demonstrate the leaf punching and sinking portion of the instructions. Afterward, if students feel confident that they can do the procedure, they may skip directly to Testing Conditions for Photosynthesis. To save even more time, do the demonstration the day before the lab day.

Shortcuts

It is possible to run the trials while leaving the leaf disks in the syringe with about 10–15 mL of the sodium bicarbonate solution. The syringe should be placed upright (as in a test tube rack) near the light source, and the leaf disks will rise just as they would in a cup.

Depending on the students' abilities, you may want to determine ahead of time what condition to have them test. Alternatively, you may lead a class discussion prior to the exercise and have students select a condition to test.

Suggested Factors for Testing (Question 3)

Different light intensities can be tested by moving the light source closer to or farther from the trial cups. Or you could try testing differing quantities of naturally occurring light (e.g., a sunny windowsill versus a shady corner).

Different wavelengths of light can be obtained by filtering the light through different colors of cellophane. An interesting comparison would be a regular versus a full-spectrum fluorescent light.

B Remove the plunger from the syringe and gently load the disks into the bottom of the syringe. You may need to use forceps to gently push the disks to the bottom of the syringe. Replace the plunger in the syringe and push it down as far as it will go without squashing the disks.

C Pull 10 mL of 2.5% sodium bicarbonate solution into the syringe. Hold the syringe vertical with the tip up. Tap the syringe to get as many air bubbles as possible to rise to the tip and be released. Gently depress the plunger to force any remaining air from the syringe.

D Cover the syringe opening and pull back on the plunger to create a vacuum in the syringe. Maintain the vacuum for at least 10 seconds while shaking the disks to suspend them in the solution. Tiny air bubbles should be seen at the edge of the disks where the air is being pulled out of the leaf disk spaces. Tap the syringe to release the air bubbles from the disks. The leaf disks should begin to sink.

E Now depress the plunger while covering the syringe opening. This will force the bicarbonate solution into the leaves.

You may have to repeat Steps D and E several times to get all the leaf disks to sink.

1. Why is a solution of sodium bicarbonate ($NaHCO_3$) forced into the leaves instead of just tap water or distilled water? (*Hint:* Review the chemical equation for photosynthesis and think about what sodium bicarbonate might supply for the leaf disks.)

   Sodium bicarbonate supplies the carbon atoms needed by the Calvin cycle (light-independent phase) of photosynthesis.

Once the leaf disks have sunk, they are ready to be used for rate of photosynthesis trials.

MEASURING THE RATE OF PHOTOSYNTHESIS

F Pour the disks from the syringe into a clear plastic cup.

G Add 2.5% sodium bicarbonate solution to the cup to a depth of 2 cm. Separate the leaf disks so that they are not overlapping. Discard any disks that do not sink.

H Place the cup under a light source.

I Observe the number of leaf disks at the surface after each minute for 25 minutes. Record your observations in Table 1.

---

2. Describe what you observe.

   Students should observe one or more disks rise after several minutes, with additional disks floating at brief intervals.

Testing Conditions for Photosynthesis

In this part of the activity you will design an experiment to use the leaf disk technique to test how varying factors can affect the rate of photosynthesis.

J Choose an environmental factor to test. Some suggestions are light intensity or wavelength, pH, and temperature. Note that these are not the only factors you might test. Check with your teacher for approval if you think of something not on this list.

3. What environmental factor have you chosen to test? Write your answer in the form of a research question.

   Answers will vary. *Example:* Does temperature affect the rate of photosynthesis?

4. What is your hypothesis?

   Answers will vary. *Example:* The best photosynthetic rate will take place in warm water.

5. Explain why you think the factor you've chosen to test affects the rate of photosynthesis.

   Answers will vary.

6. What is the independent variable in your experiment?

   Answers will vary. *Example:* temperature

7. What is the dependent variable in your experiment?

   See margin for answer.

8. What will be the standardized variables in your experiment?

   See margin for answer.

If pH is chosen as the condition, be aware that dramatic results will not be seen unless the pH values of the trials are very different. If temperature is chosen, steps will need to be taken to ensure that the trial cups remain at constant temperature.

Students could also compare the photosynthesis rates of the disks in the sodium bicarbonate solution with those in plain water, which actually contains small amounts of carbon because of the dissolving of atmospheric carbon dioxide.

Question 7 Answer

Answers will vary. *Example:* the amount of time it takes for the leaf disks to float—if they float at all. Students may also give "rate of photosynthesis" as a correct response since measuring the time for the leaf disks to float is an indirect measurement of the rate of photosynthesis.

Question 8 Answer

Answers will vary depending on what is chosen as the independent variable.

9. What will you use as a control group?

   See margin for answer.

10. Describe the procedure that you will use to test your hypothesis.

    Answers will vary.
    _____
    _____
    _____
    _____
    _____

K Run your experiment.

L Record your data in Table 1. Space has been provided for you to test a range of values for the factor you have chosen to test.

Running the experiment ▶

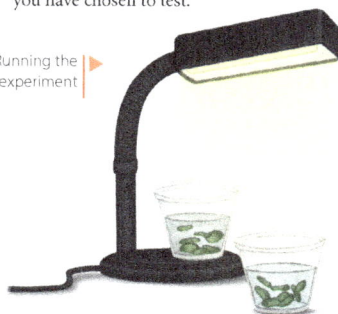

ANALYSIS

M Graph your results in the graphing area. Place your time data on the *x*-axis and the number of floating disks on the *y*-axis.

N Use a different-colored pencil to graph the data for each group. Don't forget to either label each resulting curve or provide a key for which group is represented by each color.

CONCLUSION

11. On the basis of your data, determine whether the rate of photosynthesis is affected by the factor you were testing.

    Answers will vary.

12. Was your hypothesis supported or falsified? Use data to support your answer.

    Answers will vary.
    _____
    _____
    _____

13. At what values for the factor (if a range was tested) did photosynthesis happen at the fastest rate?

    Answers will vary.

14. At what values for the factor (if a range was tested) did photosynthesis happen at the slowest rate?

    Answers will vary.

15. What can you conclude regarding the rate of photosynthesis and the factor you were testing?

    Answers will vary.
    _____
    _____
    _____
    _____
    _____

16. If your results did not support your hypothesis, what factors may have affected your results?

    Answers will vary.
    _____
    _____
    _____

17. What steps would you recommend to improve the experiment you performed?

    Answers will vary.
    _____
    _____
    _____
    _____
    _____

GOING FURTHER

18. Recall from your textbook that a chemical formula, such as the formula for photosynthesis, is a model, and models need to be tested and, if necessary, modified. What evidence did the leaf disk procedure produce that demonstrated the validity of the photosynthesis equation?

Answers will vary. *Example:* Oxygen was produced, according to the equation, and oxygen gas is less dense than water. The disks floated, so a gas was probably produced.

19. In this lab activity you measured something that you could easily observe (floating leaf disks) to make conclusions about a process that you cannot easily observe (photosynthesis). This kind of indirect observation is actually a very common technique in science. Can you think of another process that can be measured indirectly? How is that process measured?

Answers will vary. *Example:* The number of calories burned during exercise can be estimated by observing the amount of work done on a treadmill.

20. When it comes to the need for exercising wise dominion over the earth, food production certainly comes to mind. Why would it be important for farmers to understand what factors influence rates of photosynthesis? How would they apply that understanding to growing crops?

Although farmers may be thinking about plant growth rather than photosynthesis, the two concepts are obviously linked— the rate of the former is determined by the rate of the latter. Farmers need to know what factors (e.g., light intensity and duration requirements, soil pH, water demand, and temperature) influence the growth of their crops so that they can either choose crops suited to local conditions or tailor those conditions to the needs of their crops.

✝ *Applying Christian Understanding to Life*

Carrying out the Creation Mandate makes it necessary to feed a growing population. Modern farming requires the application of good science in order to maximize crop production per unit of land, which also makes it possible to conserve land for other uses, such as habitat for wildlife or open spaces for people to enjoy.

TABLE 1 Number of Floating Leaf Disks

TIME (min)	PRACTICE RUN GROUP (STEP I)	CONTROL GROUP	GROUP 1	GROUP 2	GROUP 3
0		0	0		
1		0	0		
2		0	0		
3		0	0		
4		0	0		
5		0	0		
6		0	0		
7		0	0		
8		0	0		
9		0	0		
10		0	0		
11		0	0		
12		0	0		
13		0	0		
14		1	0		
15		2	0		
16		3	0		
17		3	0		
18		5	0		
19		6	0		
20		7	1		
21		8	1		
22		9	1		
23		11	1		
24		12	1		
25		13	1		

✓ **Sample Data**

The data provided was collected when the lab activity was developed. Students' data should be similar.

GRAPHING AREA A

Floating Disks vs. Time

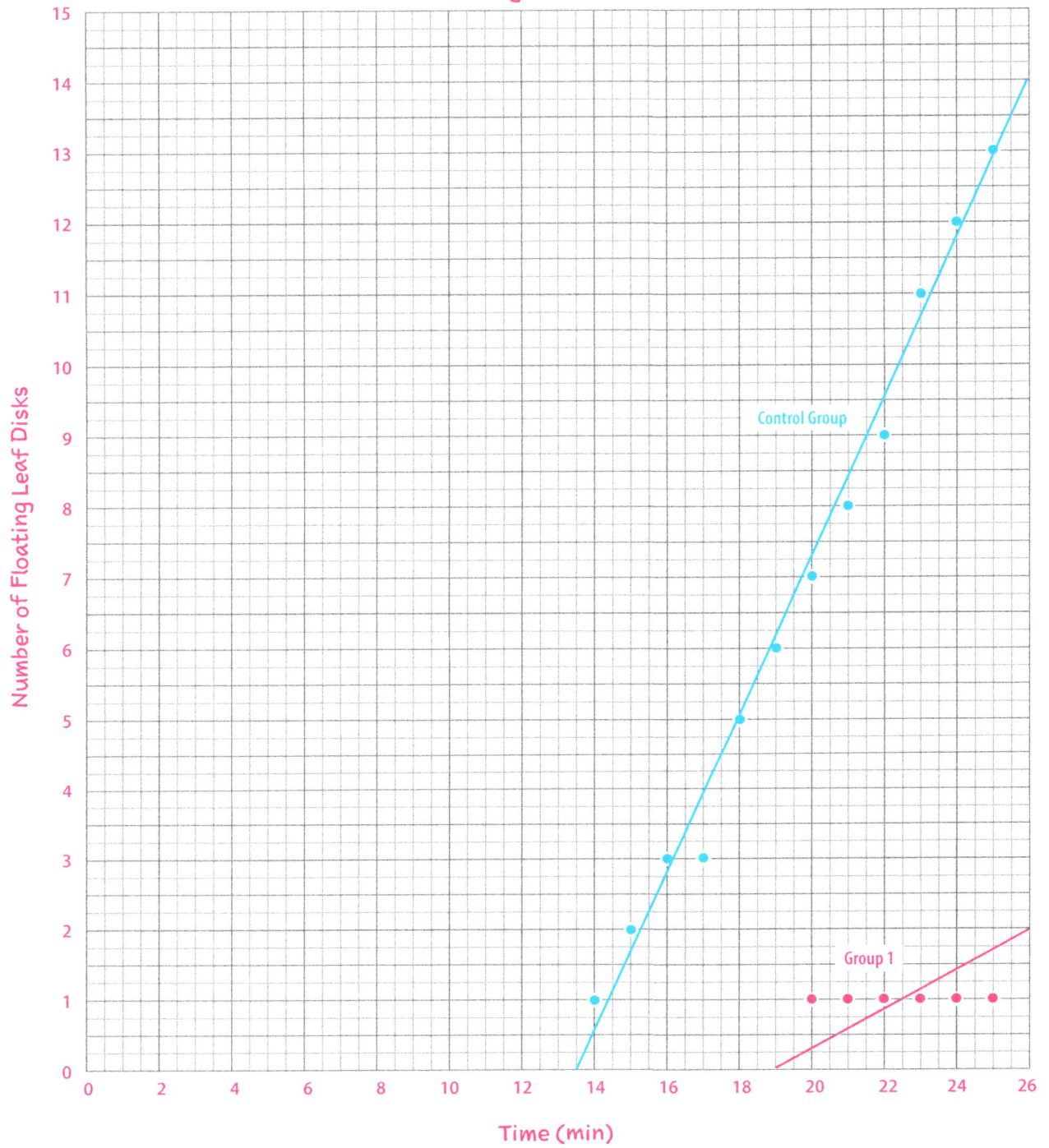

7B LAB

On the Road to Alternative Fuels

Fermentation and Biofuels

In America we can scarcely imagine what life was like before Karl Benz invented the first gas-powered automobile in 1885. Gasoline and diesel fuel, which are used to power most cars, are examples of *fossil fuels*—energy sources derived from the buried remains of plants and animals that have been changed into coal, oil, and natural gas. Because there are no new sources of these fuels being produced today, they are described as *nonrenewable*, meaning that it is possible that one day we could run out of them.

Many people today are engaged in the search for new sources of energy, and *renewable energy* is often their goal. A renewable energy source, as the name suggests, is a source that can be replaced. If a tree is cut down for firewood, for example, a new tree can be planted and grown to replace it. Many scientists and engineers are also looking for *clean energy* sources and *green energy* sources. Clean energy does not produce pollution during its production or use, while green energy sources add little or no greenhouse gases to the atmosphere.

In this lab activity you will investigate some of the issues concerning the production of a particular biofuel—*ethanol*. Ethanol is a kind of alcohol that is frequently used as an additive in gasoline. It isn't extracted from the ground as crude oil is. It's a byproduct of anaerobic cellular respiration, a means of obtaining energy used by many microorganisms, including yeast.

Yeasts use alcoholic fermentation, one form of anaerobic cellular respiration, to extract energy from carbohydrates. In the process, ethanol and carbon dioxide gas are produced as wastes. You should see from this lab activity that we need an inexpensive source of carbohydrates to produce affordable ethanol. In the first part of the activity you will test yeast's ability to ferment different carbohydrate sources, known as *feedstocks*.

Equipment

measuring spoons	resealable plastic bags, snack-size (3)
graduated cylinder, 50 mL or larger	glucose syrup
permanent marker	cornmeal
	wood shavings or grass clippings

? What are some of the economic and societal costs of using fermentation to produce biofuels?

Although many people have invented powered vehicles, Karl Benz's automobile was the first that was powered by gas. The car manufacturer Mercedes-Benz bears his name.

QUESTIONS
- What is fermentation?
- What are biofuels?
- How is fermentation used to produce biofuels?
- Are biofuels worth producing?

active dry yeast
warm water (about 40 °C)
goggles

LAB 7B OBJECTIVES

- Define *fermentation*.
- Determine which of three feedstocks is the most effective for fermentation.
- Describe the issues regarding ethanol production and use.
- Explain how worldview affects one's response to new technologies.

✔ Prelab Discussion

If you're the parent or teacher of a high schooler, you might be old enough to remember life before the advent of widespread cellphone use. Ask your students what they think life would be like without smartphones, texting, or social media. You might also describe for them the technologies that were in their infancy when you were their age. What societal factors determine whether a new technology will be widely accepted?

🧪 Equipment Notes

One 4 oz jar of dry active yeast contains approximately 36 tsp of yeast, or enough for 36 bags (12 groups of students).

Karo Light syrup can be used as glucose syrup.

Prior to the activity you may discuss with students alternative sources of cellulose, such as leaves or twigs. Any such source will be suitable for this activity as long as it is a plant part that does not contain large amounts of sugar.

Evaluating Costs and Benefits

The main idea of this lab activity is to get students to think about how worldview and a basic understanding of science shape a person's attitudes toward whether to adopt new technologies, such as alternative fuels (including ethanol). Worldview affects how a person views the costs and benefits of alternative fuels.

In the first part of the activity students will discover that glucose, that is, sugar, is the most effective feedstock for fermentation. However, during their research they will discover that, at least in the United States, the primary source of this glucose is not sugar cane or sugar beets, but corn, and this is where some important issues must be considered. Because corn starch must first be rendered to glucose prior to fermentation, it incurs added production costs. As of 2016, these added costs make ethanol more expensive than gasoline, as has been the case in most years going all the way back to 1982 according to government data. As a result, the primary push for the widespread use of ethanol comes not from market forces but from government directives and incentives. This calls into question the motives of those who are advocating for ethanol production and use. As students research the topic of ethanol production, hopefully they will see that though scientists can show us how to do a particular thing, the decision about whether such a thing is good and useful is a value judgment closely linked to a person's worldview.

Suggested Modification

This lab activity can be easily modified to be an examination of factors that affect fermentation, as was done in Lab 7A for photosynthesis.

Catalysts and Enzymes (Question 5)

To help students answer Question 5, review the Catalysts and Enzymes subsection on pages 36–37 in the Student Edition.

Students may incorrectly conclude that yeast lacks the enzymes to break down starch; it has the necessary enzymes, but the process is slow. For the purposes of this activity it is necessary only that students conclude that yeast prefers glucose as a feedstock.

PROCEDURE

Fermenting Carbohydrates

A Label each of the three bags with the feedstock that will be fermented in that bag (glucose syrup, cornmeal, or shavings/clippings).

B In each bag, combine 1 tsp of the appropriate feedstock with 1 tsp of active dry yeast.

C To each bag, add 50 mL of warm water and seal each one after removing as much air as possible. Gently mix the contents of each bag.

D Lay each bag on a flat surface and observe for 15 minutes. If a bag appears to be inflating to the point of bursting, you may release some of the gas.

1. After 15 minutes of observation, rank the bags by the amount of gas produced from most to least.

   glucose syrup, cornmeal, shavings/clippings

2. What is the relationship between the gas produced and the rate of fermentation? Explain.

   More gas produced indicates a greater rate of fermentation since more carbon dioxide is produced.

3. On the basis of your results, which of the three do you think would be the best feedstock for producing ethanol? Explain.

   Glucose syrup produced much more gas per unit of feedstock than either cornmeal or shavings/clippings. More gas indicates more fermentation taking place.

The carbohydrates found in the three feedstocks that you tested are different. Glucose syrup contains *glucose*, a simple sugar. Cornmeal is made primarily of *starch*, a long polymer made of many glucose molecules chemically bonded together. Wood shavings, grass clippings, and other plant materials are made mostly of *cellulose*, which is also a long glucose polymer, but the glucose molecules are bonded together in a different manner than those in starch. For cells to process these different carbohydrates, they need customized enzymes that help break down the molecules to release energy and other byproducts.

4. On the basis of your answer to Question 2, which carbohydrate do you think yeast prefers as an energy source?

   glucose

5. Why is yeast able to ferment some feedstocks efficiently but not others?

   Yeast has the enzymes to metabolize glucose but does not have the enzymes needed to break down starch or cellulose efficiently.

Research

Now that you've determined which of the three feedstocks ferments best, you may be wondering whether that feedstock is, in fact, the one most used for ethanol production. The answer may surprise you.

E Do some research on the questions and issues surrounding the production and use of ethanol as a biofuel and summarize your findings in a short paper or presentation. Consider the following important topics:

» What are some of the advantages and disadvantages of ethanol compared with those of fossil fuels?

» Who are the primary voices advocating the widespread use of ethanol?

» What are the primary feedstocks being used to produce ethanol?

» Are any other feedstock sources being investigated?

» How much does it cost to produce ethanol?

» How much does ethanol cost for consumers?

» Is ethanol a safe fuel additive?

» Does ethanol yield as much energy per unit of volume as gasoline?

» What are some of the hidden costs of using ethanol on a large scale? (*Hint:* Any feedstock used to make ethanol obviously can't be used for its original purpose.)

F Be prepared to present and discuss your findings when your presentation is due.

CONCLUSION

Clean, renewable energy sounds like a wonderful solution for mankind's energy needs, and perhaps one day it will be. But new technologies often bring along with them some hidden drawbacks. Many hard questions need to be asked and answered before those new technologies gain widespread acceptance. How those questions are answered often depends on one's worldview.

6. Why is there such a push today to develop alternative energies?

See margin for answer.

_____

_____

_____

7. Are the reasons you stated in Question 6 legitimate justifications for pursuing alternative energies?

Answers will vary. Whether something like finding fossil fuel

alternatives is "good" or "worthy" is a value judgment, which is

often made at least partly on the basis of one's worldview.

_____

Presentation Format

Decide ahead of time the form that students should use to present their findings, whether as a report (written or oral), display, or A/V presentation; you might allow them to choose. Consider using a rubric to evaluate the projects, taking into account the quality of the presentation, the thoroughness of the research done, and the value of the content. You can tailor the project to your students by selecting one or a few of the questions from the suggested list for them to research.

Question 6 Answer

Answers will vary. Many people today fear that fossil fuels will one day be exhausted. Others believe that biofuels are a cleaner energy source and reduce greenhouse gas emissions Many large cities are already dealing with the smog created by pollutants from gasoline engine usage and from industry.

8. How does worldview influence a person's view of the cost versus benefits of adopting certain kinds of alternative technology? What role does worldview play in determining the wide-scale adoption of certain kinds of technology?

   See margin for answer.

9. How should a biblical worldview affect our attitudes and opinions about ethanol and other fossil fuel alternatives?

   We should definitely be open to the use of renewable resources if they show promise for long-term sustainability and value. We cannot postpone human need until that point, however, so we must accept and wrestle with short-term, viable solutions to the energy needs in our country and around the world, even if that means that we cannot achieve complete dependence on renewable resources.

8A LAB

Let's Split

Mitosis and Meiosis

Like the other US states, Wisconsin has a state bird and a state flower—the American robin and the wood violet. But in 2010 Wisconsin lawmakers, in recognition of their state's status as the leading cheese producer in the country, were considering something rather unusual—an official state microbe: the bacterium *Lactococcus lactis*. This microbe loves lactose, the sugar found in milk. When *L. lactis* bacteria are placed in milk, they begin breaking down the milk's lactose for energy, creating lactic acid as a byproduct. Given the abundant supply of lactose in milk, the bacteria begin dividing, making more bacteria that make even more lactic acid. The lactic acid causes the milk to curdle, a process that is the first step in producing several delicious types of cheese, including colby, cheddar, and cottage cheese.

Life as we know it depends on cell division. Cells divide through mitosis to create new cells as old cells die, and many microorganisms, like *L. lactis*, produce new organisms through cell division. In sexual organisms a few cells divide through meiosis to form gametes, which unite to form zygotes that grow and develop into new adults. In this lab activity you will use a microscope to observe cells that have been frozen during the different phases of mitosis and meiosis.

Equipment

microscope

preserved slide of fish embryos prepared for viewing mitosis

preserved slide of fish prepared for viewing meiosis

PROCEDURE

Mitosis

Mitosis happens rapidly in animal embryos, so if a slide of an embryo is prepared properly, it will reveal all the phases of mitosis.

A Obtain a preserved slide of fish embryos.

1. Indicate and describe the type of sectioning on your slide.

   longitudinal section (l.s.); The specimen is cut lengthwise.

B Observe the slide on high dry or oil immersion power, looking for the various stages of mitosis and interphase. You should find all the phases. Check the phases as you find them.

- ☐ interphase
- ☐ prophase
- ☐ metaphase
- ☐ anaphase
- ☐ telophase
- ☐ daughter cells

?

What are the differences between mitosis and meiosis?

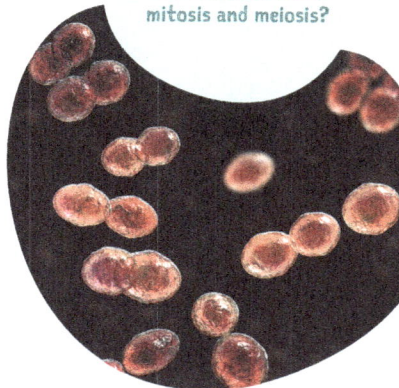

QUESTIONS

- What are the stages of mitosis?
- What are the stages of meiosis?
- What are the differences between mitosis and meiosis?

LAB 8A OBJECTIVES

- Draw the stages of mitosis.
- Explain the differences between cytokinesis in animal and plant cells.
- Identify the stages of meiosis.
- Explain the differences between mitosis and meiosis.

Equipment Notes

Any slide showing mitosis in animal cells may be used in place of the fish embryos slide.

The slide showing meiosis may contain plant or animal cells. Common examples are lily anthers and grasshopper testes.

If your students' microscopes have oil immersion objectives, some may find the higher power useful while others may find it more trouble than it's worth.

State Microbe

Although Wisconsin ultimately did not designate *Lactococcus lactis* as its state microbe, Oregon in 2013 did become the first US state to name an official state microbe: the yeast *Saccharomyces cerevisiae*.

2. How did you identify interphase?

Interphase has no observable chromosomes.

3. How did you identify prophase?

Prophase has an observable mitotic spindle.

4. How did you identify metaphase?

In metaphase chromosomes are lined up on the mitotic spindle.

5. How did you identify anaphase?

During anaphase chromosome pairs are beginning to separate, and the cell begins to split to form two new cells.

6. How did you identify telophase?

In telophase chromatids are now independent and the nucleus encloses them.

7. How did you identify the daughter cells?

The daughter cells are completely separated by their plasma membranes.

8. You may have noticed that some of the embryo cells lack chromatin material. Explain the most probable cause for this. (*Hint:* How were the eggs cut to make this type of slide?)

Some of the cells were cut in such a way that the nucleus is in another slice.

Use the next page to sketch a series of specimen drawings showing the stages of mitosis in fish embryos. Be sure that you draw typical specimens of the various stages of mitosis.

C Draw one cell in the phase indicated in each of the areas provided. Label any structures that you can identify.

PROPHASE   METAPHASE

ANAPHASE   TELOPHASE

Which Stage Is Most Frequently Seen?

While it is likely that your students will encounter interphase most frequently, they are sampling only twenty cells, and there is a very small chance that one or two students will answer Question 9 differently. This is not a problem, but those students may require help on Question 10.

D Randomly pick twenty cells from the slide and count how many of them are in interphase and in each of the four stages of mitosis.

9. Which stage is more frequently seen than the others?

   interphase

10. Why do you think this stage is more frequently seen than the others?

   Interphase takes much longer than any of the phases of mitosis.

Meiosis

While mitosis happens in most of the tissues of a plant or an animal's body, meiosis occurs only in the reproductive organs. So a slide must be made from the reproductive structures in order for meiosis to be seen.

E Obtain a preserved slide of fish meiosis. Observe the slide on high dry or oil immersion power, looking for the various stages of meiosis. You should find all the phases. Check the phases as you find them.

☐ interphase   ☐ prophase II

☐ prophase I   ☐ metaphase II

☐ metaphase I   ☐ anaphase II

☐ anaphase I   ☐ telophase II

☐ telophase I

F Choose one of the stages of meiosis I and draw it and its corresponding stage in meiosis II in the drawing areas below.

DRAWING AREA A   DRAWING AREA B

Comparing Mitosis and Meiosis

Mitosis forms two identical cells; meiosis forms gametes in preparation for sexual reproduction. The processes, while similar, have noticeable differences. Compare mitosis and meiosis by checking the correct choice in the cells in Table 1. All the information you need is in your textbook, but some of the answers may require extra thinking. For telophase assume that the two re-forming nuclei are in separate cells.

TABLE 1 Comparison of Mitosis and Meiosis

TYPE AND PHASE	STAGE CONTAINS	POSITION OF CHROMOSOMES IN THE CELL	NUMBER OF CHROMOSOMES (2n = 6)		NUMBER OF CHROMATIDS (2n = 6)	
Mitosis Prophase	☑ sister chromatids ☐ daughter chromosomes	☑ moving toward center ☐ located at center ☐ moving from center	☐ 3 ☑ 6	☐ 9 ☐ 12	☐ 6 ☑ 12	☐ 24 ☐ NA
Meiosis Prophase I	☑ sister chromatids ☐ daughter chromosomes	☑ moving toward center ☐ located at center ☐ moving from center	☐ 3 ☑ 6	☐ 9 ☐ 12	☐ 6 ☑ 12	☐ 24 ☐ NA
Mitosis Metaphase	☑ sister chromatids ☐ daughter chromosomes	☐ moving toward center ☑ located at center ☐ moving from center	☐ 3 ☑ 6	☐ 9 ☐ 12	☐ 6 ☑ 12	☐ 24 ☐ NA
Meiosis Metaphase II	☑ sister chromatids ☐ daughter chromosomes	☐ moving toward center ☑ located at center ☐ moving from center	☑ 3 ☐ 6	☐ 9 ☐ 12	☑ 6 ☐ 12	☐ 24 ☐ NA
Mitosis Telophase	☐ sister chromatids ☑ daughter chromosomes	no directional movement	☐ 3 ☑ 6	☐ 9 ☐ 12	☐ 6 ☐ 12	☐ 24 ☑ NA
Meiosis Telophase II	☐ sister chromatids ☑ daughter chromosomes	no directional movement	☑ 3 ☐ 6	☐ 9 ☐ 12	☐ 6 ☐ 12	☐ 24 ☑ NA

11. In which phase of which cellular process would you find tetrads lined up on the equatorial plate?

metaphase I of meiosis

12. Contrast the purpose of mitosis with the purpose of meiosis in macroorganisms.

Mitosis produces two identical cells in an organism, while meiosis produces gametes that unite to form a new organism.

8B LAB

The Punnett Square Dance

Inheritance Patterns

In the early twentieth century Gregor Mendel's genetic discoveries were beginning to revolutionize the field of genetics. Geneticists needed a way to predict the possible combinations of alleles in a genetic cross. Reginald Punnett was a young biology professor at Cambridge who had recently published a paper in which he had devised a tool to predict the probability of an organism expressing a particular trait: the Punnett square.

One day he was waiting with mathematician and fellow cricket player G. H. Hardy to play a cricket game that had been delayed by rain. The two professors began to discuss inheritance patterns and gene frequency in a population.

In Chapter 9 you'll learn how Hardy built on Punnett's ideas to understand population genetics. In this lab activity you will use Punnett squares to investigate the different inheritance patterns that can occur in a cross.

How can Punnett squares help scientists learn about inheritance patterns?

QUESTIONS
- What are inheritance patterns?
- How can you tell what inheritance pattern a trait has?

Equipment
none

PROCEDURE

Simple Dominance

Some people in the United States can't drink milk—they are lactose intolerant. When they drink milk, they feel bloated, get stomach cramps, and experience discomfort. This might sound like a disease to you, but this condition is perfectly normal in many places in the world. Most people in Africa and Asia are lactose intolerant, but their traditional diets have never included milk anyway. The ability to digest lactose as an adult is found primarily in people of European and Middle Eastern descent.

To digest lactose as an adult, a person must have the ability to produce *lactase*. This ability is an inherited, dominant characteristic. For the exercises that follow, use an uppercase L to represent the dominant allele and a lowercase l to represent the recessive allele.

- Identify the different basic inheritance patterns.
- Apply knowledge of inheritance patterns by making Punnett squares.

Equipment Notes

This lab activity is a review of Punnett squares. It may be assigned for homework or done in class. Because of the nature of this activity, there are no equipment or safety requirements.

Also, since this lab activity is longer than most and because it is a paper and pencil activity, you may want to assign each section separately after covering each type of inheritance in the classroom.

Review Punnett Squares

Before beginning the activity, have students review how to use Punnett squares on pages 160–64 in the Student Edition. They will gain valuable experience in filling in Punnett squares in this activity.

More on G. H. Hardy

G. H. Hardy went on to formulate Hardy's law, an important principle of population genetics. In 1943 it was recognized that Wilhelm Weinberg had independently formulated the same principle, and since then it has been referred to as the Hardy-Weinberg principle. Students will learn about this principle in Chapter 9.

A Fill in Table 1 with the correct genotype.

TABLE 1

GENOTYPE	PHENOTYPE
_LL_	a person who is lactose tolerant
_Ll_	
_ll_	a person who is lactose intolerant

1. Is there any difference between the phenotypes resulting from the two genotypes that cause a person to be lactose tolerant? Explain.

   No. They can both digest lactose.

B Jon Forrest, who is homozygous for lactase production, marries Rachel, a lactose intolerant woman. Use this information to fill out the Punnett square in the margin for their offspring and answer Questions 2 and 3.

2. If Jon and Rachel have a daughter who is heterozygous for the lactase gene, will she be able to digest lactose? Explain.

   Yes. Because the lactase gene is dominant, a heterozygous individual will be able to digest lactose.

	L	L
l	Ll	Ll
l	Ll	Ll

3. Could the Forrests have a child who is homozygous recessive for the lactase gene and thus unable to digest lactose? Explain.

   No. This couple's children will be only heterozygous for the lactase gene.

C Zack, one of the Forrests' sons, marries Ellen, who is known to be heterozygous for the lactase gene. Use this information to fill in the Punnett square in the margin and answer Questions 4–6.

4. Is there a possibility that any of Zach and Ellen's children will be lactose intolerant? If so, circle the genotype(s).

   Yes

	L	l
L	LL	Ll
l	Ll	(ll)

The *phenotypic ratio* is the ratio of all the possible phenotypes resulting from a cross. The *genotypic ratio* is the ratio of all possible genotypes resulting from a cross. Both are reduced to lowest terms. For example, if a male fruit fly heterozygous for wings (Ww) mates with a homozygous wingless female (ww), the phenotypic ratio of their offspring would be 1 winged : 1 wingless. The genotypic ratio would be 1 Ww : 1 ww.

5. What is the phenotypic ratio for lactose intolerance in Zack and Ellen's children?

   3 lactose tolerant : 1 lactose intolerant

6. What is the genotypic ratio of their children?

   1 LL : 2 Ll : 1 ll

Hugh McTaggart is lactose intolerant, but his wife Amy is lactose tolerant. Amy's mother is lactose intolerant, but Amy's father is lactose tolerant.

7. Is it possible for the McTaggarts to have a child that can digest lactose? Explain.

   Yes. Their child could inherit the dominant gene from Amy.

8. Is it possible for the McTaggarts to have a child that is lactose intolerant? Explain.

   Yes. Their child could inherit a recessive gene from both Amy and Hugh.

9. Explain how you can know Amy's genotype.

   Amy's mother is lactose intolerant, making her genotype ll, so Amy must have received a recessive allele from her mother. Since Amy can digest lactose, her genotype must be Ll.

10. How might the ability to digest lactose be helpful for a population?

    Individuals with the ability to digest lactose would be able to obtain nourishment from milk from domesticated animals.

Incomplete Dominance

Though Mendel first discovered the principles of inheritance, we now know that genetics is often much more complex than he imagined. One such complication is *incomplete dominance*.

D When a homozygous red radish plant is crossed with a homozygous white radish plant, purple radishes result.

11. The genotype of a white radish is ____, that of a red radish is ____, and that of a purple radish is ____.

    $C^w C^w$; $C^r C^r$; $C^r C^w$

⟳ Review Incomplete Dominance

Have students review the material on incomplete dominance on page 162 in the Student Edition.

✔ Allele Symbols for Incomplete Dominance

Students may use different symbols for alleles than those shown here. This is acceptable. The important thing is that, for an incomplete dominance trait, the students use a single uppercase letter to represent the trait along with different lowercase superscripted letters for each allele.

12. In the chart below, write the possible gametes that each type of radish can produce.

	C^w	C^w
C^r	C^rC^w	C^rC^w
C^r	C^rC^w	C^rC^w

TYPE OF RADISH	POSSIBLE GAMETES
White	C^w
Red	C^r
Purple	C^r, C^w

13. If the pollen from a white radish fertilizes the egg of a red radish, what will be the genotypes and the phenotypes of the offspring? Use the Punnett square in the margin to prove your answer.

genotype: C^rC^w; phenotype: purple

14. If pollen from a red radish flower fertilizes the egg of a flower on another red radish plant, what will be the genotypes and phenotypes of the offspring?

They would all be red and have the genotype C^rC^r.

15. If two purple radishes are cross-pollinated, what are the genotypic and phenotypic ratios of the F_1 generation? Prove your answer by making the proper cross on the Punnett square in the margin.

	C^r	C^w
C^r	C^rC^r	C^rC^w
C^w	C^rC^w	C^wC^w

genotypic ratio: 1 C^rC^r : 2 C^rC^w : 1 C^wC^w

phenotypic ratio: 1 red : 2 purple : 1 white

16. If a red radish and a purple radish are cross-pollinated, what will be the phenotypic and genotypic ratios? Complete a Punnett square of the cross in the margin if needed.

genotypic ratio: 1 C^rC^r : 1 C^rC^w

phenotypic ratio: 1 red : 1 purple

17. What would be the genotypic and phenotypic ratios if a white radish and a purple radish were crossed?

genotypic ratio: 1 C^rC^w : 1 C^wC^w

phenotypic ratio: 1 purple : 1 white

Let's consider a similar but more complex problem related to the tails of cats, another trait determined by incomplete dominance.

E The litter resulting from the mating of two short-tailed cats contains two kittens with long tails, six with short tails, and three without tails. In the Punnett square in the margin, diagram a cross that will show the above results.

	T^l	T^n
T^l	$T^l T^l$	$T^l T^n$
T^n	$T^l T^n$	$T^n T^n$

18. Give a key for the letters you choose to represent the alleles.

Answers will vary. *Example:* T^l = allele for long tail, T^n = allele for no tail, $T^l T^n$ = genotype for short tail

19. What are the genotypes of the parents?

$T^l T^n$

20. How does the ratio of the kittens given in the statement compare to the ratio obtained from the Punnett square? Is it close enough for you to be sure that you used the proper genotypes when you diagrammed the cross? Explain.

It does not agree exactly, but it is close enough that the ratio is accurate. The expected ratio is 1 long : 2 short : 1 tailless. The ratio given in the problem is 2 : 6 : 3.

The fur of some types of mice illustrates incomplete dominance. The fur can be black agouti, normal black, or albino (white). A mouse that is heterozygous is normal black. For Questions 21–25, use F^b to indicate the allele for black agouti fur and F^w to indicate the allele for albino white fur.

21. Determine the black agouti mouse genotype.

$F^b F^b$

22. Determine the normal black mouse genotype.

$F^b F^w$

23. Determine the albino mouse genotype.

$F^w F^w$

F Using the Punnett square in the margin, diagram a cross between a normal black mouse and an albino mouse.

	F^b	F^w
F^w	$F^b F^w$	$F^w F^w$
F^w	$F^b F^w$	$F^w F^w$

24. What are the genotypic and phenotypic ratios in the F_1 generation?

genotypic ratio: 1 $F^b F^w$: 1 $F^w F^w$

phenotypic ratio: 1 normal black mouse : 1 albino mouse

25. This cross produces a litter of fifteen mice. Eight of the mice are normal black, and seven are albino. Was the expected phenotypic ratio obtained? Explain.

Yes. The expected ratio is an equal number of normal black and albino mice. The results were close to this expected ratio.

This cow has a roan coat, which resulted from codominance.

	C^R	C^W
C^R	$C^R C^R$	$C^R C^W$
C^W	$C^R C^W$	$C^W C^W$

Codominance

When a white bull (genotype $C^W C^W$) crosses with a red cow (genotype $C^R C^R$), a roan cow (genotype $C^R C^W$) results.

26. A red bull mates with a red cow. Could their offspring be a roan? Explain.

   No. Red cattle are homozygous, so a red bull and a red cow could have only red offspring.

Two roan cattle mate. Use the Punnett square in the margin to diagram the cross.

27. Give the genotypic and phenotypic ratios for the offspring.

   genotypic ratio: 1 $C^R C^R$: 2 $C^R C^W$: 1 $C^W C^W$

   phenotypic ratio: 1 red : 2 roan : 1 white

Cattle typically have only one or two calves at a time, so it's difficult to actually see this phenotypic ratio in real life. However, these principles can be used to solve other real problems.

28. Ned Wright's roan cow is pregnant, and the rancher believes that she has mated with his red bull, but she may have mated with his neighbor's roan bull. The cow gives birth to twins, one roan and one white. Which bull sired the calves? Explain.

   The roan bull is the sire. The offspring of a red bull and a roan cow could be only red or roan.

29. A farmer with a red cow wants to have roan calves. What color(s) of bull can he cross his cow with in order to have roan calves? If he can get roan calves by crossing his cow with more than one color of bull, which color of bull is most likely to give him roan calves? Draw a Punnett square in the margin if needed.

   The farmer could possibly get roan calves by crossing his red cow with either a roan bull or a white bull, but a cross with the roan bull could give him red calves. So he should cross her with the white bull since he would be certain to get roan calves.

Multiple Alleles

Sometimes there will be more than one pair of alleles possible at a single locus. Three or more alleles, rather than just two, may be possible.

A gene that controls fur color in rabbits has four different alleles. The wild-type allele, symbolized by R, gives the rabbit a dark color; the exact color depends on several other genes that we will not consider here. The chinchilla allele, denoted by r^{ch}, gives the rabbit a gray look. It is recessive to the wild-type allele but dominant to the other alleles. The Himalayan allele, denoted by r^h, gives a rabbit white fur with dark ears, nose, feet, and tail. It is recessive to the wild-type and chinchilla allele. The albino allele, denoted by r, gives the rabbit an entirely white coat. This allele is recessive to all the other alleles.

G Using the information above, fill in Table 2 with possible genotypes of each phenotype.

TABLE 2

	PHENOTYPE	POSSIBLE GENOTYPE
	wild-type	RR, Rr^{ch}, Rr^{h}, Rr
	chinchilla	$r^{ch}r^{ch}, r^{ch}r^{h}, r^{ch}r$
	Himalayan	$r^{h}r^{h}, r^{h}r$
	albino	rr

30. What is the maximum number of color types that could occur in one litter?

A maximum of three different color types could occur in a litter at one time.

31. A male chinchilla rabbit with the genotype of $r^{ch}r^{h}$ is bred with a female albino rabbit. What will the genotypic and phenotypic ratios be? Use the Punnett square in the margin if you need to.

genotypic ratio: 1 $r^{ch}r$: 1 $r^{h}r$

phenotypic ratio: 1 chinchilla : 1 Himalayan

	r^{ch}	r^{h}
r	$r^{ch}r$	$r^{h}r$
r	$r^{ch}r$	$r^{h}r$

✓ **Himalayan Rabbit Fur**

The Himalayan allele is affected by temperature. Direct students to page 169 in the Student Edition for further information.

Review Polygenic Inheritance

Have students review the section on polygenic inheritance on page 163 in the Student Edition.

Question 34

Students might feel overwhelmed by the polygenic inheritance Punnett square. If you anticipate this happening, do Question 34 as a demonstration.

32. A male wild-type rabbit is crossed with a female Himalayan rabbit, and the resulting litter includes four wild-type rabbits, two Himalayan rabbits, and two albino rabbits. What are the genotypes of the parents?

Rr, r^hr

Polygenic Inheritance

Up to this point we have worked only with traits that are determined by one gene. But this is actually rarely the case. Most traits are the results of interactions between multiple genes.

Labrador retrievers can be black, chocolate (brown), and yellow. The coloration is a result of the interactions of two genes. The first gene, the *E gene*, occurs in two alleles, E and e. The dominant E allele codes for the dog to be either black or chocolate. The recessive e allele codes for the dog to be yellow. So a yellow lab is homozygous recessive for this gene. The second gene, called the *B gene*, determines whether a dog with a dominant E gene will be black or chocolate. It also comes in two alleles. The dominant B allele codes for black fur while the recessive b allele codes for chocolate fur.

33. What is the genotype of a chocolate Labrador puppy whose mother was a yellow lab? How do you know this?

bbEe; Since the puppy has chocolate fur, it must be homozygous recessive for the B gene. Since it is not a yellow lab, it must have at least one dominant allele for the E gene; since its mother was homozygous recessive for the E gene, the puppy must be heterozygous.

34. A male black lab with a genotype BBEe is crossed with a chocolate female (bbEe). What will the genotypic and phenotypic ratios of the offspring be? Use the Punnett square below.

	BE	BE	Be	Be
bE	BbEE	BbEE	BbEe	BbEe
bE	BbEE	BbEE	BbEe	BbEe
be	BbEe	BbEe	Bbee	Bbee
be	BbEe	BbEe	Bbee	Bbee

genotypic ratio: 1 BbEE : 2 BbEe : 1 Bbee

phenotypic ratio: 3 black : 1 yellow

35. A yellow lab is crossed with a black lab. The resulting litter contains three black puppies, two chocolate puppies, and four yellow puppies. The mother of the yellow parent was homozygous dominant for the B gene. What are the genotypes of the two labs?

BbEe and Bbee

36. Two yellow labs, both heterozygous for the B gene, are crossed. What are the genotypic and phenotypic ratios of the offspring?

genotypic ratio: 1 BBee : 2 Bbee : 1 bbee

phenotypic ratio: 1 yellow : 0 black : 0 chocolate

Sex-Linked Traits

In 1911 Thomas Hunt Morgan published a paper suggesting that some genes were carried on chromosomes that we now call sex chromosomes. His discovery of sex-linked traits was based on his studies of inheritance in common fruit flies. The first sex-linked trait he discovered was one that caused the flies' eyes to be white instead of the usual red.

H Fill in the Punnett square below for the cross of a white-eyed male (genotype X^rY) with a wild-type female (X^RX^R).

37. What are the genotypic and phenotypic ratios for the offspring of this cross? Be sure to include the sex of the offspring in the phenotypic ratios.

	X^r	Y
X^R	X^RX^r	X^RY
X^R	X^RX^r	X^RY

genotypic ratio: 1 X^RX^r : 1 X^RY

phenotypic ratio: 1 wild-type female : 1 wild-type male

The fly on the right is a wild-type individual. The fly on the left has white eyes because of a sex-linked mutation.

I One of the female offspring from the previous step mates with a wild-type male. Fill in the Punnett square in the margin for the cross.

38. What are the genotypic and phenotypic ratios for the offspring of this cross?

genotypic ratio: 1 X^RX^R : 1 X^RX^r : 1 X^RY : 1 X^rY

phenotypic ratio: 2 wild-type females : 1 wild-type male : 1 white-eyed male

	X^R	Y
X^R	X^RX^R	X^RY
X^r	X^RX^r	X^rY

Have students review the section on sex-linked traits on page 164 in the Student Edition.

	X^r	Y
X^R	X^{R}X^r	X^{R}Y
X^r	X^{r}X^r	X^{r}Y

J Another female from the original cross mates with a white-eyed male. Fill in the Punnett square in the margin for the cross.

39. What are the genotypic and phenotypic ratios for the offspring of this cross?

genotypic ratio: 1 X^{R}X^r : 1 X^{r}X^r : 1 X^{R}Y : 1 X^{r}Y

phenotypic ratio: 1 wild-type female : 1 white-eyed female :

1 wild-type male : 1 white-eyed male

Two flies mate. When their offspring emerge, there are thirty-one white-eyed males and twenty-nine wild-type females. Use this information to answer Questions 40–42. Draw a Punnett square in the margin if needed.

40. What is the phenotypic ratio of these offspring?

1 wild-type female : 1 white-eyed male

41. What was the genotype and phenotype of the father?

X^{R}Y, a wild-type male

42. What was the genotype and phenotype of the mother?

X^{r}X^r, a white-eyed female

9A LAB

Fix It!

Modeling Genetic Drift

Have you ever worn a warm, comfortable mohair sweater? Mohair is a fiber produced from the wool of Angora goats, believed to be descended from the markhor, a wild goat native to Central Asia. As you can see in the photos on the right, Angora goats look a lot different from their wild cousins! How did this happen? For many centuries humans have been selectively breeding (i.e., artificially selecting) goats domesticated from wild stock. As a result, there are now many varieties of domestic goats raised for meat, milk, and wool or just for getting rid of the poison ivy in your yard.

Charles Darwin and other early evolutionists imagined that natural selection worked much like artificial selection, slowly acting within populations to gradually produce fitter organisms. But later scientists recognized that the gene pools of large populations of organisms tended to remain stable over time. In other words, the change in allele frequency needed to produce new traits for natural selection to act upon was unlikely to occur in large populations. Scientists needed to propose new mechanisms that could produce the necessary change in allele frequencies, also known as genetic drift. In this lab activity you will be simulating genetic drift in a population using beads to represent the alleles for a gene.

Equipment

beads, red-colored (100)   small container
beads, white-colored (100)

PROCEDURE

Part 1: Modeling Genetic Drift

A Thoroughly mix together 50 red beads and 50 white beads in a small container. Each color of bead represents an allele for a particular gene. Your container of beads represents a gene pool in which each allele has a frequency of 0.5 (50%).

B Without looking into the container, randomly select 8 pairs of beads. These beads represent the genotypes of Generation 1, that is, the 8 descendants of the original population that will found a new population. Record the frequency of the red and white alleles in this new population in the *Generation 1* row of Table 1 (the allele frequency for Generation 0 has been filled in for you). For example, if 10 of the 16 selected alleles are red, the new frequency would be $10/16 \times 100\% = 62.5\%$ red and 37.5% white.

?

How do populations acquire new traits?

A wild markhor

A domesticated Angora goat

QUESTIONS

- What is genetic drift?
- How does genetic drift fix an allele within a population?
- Can genetic drift sufficiently explain the production of new phenotypes within a population of organisms?

- Explain how genetic drift works.
- Explain why genetic drift may be an inadequate means of fixing new traits within populations.

Introduction to Lab 9A

Throughout their years in high school, college, and beyond, students will hear evolution presented as if it were a well-established fact. The truth is that even many secular evolutionary biologists recognize that there are serious shortcomings in the ability of current evolutionary thought to explain just how new traits can be produced and fixed within populations. This lab activity is a simulation that helps students understand population genetics, especially its purported role in evolution.

Equipment Notes

If you can't find colored beads, try marbles, different varieties of dried beans, or even colored candies—if you can keep your students from eating the supplies before the activity is done! It is essential, though, that the particles all feel similar to the touch so that students can't discriminate by size or texture.

Material to Review

Students should review the material on genetic limits in Section 9.1 on pages 178–79 in the Student Edition.

Question 1 Answer

Answers will vary. Sample data in Table 1 shows the red allele being fixed after seven generations.

Question 3 Answer

Since each allele has the same frequency in the initial setup, each allele starts with an equal chance (one in two, or ½) of becoming the fixed allele in a particular group. There is a one-in-four chance (½ × ½) that the same allele will be fixed for any two groups, and that chance decreases exponentially as the number of groups increases. Therefore, the greater the number of groups, the less we would expect all groups to fix the same allele.

Question 4 Answer

Groups in which red became fixed will have frequencies that approach 100%; the frequency for red will approach 0% for groups in which white became fixed. Also, gene frequency in one generation only predicts, not determines, the frequency in the next generation, causing trend lines to differ even for groups that have the same color become fixed.

Question 6 Answer

This model assumes that no individuals move into or out of the population. It also examines the effect of only one gene variation and does not consider how that variation might work in tandem with other variations. Besides not considering competitive advantage, this model also assumes that neither allele is deleterious; both alleles are assumed to be neutral with regard to advantage.

C Assume that the bead alleles code for color in a simple dominance pattern, with red being the dominant allele. How many of the founders have the red phenotype? How many are white? Record your answers in Table 1.

D The 8 founding individuals that you selected in Step B provide the gene pool for the next generation. Refill your container with 100 beads with the same proportion of red and white beads as for Generation 1. Again, randomly select 8 pairs of beads, representing the genotypes of the next generation (Generation 2). Record the frequency of red and white alleles for this generation in the *Generation 2* row of Table 1.

E How many individuals in Generation 2 are red? How many are white? Record your answers in Table 1.

F Repeat this process, each time recording the allele frequency and number of each phenotype for each generation, until only one kind of allele is present, or until you reach 10 generations.

1. When an allele is eliminated from a population, the remaining allele is said to be *fixed*. Did one of the two alleles become fixed? If so, which one?

   See margin for answer.

2. If an allele became fixed, how many generations did the process take? If neither allele became fixed, what was the lowest frequency reached by one allele?

   Answers will vary.

G Graph the frequency of the red allele in each generation in Graphing Area A. Place the generation number on the *x*-axis and the allele frequency in percent on the *y*-axis.

3. Compare your results with other groups in your class. Did the same allele become fixed for each group? Is this the result that you would expect, given the experimental design? Explain.

   See margin for answer.

4. Did the graphs for the frequency of the red allele look the same for every group? Explain.

   See margin for answer.

5. For this simulation, nothing was said about whether one allele or the other gave an organism a competitive advantage. Suppose the red allele actually gave an organism a competitive advantage. How would that affect the outcome of the experiment?

   If the red allele offered a competitive advantage, then it would be more likely to become fixed and less likely to be lost within the population.

6. What are some other factors that could affect genetic drift in a population that were not considered in this exercise?

   See margin for answer.

Part 2: Genetic Drift Simulations

Whew! Handling all those beads was a chore, wasn't it? What if you wanted to run the experiment again, but instead of looking at a population of 8 individuals you want to see whether you get the same result for a population of hundreds of organisms? You're going to need a lot more beads!

But don't worry. Scientists and students nowadays have it easier than their counterparts back in the early days of population genetics research. Computers can now simulate things like genetic drift for large populations over many generations. In this section you'll try out one of those simulations.

H Do an internet search using the keywords "genetic drift simulation." You should find several sites that have simple genetic drift simulators. Running the simulation will produce a graph like the one you created in Part 1, showing the trend toward fixing or losing an allele within a population.

I Set the population size to 8 (like the population in Part 1) and the allele frequency to 0.5 and run the simulation. Some simulators allow you to test multiple populations at the same time.

7. What happens to the allele being tested? Is it fixed or lost? Over many generations or just a few?

   See margin for answer.

8. Run the simulation again with the same settings. Do you get the same results? Explain.

   The simulation will produce slightly different results every time it is run. Since the allele does not confer advantage, it has an equal chance of being either fixed or lost.

9. Now try changing the population size. Run the simulation with populations of 100, 500, and 1000. How is the allele frequency affected?

   The allele frequency changes more slowly with each increase in population size, and the allele becomes less likely to be fixed or lost.

10. Suppose the initial frequency of a desired allele is less than 0.5. Set the initial frequency to 0.1 and run the simulation for different-sized populations. What happens to the allele frequency?

    The allele frequency usually declines. A lower initial frequency makes it more likely that an allele will be lost.

11. Make a prediction about how the allele frequency will be affected if the initial frequency is greater than 0.5.

    Answers will vary.

12. Now try running the simulation with the initial frequency set at 0.9. How is the allele frequency affected?

    The allele frequency usually increases. A higher initial frequency makes it more likely that an allele will become fixed.

13. Using what you have learned from Parts 1 and 2, write a summary of how population size and initial allele frequency affect whether an allele becomes fixed in a population of organisms. Why is this important in a discussion of natural selection?

    See margin for answer.

14. Both creationists and evolutionists recognize that for natural selection to work, new genetic information within a population is necessary. Did the simulations that you used show a gain of new genetic information or a loss of genetic information? Explain. (*Hint:* Remember that the simulations are all based on two or more alleles for a gene. If one increases in frequency, the others must necessarily decrease.)

    These simulations demonstrate a net loss of genetic information. Even if an allele is fixed, it means that one or more other alleles for that gene are lost.

Genetic Drift Simulators

There are many free simulators available online. Some are simple, allowing adjustment of initial allele frequency and population size only and testing only one population at a time. Others allow the user to set more parameters and can test multiple populations simultaneously. Be sure to try them out before lab day! As an option, you can decide ahead of time which site you want students to use and direct them to it. To view some simulators that are available for download, view the links *Genetic Drift Simulator 1, 2,* or *3,* each of which is available as a digital resource.

Question 7 Answer
Answers will vary. The allele should be fixed or lost fairly rapidly, often in ten generations or less.

Question 13 Answer
An allele is more likely to become fixed if a population is small and the initial frequency of the allele is greater than 0.5. Increasingly less frequent alleles and larger population sizes both work to make fixing an allele less likely. For natural selection to occur, new alleles must become fixed, suggesting that small populations and a high initial frequency of an allele are necessary.

Evaluate Contrary Thinking: Evolution's Thorny Problem

Discussion on the history of evolutionary theory can be valuable. As our understanding of the natural world increases, more and more shortcomings of the current evolutionary paradigm, called the *modern synthesis* (see Chapter 10), become evident, to the point that even a growing number of secular scientists are now concluding that mutation and natural selection are inadequate to explain the observable diversity of living things. Theories such as genetic drift, intended to address some of the weaknesses of Darwinism, are often demonstrated to be not up to the task. For additional information, see *Answers in Genesis*. The AIG link *Population Genetics* is available as a digital resource.

Seed banks save heirloom varieties of seeds so that genetic information isn't lost through selective breeding.

15. Does selective breeding, such as for the Angora goat, affect genetic information? Explain.

See margin for answer.

16. How do evolutionists suggest that new information is produced within a population?

through mutation

17. How does a loss of genetic information fit within a biblical worldview?

In the beginning God pronounced everything that He created to be "very good." This included the gene pools of all created organisms. All of creation was brought under the Curse by Adam's sin, and the effects of this included the loss of genetic information, which means that there was a loss of genetic diversity, making organisms less capable of dealing with change in their environment.

GOING FURTHER

So far we have considered only the way that initial allele frequency and population size affect genetic drift. Some simulators also allow you to set values for migration (organisms moving into or leaving a population), bottlenecks (a sudden and dramatic decrease in a population), mutation (the rate at which the alleles being tested mutate), and fitness (an indication of the relative fitness of each possible phenotype that can be produced by combinations of the tested alleles).

18. Try experimenting with the other simulation settings and write a brief summary of what you discover.

Answers will vary.

TABLE 1 Frequency of Red and White Alleles in Bead Populations

GENERATION	RED ALLELE FREQUENCY (%)	WHITE ALLELE FREQUENCY (%)	NUMBER OF RED PHENOTYPE (RR or Rr)	NUMBER OF WHITE PHENOTYPE (rr)
0	50	50		
1	25	75	4	4
2	25	75	4	4
3	69	31	5	3
4	75	25	8	0
5	81	19	8	0
6	94	6	8	0
7	100	0	8	0
8				
9				
10				

✔ **Sample Data**

The data provided in Table 1 and graphed on page 90 was collected when the lab activity was developed. Students' data should be similar.

Change in Red Allele Frequency

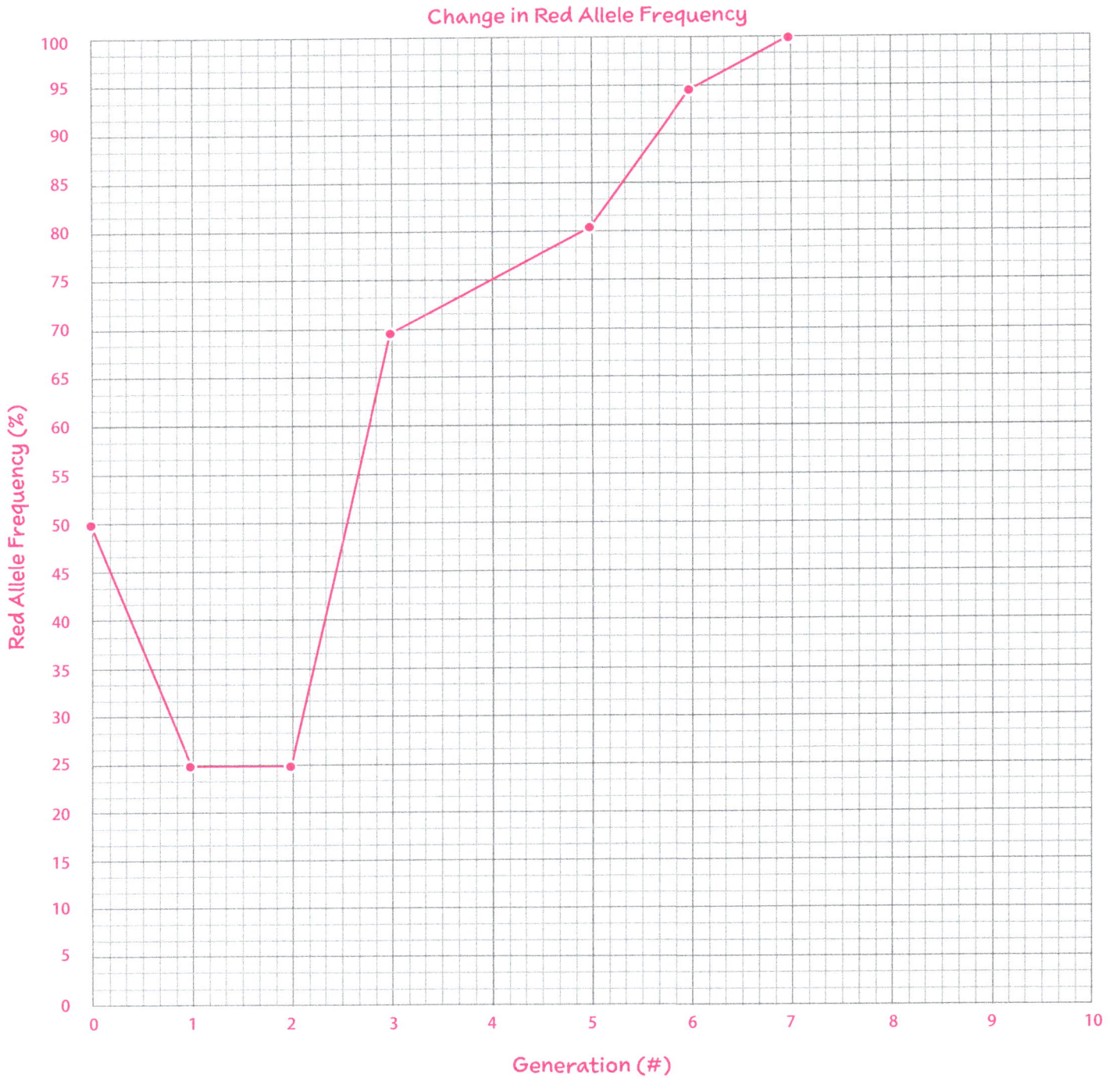

Red Allele Frequency (%) vs Generation (#)

9B LAB

Whodunit?

DNA Fingerprinting

A crime has occurred, and the culprit has been caught and punished. Justice has been served! Or has it?

DNA fingerprinting is an incredibly useful tool in forensics. It is used to analyze evidence in criminal investigations, resolve paternity cases, and identify victims of crimes or disasters. The idea behind this technique is that each individual has unique DNA sequences. Though there are several DNA sequences that all people and even some animals share, everyone has some unique DNA sequences in noncoding regions. These sequences can show the difference between individuals, just like fingerprints from a person's finger. What might surprise you is that DNA fingerprinting has shown a large number of people, already convicted of a crime, to be innocent rather than guilty.

The process of DNA fingerprinting involves a series of techniques that produce a pattern of DNA that can be analyzed. This pattern looks like a bar code. To get this pattern, DNA must be isolated and purified in a way that is similar to the technique that you learned in Lab 6B. In this lab activity you will simulate the steps of DNA fingerprinting to analyze some "DNA samples."

How is DNA fingerprinting used to serve justice?

Equipment
none

THE CRIME

The day was going to be perfect! The sophomore class had developed a flawless recipe for Super Chocolate Brownies. They were entering their recipe, along with samples, into the school bakeoff competition, the tiebreaking event in this year's school spirit week. But when Mr. Sokolata, the sophomore class advisor, entered his classroom after lunch, he was shocked to find that the freshly baked brownies, which he had left on his desk to cool, had been mostly eaten, and the recipe card was missing! Who could have done such a thing?

No one saw what happened, but on the basis of an abundance of circumstantial evidence, a student named Chip Lee was charged with theft that very afternoon and expelled from school. The case didn't sit well with Mrs. Gumschuh, the biology teacher. She decided to look into the case. Happily, Mr. Sokolata had no afternoon classes in his room, and the crime scene was still mostly undisturbed when Mrs. Gumschuh stopped by after school.

QUESTIONS
- What is DNA fingerprinting?
- How is DNA fingerprinting done?
- How is DNA evidence used to help solve crimes?

Whodunit? 91

LAB 9B OBJECTIVES
- Outline the process of DNA fingerprinting.
- Explain how DNA fingerprinting is used to help solve criminal cases.

Introduction to Lab 9B

DNA fingerprinting has become a standard tool for criminal investigators. This simple paper-and-pencil lab activity will familiarize students with the basics of DNA fingerprinting without the need to purchase expensive electrophoresis kits. Because this lab activity is relatively brief and requires no equipment, it can be assigned as homework to save class time.

Prelab Discussion

It might be helpful to probe students' prior understanding of DNA fingerprinting. The term is commonplace, but students might have misconceptions about how the process works.

Equipment Notes

For teachers wanting to delve further into this topic, DNA electrophoresis kits are available from most science equipment suppliers. For this exercise no equipment is necessary, and there are no safety concerns.

Trick DNA Evidence

There are known instances of criminals who have purposefully left falsified DNA evidence at crime scenes for investigators to collect, including DNA from other individuals or synthesized DNA.

THE EVIDENCE

Here's what she found. The baking pan, a few partially eaten brownies still in it, remained on the desk where Mr. Sokolata had left it. At a student desk near the pan of brownies was a senior class T-shirt with some hairs on the collar. Mrs. Gumschuh learned from Mr. Stern, the dean of students, that no sign had been found of forced entry into the classroom from outside the building. Mrs. Gumschuh collected samples of the hair from the T-shirt and swabbed the brownie pan for saliva to test for DNA. She also collected cheek swabs from each of the four original suspects. Read on.

Suspect Data

After interviewing students, Mr. Stern had narrowed the list of suspects down to four individuals. Circumstantial evidence pointed to the following persons.

SUSPECT 1: I. K. "BEN" ORDELIJK, CUSTODIAN

Mr. Ordelijk is in charge of cleaning the classrooms and has keys to every room at the school, so he definitely had access to the brownies, even if the door hadn't been open. He's also been telling other staff members for weeks that he's been craving sugar since starting a new low-carb diet.

SUSPECT 2: CHIP LEE, STUDENT

Chip has been an unwilling visitor to Mr. Stern's office on several occasions and has been known to try to talk his way out of consequences for his actions. Chip was seen entering the classroom by Mr. Ordelijk, though he claims he was retrieving a notebook. His addiction to brownies (especially those with chocolate chips!) is legendary. He had mentioned to some students earlier that Mr. Sokolata often left his door open during lunch and had reportedly quipped, "Boy, Mr. Sokolata better not leave those brownies where I can get to them!" Chip usually eats lunch with the same group of students every day, but on the day of the crime he arrived late and had dark-colored crumbs on his clothing.

SUSPECT 3: BROCK LEE, STUDENT

Brock is Chip's twin brother and Mr. Sokolata's classroom aid. He stops by Mr. Sokolata's room every day at lunch to restock supplies, but it's common knowledge that he prefers doughnuts to brownies.

SUSPECT 4: PATSY FAHLGUY, SENIOR CLASS PRESIDENT

Patsy had vowed that the senior class would win the spirit week competition at any cost. She had all the seniors wear their class T-shirts to show team spirit.

DNA Analysis

Mrs. Gumschuh isolated DNA from the hair sample, saliva sample, and cheek cell samples from each of the suspects. Next, she cut the DNA with a restriction enzyme. The restriction enzyme Mrs. Gumschuh used always cuts at the sequence ATTA between the two Ts.

1. The restriction enzyme used in this step cut DNA at the same specific site every time. Why is this essential for the procedure?

   If the enzyme cut at random sites each time, you would never

   get a consistent fingerprint for the same DNA sample.

✔ No Complementary Strands

To avoid confusion, only one DNA strand from each sample is shown. The complementary strands have been omitted.

A DNA sequences for each of the samples are listed below. Draw a line through each cut site in each DNA sample. The saliva DNA sample has been done for you as an example.

SALIVA DNA SAMPLE:

A T G T A G A C T G G A C C A T A T|T A C

G A T|T A G G C A C T C A T|T A G C C G T

A C A G T A C T C A C C

HAIR DNA SAMPLE:

A T C T C G T G A C A T|T A C C T T G T A T

C G A T|T A G C A A T|T A A G G A T C C T

G C A G T A G C A C C

SUSPECT 1 DNA SAMPLE:

A T|T A A C G G G T A T C T T C G G A T|T A

C G G A G A C T A A G T G C C T A G A T|T

A C G A A G C T A C C

SUSPECT 2 DNA SAMPLE:

A T C A G C A T G T G T T C A A T|T A G C C

G A G A T|T A A G G C C A C T G G A G T A

C T A C G G C C A C C

SUSPECT 3 DNA SAMPLE:

A T G T A G A C T G G A C C A T A T|T A C

G A T|T A G G C A C T C A T|T A G C C G T

A C A G T A C T C A C C

SUSPECT 4 DNA SAMPLE:

A T C T C G T G A C A T|T A C C T T G T A T

C G A T|T A G C A A T|T A A G G A T C C T

G C A G T A G C A C C

2. What do all the DNA sequences have in common? Explain this similarity.

The DNA sequences are all the same length. It is important to analyze analogous DNA segments when doing DNA fingerprinting.

Now you will need to count the number of base pairs in the fragments (created by the cut sites) of each DNA sample. For example, the saliva sample has four fragments, with the first fragment having 18 base pairs (bp), then 6 bp, 11 bp, and 19 bp.

3. Count the fragment lengths in the hair DNA sample and each suspect DNA sample.

Hair: 12 bp, 14 bp, 7 bp, 21 bp

Suspect 1: 2 bp, 18 bp, 22 bp, 12 bp

Suspect 2: 17 bp, 10 bp, 27 bp

Suspect 3: 18 bp, 6 bp, 11 bp, 19 bp

Suspect 4: 12 bp, 14 bp, 7 bp, 21 bp

The DNA fragments are sorted by loading each sample of DNA into the well of a gel. An electric current moves the negatively charged DNA fragments toward the positive end of the gel, sorting them by size with the smaller pieces moving the farthest. This creates a unique pattern, or DNA fingerprint.

B Use Table 1 to sort the DNA fragments by drawing a dash in the gel box that corresponds to the fragment length. The saliva sample is done for you as an example.

Next, Mrs. Gumschuh stained the DNA gel with a fluorescent dye.

4. Why do you think this step is essential in the DNA fingerprinting sequence?

The DNA bands are not visible in the gel until it has been stained.

EVALUATING THE EVIDENCE

Now let's analyze the evidence to find out whodunit!

5. Whom does the hair sample belong to?

Patsy Fahlguy

6. Whom does the saliva sample belong to?

Brock Lee

7. Who committed the crime?

Brock Lee

8. How do you know that this person committed the crime?

His DNA sample matches the DNA swabbed from the brownie pan, indicating that he is the one who ate the brownies.

9. Using the evidence above, explain how DNA fingerprinting can be used to clear a person as a suspect even though circumstantial evidence places that person at the scene of the crime.

See margin for answer.

Question 9 Answer

Although Chip was seen in Mr. Sokolata's classroom and the DNA from the hair puts Patsy there as well, the DNA from their cheek swabs did not match the DNA from the brownie pan. (Students may mention that Mr. Ordelijk's DNA does not match either, but it was not established that he was in the classroom, though he had access.)

10. Why were Chip's and Brock's DNA fingerprints not identical even though they are twins?

Identical twins have matching DNA, but fraternal twins do not. It was not stated that Chip and Brock are identical twins.

11. What are some limitations to using this technique as evidence in a trial?

(1) There may not be a good source of DNA to take a sample from, or the sample might be too small. (2) In a crime scene the DNA may be contaminated from the outside environment or have more than one person's DNA mixed together (e.g., in blood). (3) The crime scene could be old and the DNA may have degraded or broken down. (4) The evidence also needs to be corroborated with other evidence. For example, if your hair is found at a crime scene because you were there earlier in the day before the crime was committed, it could be misleading.

GOING FURTHER

Eating a pan of brownies and getting expelled is not the worst thing that could happen to a person. Many people have been convicted of far worse crimes and sent to prison, often for many years, on the basis of less than completely convincing evidence. The Innocence Project is a nonprofit organization that seeks to exonerate people wrongfully convicted of crimes. As of 2021 they have gotten nearly 200 convictions overturned; some of the exonerated were even on death row. Do an internet search using the keywords "Innocence Project" and answer the following questions.

12. What is the primary way in which the Innocence Project exonerates people wrongfully convicted of crimes?

The Innocence Project primarily uses DNA evidence to get convictions overturned.

13. The Innocence Project is not always successful in getting convictions overturned. Why not?

About 29% of Innocence Project cases are closed because the DNA evidence is either missing or has been destroyed.

14. How does the work of the Innocence Project fit with a biblical worldview?

See margin for answer.

✝ *Justice*

Most people want to live in a just society, one in which wrongdoers are apprehended and held accountable for their actions. However, the Fall affected our ability to carry out perfect justice. As a consequence, our pursuit of justice sometimes leads to perpetrations of injustice. A scientific tool such as DNA fingerprinting can help avoid wrongful convictions and ensure that it is in fact the truly guilty to whom punishment is meted out.

Question 14 Answer

Many places in Scripture teach us to do justly and to love our neighbors. Overturning wrongful convictions not only serves justice and demonstrates compassion but also often leads to the apprehension and conviction of the actual perpetrators of some crimes.

TABLE 1 Electrophoresis Gel Box

	SALIVA	HAIR	SUSPECT 1	SUSPECT 2	SUSPECT 3	SUSPECT 4
30						
29						
28						
27				▬		
26						
25						
24						
23						
22			▬			
21		▬				▬
20						
19	▬				▬	
18	▬		▬		▬	
17				▬		
16						
15						
14		▬				▬
13						
12		▬	▬			▬
11	▬				▬	
10				▬		
9						
8						
7		▬				▬
6	▬				▬	
5						
4						
3						
2			▬			
1						

10A LAB

In Darwin's Own Words

Examining On the Origin of Species

You have been learning about Charles Darwin's ideas and how they were shaped by his culture and times. Let's look at the introduction to *On the Origin of Species*. You will read Darwin's work and analyze his words by answering some questions. As you read, look for Darwin's three themes: variation, the struggle for survival, and natural selection in species. Be ready to evaluate his ideas from a biblical perspective.

Equipment
none

INTRODUCTION FROM ON THE ORIGIN OF SPECIES

When on board H.M.S. 'Beagle,' as naturalist, I was much struck with certain facts in the distribution of the inhabitants of South America, and in the geological relations of the present to the past inhabitants of that continent. These facts seemed to me to throw some light on the origin of species—that mystery of mysteries, as it has been called by one of our greatest philosophers. On my return home, it occurred to me, in 1837, that something might perhaps be made out on this question by patiently accumulating and reflecting on all sorts of facts which could possibly have any bearing on it. After five years' work I allowed myself to speculate on the subject, and drew up some short notes; these I enlarged in 1844 into a sketch of the conclusions, which then seemed to me probable: from that period to the present day I have steadily pursued the same object. I hope that I may be excused for entering on these personal details, as I give them to show that I have not been hasty in coming to a decision.

My work is now nearly finished; but as it will take me two or three more years to complete it, and as my health is far from strong, I have been urged to publish this Abstract. I have more especially been induced to do this, as Mr. Wallace, who is now studying the natural history of the Malay Archipelago, has arrived at almost exactly the same general conclusions that I have on the origin of species. Last year he sent to me a memoir on this subject, with a request that I would forward it to Sir Charles Lyell, who sent it to the Linnean Society, and it is published in the third volume of the journal of that Society. Sir C. Lyell and Dr. Hooker, who both knew of my work—the latter having read my sketch of 1844—honoured me by thinking it advisable to publish, with Mr. Wallace's excellent memoir, some brief extracts from my manuscripts.

What is *On the Origin of Species* all about?

QUESTIONS
- What are Darwin's key points?
- What does Darwin say that is compatible with a biblical worldview?
- What does Darwin say that is incompatible with a biblical worldview?

▲ This photo of Charles Darwin was taken in 1854 when he was working to publish *On the Origin of Species*.

- Identify the main themes of Darwin's On the Origin of Species.
- Analyze Darwin's ideas from a biblical worldview.

✔ Introduction to Lab 10A

This activity integrates three subjects: science, literature, and history. It is important for students to see how these subjects overlap by having them examine a selection from some original source material. This activity can be done in groups or on an individual basis, or it can even be assigned for homework. If you choose the latter, be sure to discuss students' work with them in a class setting.

This activity is structured with a *before-reading* section to orient students, *during-reading* features that clarify and illustrate the text, and *after-reading* questions that lead students to analyze Darwin's words.

💻 Accessing Darwin's Work

To see the other chapters from *On the Origin of Species*, view the link *On the Origin of Species*, which is available as a digital resource.

💻 Darwin: The Voyage That Shook the World

Creation Ministries International has produced a video on Darwin's voyage on HMS *Beagle* and his publication of *On the Origin of Species*. This documentary will help your students view Darwin's life and work through the lens of Scripture.

✔ Vestiges of Creation

Darwin is here referencing the 1844 work entitled *Vestiges of the Natural History of Creation* by Robert Chambers. This book formed a big story for a naturalistic view of origins, including the evolution of stars and ideas on the evolution of species that predated the book. The publication was at first well received by Victorian society until theologians began to criticize its countercultural themes. Darwin felt that this work prepared the way for his ideas of natural selection.

This Abstract, which I now publish, must necessarily be imperfect. I cannot here give references and authorities for my several statements; and I must trust to the reader reposing some confidence in my accuracy. No doubt errors will have crept in, though I hope I have always been cautious in trusting to good authorities alone. I can here give only the general conclusions at which I have arrived, with a few facts in illustration, but which, I hope, in most cases will suffice. No one can feel more sensible than I do of the necessity of hereafter publishing in detail all the facts, with references, on which my conclusions have been grounded; and I hope in a future work to do this. For I am well aware that scarcely a single point is discussed in this volume on which facts cannot be adduced, often apparently leading to conclusions directly opposite to those at which I have arrived. A fair result can be obtained only by fully stating and balancing the facts and arguments on both sides of each question; and this cannot possibly be here done.

I much regret that want of space prevents my having the satisfaction of acknowledging the generous assistance which I have received from very many naturalists, some of them personally unknown to me. I cannot, however, let this opportunity pass without expressing my deep obligations to Dr. Hooker, who for the last fifteen years has aided me in every possible way by his large stores of knowledge and his excellent judgment.

In considering the origin of species, it is quite conceivable that a naturalist, reflecting on the mutual affinities of organic beings, on their embryological relations, their geographical distribution, geological succession, and other such facts, might come to the conclusion that each species had not been independently created, but had descended, like varieties, from other species. Nevertheless, such a conclusion, even if well founded, would be unsatisfactory, until it could be shown how the innumerable species inhabiting this world have been modified so as to acquire that perfection of structure and coadaptation which most justly excites our admiration. Naturalists continually refer to external conditions, such as climate, food, etc., as the only possible cause of variation. In one very limited sense, as we shall hereafter see, this may be true; but it is preposterous to attribute to mere external conditions, the structure, for instance, of the woodpecker, with its feet, tail, beak, and tongue, so admirably adapted to catch insects under the bark of trees. In the case of the mistletoe, which draws its nourishment from certain trees, which has seeds that must be transported by certain birds, and which has flowers with separate sexes absolutely requiring the agency of certain insects to bring pollen from one flower to the other, it is equally preposterous to account for the structure of this parasite, with its relations to several distinct organic beings, by the effects of external conditions, or of habit, or of the volition of the plant itself.

The author of the *Vestiges of Creation* would, I presume, say that, after a certain unknown number of generations, some bird had given birth to a woodpecker, and some plant to the mistletoe, and that these had been produced perfect as we now see them; but this assumption seems to me to be no explanation, for it leaves the case of the coadaptations of organic beings to each other and to their physical conditions of life, untouched and unexplained.

▲

Darwin used the relationship of mistletoe, a parasitic plant, to the birds that spread its seeds to demonstrate that ecological relationships are the product of natural selection.

It is, therefore, of the highest importance to gain a clear insight into the means of modification and coadaptation. At the commencement of my observations it seemed to me probable that a careful study of domesticated animals and of cultivated plants would offer the best chance of making out this obscure problem. Nor have I been disappointed; in this and in all other perplexing cases I have invariably found that our knowledge, imperfect though it be, of variation under domestication, afforded the best and safest clue. I may venture to express my conviction of the high value of such studies, although they have been very commonly neglected by naturalists.

From these considerations, I shall devote the first chapter of this Abstract to Variation under Domestication. We shall thus see that a large amount of hereditary modification is at least possible, and, what is equally or more important, we shall see how great is the power of man in accumulating by his Selection successive slight variations. I will then pass on to the variability of species in a state of nature; but I shall, unfortunately, be compelled to treat this subject far too briefly, as it can be treated properly only by giving long catalogues of facts. We shall, however, be enabled to discuss what circumstances are most favourable to variation. In the next chapter the Struggle for Existence among all organic beings throughout the world, which inevitably follows from their high geometrical powers of increase, will be treated of. This is the doctrine of Malthus, applied to the whole animal and vegetable kingdoms. As many more individuals of each species are born than can possibly survive; and as, consequently, there is a frequently recurring struggle for existence, it follows that any being, if it vary however slightly in any manner profitable to itself, under the complex and sometimes varying conditions of life, will have a better chance of surviving, and thus be naturally selected. From the strong principle of inheritance, any selected variety will tend to propagate its new and modified form.

This fundamental subject of Natural Selection will be treated at some length in the fourth chapter; and we shall then see how Natural Selection almost inevitably causes much Extinction of the less improved forms of life and induces what I have called Divergence of Character. In the next chapter I shall discuss the complex and little known laws of variation and of correlation of growth. In the four succeeding chapters, the most apparent and gravest difficulties on the theory will be given: namely, first, the difficulties of transitions, or understanding how a simple being or a simple organ can be changed and perfected into a highly developed being or elaborately constructed organ; secondly the subject of Instinct, or the mental powers of animals; thirdly, Hybridism, or the infertility of species and the fertility of varieties when intercrossed; and fourthly, the imperfection of the Geological Record. In the next chapter I shall consider the geological succession of organic beings throughout time; in the eleventh and twelfth, their geographical distribution throughout space; in the thirteenth, their classification or mutual affinities, both when mature and in an embryonic condition. In the last chapter I shall give a brief recapitulation of the whole work, and a few concluding remarks.

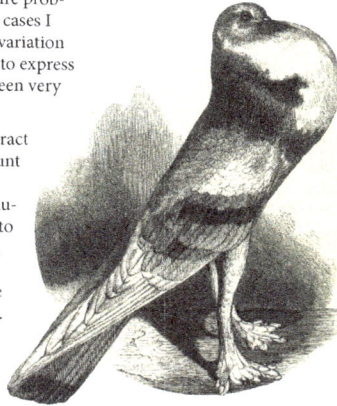

Darwin observed much variation in domesticated species. Among his favorite animals to observe were pigeons. He even began to breed pigeons himself around the time *On the Origin of Species* was published.

🄸 *Immutability of Species*

When Darwin refers to the independent creation of species and the immutability of species, he is referring to the then-common idea that species do not change over time. This idea is what the Student Edition calls *fixity of species*.

No one ought to feel surprise at much remaining as yet unexplained in regard to the origin of species and varieties, if he makes due allowance for our profound ignorance in regard to the mutual relations of all the beings which live around us. Who can explain why one species ranges widely and is very numerous, and why another allied species has a narrow range and is rare? Yet these relations are of the highest importance, for they determine the present welfare, and, as I believe, the future success and modification of every inhabitant of this world. Still less do we know of the mutual relations of the innumerable inhabitants of the world during the many past geological epochs in its history. Although much remains obscure, and will long remain obscure, I can entertain no doubt, after the most deliberate study and dispassionate judgment of which I am capable, that the view which most naturalists entertain, and which I formerly entertained—namely, that each species has been independently created—is erroneous. I am fully convinced that species are not immutable; but that those belonging to what are called the same genera are lineal descendants of some other and generally extinct species, in the same manner as the acknowledged varieties of any one species are the descendants of that species. Furthermore, I am convinced that Natural Selection has been the main but not exclusive means of modification.

Charles Darwin.

ANALYZING DARWIN'S WORDS

1. On what did Darwin base his conclusions about where species come from?

Darwin based his conclusions on his observations of the natural world while on a circumnavigation of the world, as well as on the opinions and ideas of other scientists and philosophers.

2. Name three of the other scientists Darwin mentions and describe how they influenced his work.

Answers may include Albert Wallace, Charles Lyell, and a Dr. Hooker who reviewed his work and gave feedback on his manuscripts. Albert Wallace was an entomologist whose work Darwin viewed as a confirmation of his theory, and Lyell was a geologist who encouraged Darwin to publish his work.

3. What does Darwin plan to discuss in Chapter 1? How does this apply to the origin of species?

Darwin plans to discuss variations in animals through selective breeding. If species can change through breeding, this automatically negates the notion of fixity of species.

4. What does Darwin plan to discuss in Chapter 2? Who inspired his ideas in this chapter?

   Darwin plans to discuss the struggle for survival. He is basing his

   ideas on Thomas Malthus's teachings about the limitations that

   resources put on human populations.

5. What does Darwin plan to discuss in Chapter 4?

   natural selection

6. How does Darwin say natural selection affects a population?

   Darwin believed that natural selection affects a population by

   causing less effective organisms to become extinct. This would

   allow variations of certain traits to thrive in a population, enabling

   individuals to better survive and reproduce.

7. What is Darwin's position on fixity of species?

   Darwin was a denier of fixity of species.

8. Name at least three things that you can agree with Darwin about.

   Answers will vary. Students may mention his observation of

   change in nature, his observation of the fight for survival, and

   his denial of fixity of species.

9. Name one thing that you disagree with Darwin about.

   Answers will vary. Students may mention his basic premise that

   natural selection is a sufficient means to produce new species.

10. Name one weakness of Darwin's work.

    Answers will vary. Darwin acknowledged that his work was

    imperfect, including his lack of references and authorities, his

    inclusion of possible errors, and his general statements with

    limited use of examples.

11. Name one strength of Darwin's work.

    Answers will vary. Darwin mentioned that his ideas were the

    result of much scientific observation over an extended period

    of time and named several academic peers with whom he had

    consulted.

🔵 💻 Darwin and Religion

Darwin was a theist, and his views of God were close to the tenets of deism. He was involved in his local Unitarian church and had initially intended to become a clergyman. But later in life he neglected church attendance and described himself as an agnostic.

To give students a better idea of Darwin's personal thoughts and feelings, visit the Darwin Correspondence Project; this link is available as a digital resource where you can find a database of over 8000 personal letters that Darwin wrote to friends and family.

12. Darwin and Chambers believed in God, but they viewed Him as someone who worked in the world through the laws of science without being personally involved. Comment on this view of God from a biblical perspective.

God does provide for His creation through natural laws, but He is also intimately involved in His creation. The Bible tells us that He clothes the plants, knows the number of hairs on our heads, and feeds the birds (Matt. 6:25–33; Luke 12:7). He loves His creation and has stepped in at special times in history to do miracles.

13. What was Darwin's source of reliable truth for his ideas about the origin of life?

Darwin's authoritative source was ultimately himself, his own reasoning about other scientists' ideas, and his observations of the natural world.

14. Analyze the authority for Darwin's ideas from a biblical perspective.

Though we should seek to interact with and learn about the world, our ultimate authority is God's Word. We must make sure that our ideas are in harmony with God's revelation.

GOING FURTHER

A Consider reading other chapters from this work that interest you and share your thoughts with someone else.

B Another of Darwin's works, *The Descent of Man*, was published in 1871. In this book Darwin takes his ideas on the evolution of animals and applies them to mankind. This book contains some very controversial ideas related to social Darwinism, including the notion that weak individuals in the human population should not be allowed to reproduce so that they don't weaken the human race. Social Darwinism began to affect public thinking in the 1870s. The moral implications of social Darwinism are completely contrary to the central principles of Scripture. Read a selection from *The Descent of Man* and share your thoughts with someone else.

10B LAB

Worldview Sleuthing

Evaluating Worldview in Popular Science Literature

Do you or your teacher have a subscription to *National Geographic*? *Scientific American*? *Science News*? *Popular Science*? These periodicals, like many others, can give you interesting information in ways that are easy to understand. Television programs like *NOVA* and *Nature* serve a similar purpose.

Have you ever read something in these periodicals or seen something on these programs that made you uncomfortable? Probably. But why? We feel uncomfortable when someone says or writes something that contradicts what we believe, especially when it is one of our core beliefs. In this lab activity you will read three snippets from popular science magazines and analyze them from a biblical perspective.

Equipment

current issues of science periodicals or digital copies of online articles

PROCEDURE

Evolution is not just a theory; it forms a philosophical framework that shapes how people view life. This extends to ideas about extraterrestrial life and conservation here on planet Earth. Evolution even affects people's ideas about culture and technology and the future of mankind. You'll look at excerpts from *National Geographic* and *Science News* on these topics.

Extraterrestrial Life

The first quote comes from an article about the history of the search for extraterrestrial life. Recent efforts to find alien life-forms focus on exoplanets in the "Goldilocks" zone of their host star—planets that have just the right conditions for life to evolve.

"The universe is vast and old, so advanced civilizations should have matured enough by now to send emissaries to Earth. Yet none have. Fermi suspected that it wasn't feasible or that aliens didn't think visiting Earth was worth the trouble. Others concluded that they simply don't exist. Recent investigations indicate that harsh environments may snuff out nascent life long before it evolves the intelligence necessary for sending messages or traveling through space."[1]

[1] "To An Ancient Question, No Reply" by Tom Siegfried, *Science News*, April 30, 2016, pp. 24–25

Should I believe everything I read?

QUESTIONS

- What controversial issues do people write about in popular science periodicals?
- How can I identify a science writer's worldview?
- How can I tell what to believe about what I read?

An organization called SETI (Search for Extraterrestrial Intelligence) uses a group of telescopes called the Allen Array to survey the universe for signs of extraterrestrial life.

- Identify places in popular science literature in which a writer's worldview is apparent.
- Evaluate ideas in popular science literature.
- Rebuild secular, naturalistic ideas in science literature according to a biblical worldview.

Introduction to Lab 10B

This activity integrates science and popular literature. It is important for students to be able to critically evaluate what they read. This activity can be done in groups or on an individual basis, or it can even be assigned for homework. If you choose the latter, be sure to discuss students' work with them in a class setting.

Prelab Discussion

To start students thinking about evaluating worldview in popular science literature, you could show them a clip from a NOVA, Nature, or National Geographic program and discuss it, identifying the show's worldview and critically evaluating its content on the basis of its worldview. The links *NOVA* and *Nature*, each of which is available as a digital resource, along with a database of other recent programs.

1. What words or phrases in the article suggest that the writer of this article is not writing from a biblical worldview?

   There is nothing here that clearly contradicts a biblical worldview. Students may mention the description of the universe being "vast and old," which does contradict a recent-creation point of view. They may also point out that the article assumes that intelligence would evolve, which the Bible does not suggest or support.

2. What do scientists assume about how life should evolve on other planets? Is that a good assumption? Explain.

   See margin for answer.

   _____

   _____

   _____

   _____

3. How do evolutionists relate intelligence and communication to life?

   Evolutionists assume that other life-forms have the intelligence and ability to communicate over the long distances of space.

   _____

Bison grazing in Yellowstone National Park

Conservation

The second excerpt comes from an issue of *National Geographic* dedicated to the history and conservation of Yellowstone National Park. The quote below refers to a conversation between the writer and Dave Hallac, chief of the Yellowstone Center for Resources, which oversees programs that maintain the ecological health of the park.

"Dave Hallac stood in his office at the Yellowstone Center for Resources, the park body charged with science and resource management, in a rambling old clapboard building amid the formidable stone structures of Mammoth Hot Springs. Hallac ticked through a list of inter-related concerns, nagging issues in Yellowstone familiar to us both: bison management, elk migration, grizzly bear conservation, private land development in the region surrounding the park, human population growth driving that development, invasive species and their impacts on native species, water use, climate change, and finally the overarching problem that exacerbates all these others—an absence of coordinated, transboundary management. 'We go around telling everybody this is the most intact ecosystem in the lower 48,' Hallac said. Well, if it's that important, that special, it's time for us to do a lot better when it comes to protecting it."[2]

[2] "Living With the Wild" by David Quammen, *National Geographic*, May 2016, p. 133

Question 2 Answer

Many scientists assume that life should evolve on other planets just as (they believe) life evolved on Earth. That assumption is valid in an evolutionary worldview because nothing about our world is unique—it's just a random occurrence. A biblical worldview holds that Earth is unique in its design and in its inhabitants and would lead someone to expect life as we know it nowhere else in the universe.

4. Is there anything in this article that is at odds with a biblical worldview? Explain.

   There are no themes in this article that contradict a biblical worldview. Issues of conservation, invasive species, the needs of a growing human population, and climate change are not contrary to a biblical worldview.

5. Most of the conservationists at work in Yellowstone are evolutionists. Comment on the ability of evolutionists to do good science from a biblical perspective.

   Despite having a worldview that differs from a biblical one, evolutionists often engage in good scientific work where their worldview overlaps with a biblical one. This would include wisely using Earth's resources in ways that provide for a growing human population.

6. Is a growing human population a good thing or a bad thing from a biblical worldview? Why?

   A growing human population is in direct obedience to God's Creation Mandate given in Genesis 1:28.

7. Conservationists at Yellowstone National Park are working to protect the park but still keep it accessible to a growing number of visitors. How does this show the interaction between the two parts of the Creation Mandate?

   To provide for a growing population, we must wisely use God's resources with a mind for the future.

8. Why do evolutionists value conservation? Analyze their reasons in light of their other beliefs.

   Some evolutionists value conservation because they view man as one of the top forms of life, a concept that is parallel to the biblical idea of man being the head of God's creation because he is made in God's image. Some environmentalists value conservation because it feels right to them. They want to enjoy the world and give it to others because that feels right too. But since evolution ignores what is immaterial, it cannot justify assigning value or morality to any idea.

✝ **Protecting Wild Places**

There are many reasons why a Christian can value the work of conservation. We can conserve land because it has resources that we can wisely use, but we can also protect land just because its natural state is beautiful. Appreciating beauty is part of the expression of the image of God in people.

Be aware of any student in your class who may have difficulty discussing death, especially if he is dealing with his own illness or the illness or death of a family member or friend.

Death

A final quote comes from an article about peoples' near-death experiences. Death is a common human experience, and evolution relies on death for populations to change over time. Humans are the most evolved species, according to this worldview, so in the future they should be able to find a way to fight against the effects of death.

"Death is 'a process, not a moment,' writes critical-care physician Sam Parnia in his book *Erasing Death*. It's a whole-body stroke, in which the heart stops beating but the organs don't die immediately. In fact, he writes, they might hang on intact for quite a while, which means that 'for a significant period of time after death, death is in fact fully reversible.'

"How can death, the very essence of forever, be reversible? What is the nature of consciousness during that transition through the gray zone? A growing number of scientists are wrestling with such vexing questions.

"In Seattle biologist Mark Roth experiments with putting animals into a chemically induced suspended animation, mixing up solutions to lower heartbeat and metabolism to near-hibernation levels. His goal is to make human patients who are having heart attacks 'a little bit immortal' until they can get past the medical crisis that brought them to the brink of death."[3]

[3] "The Crossing" by Robin Marantz Henig, *National Geographic*, April 2016, p. 36

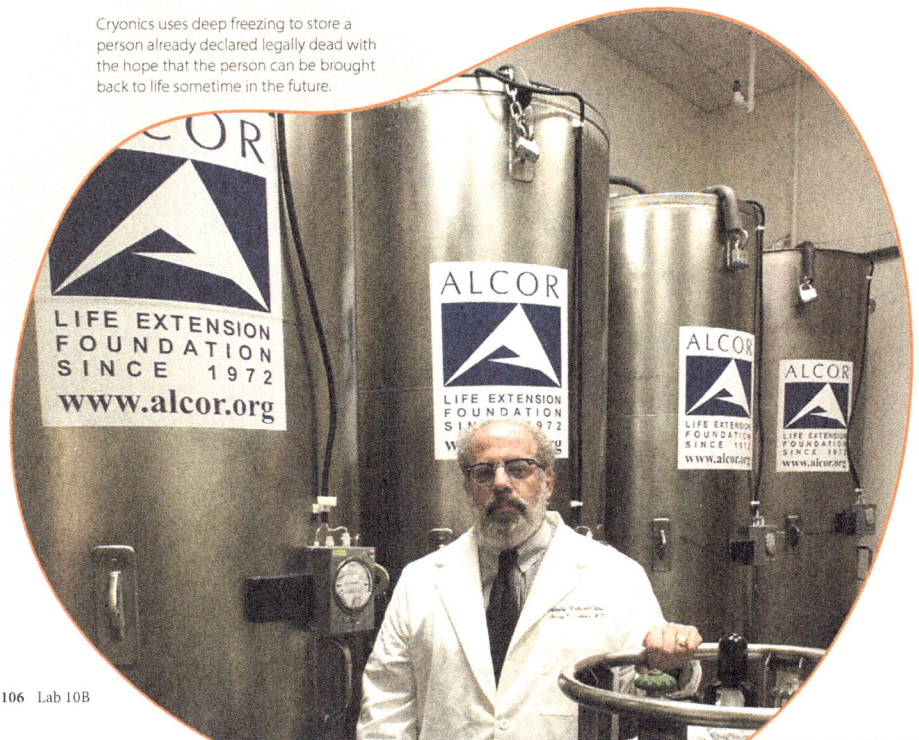

Cryonics uses deep freezing to store a person already declared legally dead with the hope that the person can be brought back to life sometime in the future.

9. Is it wrong to try to extend people's lives through medical intervention? Explain.

It is not necessarily wrong to extend people's lives to allow them to heal, but many families must make difficult decisions about life support when a family member is unable to independently breathe or has been declared brain-dead.

10. If death is such a big player in evolution, why do you think that people without a biblical worldview fear it and try to work against it? Why not let nature do its work?

See margin for answer.

11. How might scientists and doctors operate within a biblical worldview even though they might not be Christians?

In desiring to preserve the human species, evolutionists are valuing human life. Valuing human life is harmonious with a Christian worldview, which views people as image-bearers of God.

12. From a biblical perspective, will people and technology ever advance to the degree that death can be completely overcome? Explain.

Death is God's penalty for sin, and technology is not the solution to sin. Humans will never be able to "cure" death. Jesus alone can save us from the deadly effects of sin and restore to us the abundant life that God wants us to enjoy (John 3:16).

Question 10 Answer
From a naturalistic point of view, death is a natural part of life, and yet it can still feel like a very unnatural part of human life. People were created to live forever, and because of this people instinctively want to preserve life. People are also fearful of dying themselves because they don't know what will happen afterward. God has put a sense of eternity into our hearts (Eccles. 3:11).

Additional Articles for Students to Review

For a website that links to summaries of research articles written in a style that will be easy for high schoolers, see the link *Science Articles*, which is available as a digital resource.

GOING FURTHER

Find some articles of your own to do some worldview sleuthing. You can check *National Geographic*, *Smithsonian*, *Popular Mechanics*, and *Scientific American*, looking either at physical copies or exploring online.

13. What topic did you read about?

    See margin for answer.

14. What is the viewpoint of the writer on this topic?

    Answers will vary. An evolutionary worldview will be pervasive in science articles and may include an emphasis on preserving resources in a way that is potentially harmful to people or remiss regarding their needs. Many common themes involve the age of Earth, the origin of Earth, the origin of life, and the place of humans within the universe.

15. Analyze this viewpoint from a biblical worldview.

    Answers will vary. Most science topics can be analyzed in light of the Creation Mandate, the second great commandment, and the first chapter in Genesis, all of which are related to God's creation of the world and the image of God in people.

11A LAB

The Key Concept

Using Dichotomous Keys

Edmund and Tenzing were out backpacking in the Colorado foothills. They both enjoy the taste of wild onions that grow in the area, such as Geyer's onion (*Allium geyeri*). They also found another plant that looked very much like an onion but lacked the strong wild-onion smell. They decided to avoid that plant since they know that most species in the genus *Allium* smell like onions. The two adventurers had performed a simplified version of classification, and though they might not have realized it, their lives had depended on accurately identifying the unfamiliar plant as a member of the genus *Allium* or not. The unknown onion look-alike happened to be a meadow death camas (*Toxicoscordion venenosum*), a plant whose common name correctly suggests that it is not a suitable trailside treat!

Like our intrepid hikers, biologists often need to identify unknown organisms, and a commonly used tool for the task is a dichotomous key. It's a series of paired statements or characteristics about the specimen being identified. Only one statement from each pair can be true, and selecting the correct statement either identifies the organism or refers the user to another pair of statements to consider. Today you will use a simplified dichotomous key to sort out some finned friends.

Tasty!

Why is being able to identify organisms important?

Deadly!

Equipment

Key for Selected Fishes of North America (pages 119–20)

PROCEDURE

Use the key on pages 119–20 to identify the fish on pages 112–18.

A Start by reading the first pair of statements, then decide which of the two statements describes your specimen. The statement you choose will either give you the identity of your fish or direct you to another pair of statements.

B Record each number from the key that you use to identify each organism in Table 1.

C When you have finally identified your fish, record its scientific and common names in Table 1.

D Repeat Steps A through C for additional specimens. Specimen 1 has been completed for you.

QUESTIONS
- What is a dichotomous key?
- How does a dichotomous key work?
- How can a dichotomous key help identify an unknown organism?

LAB IIA OBJECTIVES

- Define *dichotomous key*.
- Explain the use of a dichotomous key.
- Identify organisms using a dichotomous key.

Prelab Discussion

Ask students to think of situations that might require being able to accurately identify unfamiliar organisms. If they found themselves in such a situation, how would they go about trying to identify those organisms?

Equipment Notes

The key used in this activity is a very simplified version of what one would expect to use in the field. Have a selection of actual keys on hand for students to examine and perhaps even try out.

Keying Specimens

Guide students through the process of keying a specimen. Adjust the number of specimens that your students should identify according to their abilities and amount of class time available, or assign unkeyed specimens as homework.

TABLE 1

Specimen Number	Numbers from Key	Common Name	Scientific Name
1	1, 2, 4, 5, 8, 9, 10	rainbow trout	*Oncorhynchus mykiss*
2	1, 2, 4, 5, 6, 7	channel catfish	*Ictalurus punctatus*
3	1, 2, 3	bowfin	*Amia calva*
4	1, 2, 4, 11, 18, 19	blue-spotted sunfish	*Enneacanthus gloriosus*
5	1, 2, 4, 11, 12, 13, 14, 15	silver redhorse	*Moxostoma anisurum*
6	1, 2, 4, 11, 12, 16, 17	muskellunge	*Esox masquinongy*
7	1, 2, 4, 11, 18, 23, 24, 26, 27	striped bass	*Morone saxatilis*
8	1, 2, 4, 5, 8, 9	Bonneville cutthroat trout	*Oncorhynchus clarkii utah*
9	1, 2, 4, 11, 18, 23, 24, 26, 27	white bass	*Morone chrysops*
10	1, 2, 4, 11, 18, 19, 20, 21	redbreast sunfish	*Lepomis auritus*
11	1, 2, 3	longnose gar	*Lepisosteus osseus*
12	1, 2, 4, 5, 8	lake whitefish	*Coregonus clupeaformis*
13	1, 2, 4, 11, 12, 13, 14, 15	river carpsucker	*Carpiodes carpio*
14	1, 2, 4, 5, 6, 7	black bullhead	*Ameiurus melas*
15	1, 2, 4, 11, 18, 23, 24, 25	smallmouth bass	*Micropterus dolomieu*
16	1, 2, 4, 11, 18, 19, 20, 21	pumpkinseed	*Lepomis gibbosus*
17	1, 2, 4, 11, 12, 13	American shad	*Alosa sapidissima*
18	1, 2, 4, 11, 12, 16, 17	northern pike	*Esox lucius*
19	1, 2, 4, 11, 18, 23, 24, 25	largemouth bass	*Micropterus salmoides*
20	1, 2, 4, 11, 18, 19, 20, 22	black crappie	*Pomoxis nigromaculatus*

Specimen Number	Numbers from Key	Common Name	Scientific Name
21	1, 2, 4, 11, 12, 16	redfin pickerel	*Esox americanus*
22	1, 2, 4, 5, 6	brindled madtom	*Noturus miurus*
23	1, 2, 4, 11, 12, 13, 14	common carp	*Cyprinus carpio*
24	1, 2, 4, 11, 18, 23	yellow perch	*Perca flavescens*
25	1, 2, 4, 5, 8, 9, 10	brook trout	*Salvelinus fontinalis*
26	1, 2, 4, 11, 18, 23, 24, 26	white perch	*Morone americana*
27	1	Atlantic sturgeon	*Acipenser oxyrinchus*
28	1, 2, 4, 11, 18, 19, 20, 22	white crappie	*Pomoxis annularis*

GOING FURTHER

1. Like most science tools, a dichotomous key has some limits to its usefulness. What are some things you can think of that might hinder the effectiveness of a dichotomous key as a tool for identifying organisms?

   See margin for answer.

   _____

   _____

   _____

   _____

2. One of the first tasks that God assigned to Adam in Eden was the giving of names to all the animals that God brought to him (Gen. 2:19–20). On the basis of what you learned about speciation in Chapter 10, do you think that Adam had a tougher job than modern biologists? Or was his easier? Explain.

   Adam was responsible for naming the different kinds of animals, which at that time had not had enough time to speciate into the animals we know today, so in that respect Adam may have had an easier job. But some animal kinds were lost in the years following the Flood, so the exact number of kinds that Adam had to identify cannot be known. Also, Adam's intellect had not yet suffered from sin's curse, so he certainly would have been up to the task of naming a great number of animal kinds.

Question 1 Answer
Answers will vary. Among the limits on the usefulness of a key are the skill of the user in correctly identifying distinguishing characteristics, the relative ease or difficulty in identifying those characteristics, the accuracy of the descriptions included in the paired statements, the similarity of the organisms being keyed, and whether the specimen being keyed is typical of the species.

✝ The First Act of Classification

It is easy to fall into the trap of thinking of biblical characters, such as Adam, as somehow less intelligent or capable than "modern" humans. This misconception may even be reinforced by illustrated children's books depicting Bible characters in primitive clothing and using simple tools. It should be remembered that Adam was formed without sin and possessed an unblemished intellect. Even without the radiation of variety in animal forms that would have followed the initial Creation Week events, there would have been a large number of different kinds of animals. Adam in essence performed the first act of classification by giving names to all the animal kinds that God created—an intellectually demanding and by no means small feat.

2

4

1

3

7

5

8

6

10

9

12

11

15

13

16

14

20

17

19

18

21

23

22

24

26

28

25

27

Key to Selected Fishes of North America

1. (a) Body is covered with bony scutes .... Atlantic sturgeon (*Acipenser oxyrinchus*)

   (b) Body is covered with scales or is scaleless .... go to 2

2. (a) Has a heterocercal caudal fin .... go to 3

   (b) Has a homocercal caudal fin .... go to 4

3. (a) Has a short snout and long dorsal fin .... bowfin (*Amia calva*)

   (b) Has a long snout and short dorsal fin .... longnose gar (*Lepisosteus osseus*)

4. (a) Has an adipose fin .... go to 5

   (b) Does not have an adipose fin .... go to 11

5. (a) Has barbels .... go to 6

   (b) Does not have barbels .... go to 8

6. (a) Caudal fin is rounded .... brindled madtom (*Noturus miurus*)

   (b) Caudal fin is forked .... go to 7

7. (a) Caudal fin is deeply forked .... channel catfish (*Ictalurus punctatus*)

   (b) Caudal fin is slightly forked .... black bullhead (*Ameiurus melas*)

8. (a) Caudal fin is deeply forked .... lake whitefish (*Coregonus clupeaformis*)

   (b) Caudal fin is slightly or moderately forked .... go to 9

9. (a) Has a prominent red "slash" mark under jaw ...... Bonneville cutthroat trout (*Oncorhynchus clarkii utah*)

   (b) Has no red "slash" mark under jaw .... go to 10

10. (a) Has black spots on pectoral, pelvic, and anal fins .... rainbow trout (*Oncorhynchus mykiss*)

   (b) Has pectoral, pelvic, and anal fins red with no spots .... brook trout (*Salvelinus fontinalis*)

11. (a) Has one dorsal fin with only soft rays .... go to 12

   (b) Has one or two dorsal fins with spiny and soft rays .... go to 18

12. (a) Leading edge of dorsal fin is at or near midpoint of body .... go to 13

   (b) Dorsal fin is set far back on body .... go to 16

13. (a) Mouth is terminal .... American shad (*Alosa sapidissima*)

   (b) Mouth is subterminal or inferior .... go to 14

14. (a) Has barbels .... common carp (*Cyprinus carpio*)

   (b) Does not have barbels .... go to 15

15. (a) First rays of dorsal fin are nearly equal in length .... silver redhorse (*Moxostoma anisurum*)

   (b) First rays of dorsal fin are greatly elongated .... river carpsucker (*Carpiodes carpio*)

16. (a) Has no spots on fins .... redfin pickerel (*Esox americanus*)

   (b) Has spots on fins .... go to 17

17. (a) Has irregular vertical barring on sides .... muskellunge (*Esox masquinongy*)

   (b) Has light spots on olive green sides .... northern pike (*Esox lucius*)

18. (a) Spiny dorsal fin and soft dorsal fin are fused .... go to 19

   (b) Spiny dorsal fin and soft dorsal fin are separate or nearly separate .... go to 23

19. (a) Caudal fin is rounded with no fork . . . . blue-spotted sunfish (*Enneacanthus gloriosus*)

(b) Caudal fin is forked . . . . go to 20

20. (a) Fins are reddish-orange with no spots . . . . go to 21

(b) Fins are heavily spotted . . . . go to 22

21. (a) Has white vermiculated markings on operculum . . . . pumpkinseed (*Lepomis gibbosus*)

(b) Has black, greatly lengthened anterior margin of operculum . . . . redbreast sunfish (*Lepomis auritus*)

22. (a) Sides are silvery white with black vertical barring . . . . white crappie (*Pomoxis annularis*)

(b) Sides are white with heavy black spotting . . . . black crappie (*Pomoxis nigromaculatus*)

23. (a) Sides are yellow with eight or nine prominent, black vertical bars . . . . yellow perch (*Perca flavescens*)

(b) Sides are silver or olive . . . . go to 24

24. (a) Sides are olive . . . . go to 25

(b) Sides are silver . . . . go to 26

25. (a) Posterior margin of mouth extends past eye . . . . largemouth bass (*Micropterus salmoides*)

(b) Posterior margin of mouth does not extend past eye . . . . smallmouth bass (*Micropterus dolomieu*)

26. (a) Dorsal surfaces are mottled . . . . white perch (*Morone americana*)

(b) Dorsal surfaces are striped . . . . go to 27

27. (a) Anterior margin of anal fin is squared . . . . striped bass (*Morone saxatilis*)

(b) Anterior margin of anal fin is rounded . . . . white bass (*Morone chrysop*)

Key Terminology

barbel—a fleshy sensory structure on the snout of a fish; a "whisker"

dorsal—back, top

heterocercal—a tail with unequally sized lobes (The dorsal lobe is usually larger, and the vertebral column may extend into it.)

homocercal—a tail with equal-sized lobes

inferior—beneath the snout

posterior—rear

scute—a bony plate

soft ray—a soft, flexible bone that supports a fin

spiny ray—a stiff, sharp bone that supports a fin

subterminal—slightly below the snout

terminal—at the end of the snout

vermiculated—wormlike

ventral—underside, bottom

Additional fish external anatomy terms can be found on pages 402–3 in your textbook.

11B LAB

All Myxed Up

A Case Study in Classification

How does taxonomy enhance our understanding of living things? People classify things all the time, and usually for a very good reason: categorized things are easier to handle. Think of a mechanic doing a tune-up on the family car. If the mechanic needs a particular wrench to loosen the oil pan plug, the job could be done more efficiently if all the wrenches are sorted by size in one drawer of the tool chest rather than tossed together willy-nilly.

Taxonomy is the science of classifying living things, and taxonomists classify living things for much the same reason as mechanics classify tools. Living things are easier to study and discuss if they have been grouped into categories on the basis of shared characteristics. Exactly how taxonomy has done this, however, has changed over the years, and the reason why it has changed has significance for scientists—and biology students—who believe in the truth of God's Word.

In this lab activity you will read portions of Carl Linnaeus's *Systema Naturae* along with the abstract from a modern scientific paper. An *abstract* is a paragraph that sums up a journal article or research paper, including the research question and the answer that the researcher found. The abstract that you will read is from a paper on the classification of myxozoans.

Myxozoans are microscopic marine and freshwater parasites whose main primary hosts are fish. The first myxozoan was identified in 1825, but myxozoans were not taxonomically classified until 1881. Since then, the taxonomy of myxozoans has undergone at least four major revisions, along with many minor revisions, and yet there is still disagreement on how they should be classified. You may not understand all the terminology in the following passages, but don't be too anxious about the specifics—we're mainly interested in the central ideas. Your goal is to consider what classification is good for and what it's *not* good for.

Why do scientists classify things?

?

Equipment

PROCEDURE

Carl Linnaeus and Systema Naturae

Carl Linnaeus is considered the Father of Modern Taxonomy. In 1735 he published *Systema Naturae*, the first thorough attempt to classify living things. In it Linnaeus introduced the system of *binomial nomenclature*, which identifies the genus and species of the organism and is still used today for naming all life-forms. What motivated Linnaeus to write his best-known work? Let's read through some of his own words from the introduction to the first edition of *Systema Naturae*. (*Note:* The paragraphs were numbered in the original. Portions of the original are omitted here, so some numbers in the sequence are missing.)

QUESTIONS
- What is taxonomy?
- How has the science of taxonomy changed?

- Interpret scientific literature on classification.
- Explain how the science of taxonomy has changed over the years.
- Identify the worldviews of the authors of the works being read.
- Explain how the works of the authors were affected by their worldviews.
- Evaluate worldviews according to the current taxonomic status of myxozoans.

✓ Introduction to Lab 11B

The science of taxonomy has changed greatly since the publication of *Systema Naturae* in 1735—and not necessarily for the better. The main objective of this lab activity is to help students grasp that taxonomy, as it is currently practiced, is an effort to force fit God's creatures into an evolutionary narrative. As might be expected, those efforts have produced highly plastic and less than satisfactory results.

⁉ Reading Level

The reading passages for this lab activity were written by scientists for scientists, so the reading level is high. You might decide to read the passages aloud with your students to help shepherd them through the tricky vocabulary and meaning.

Observations on the Three Kingdoms of Nature

1. If we observe God's works, it becomes more than sufficiently evident to everybody that each living being is propagated from an egg and that every egg produces an offspring closely resembling the parent. Hence no new species are produced nowadays.

4. As there are no new species; as like always gives birth to like; as one in each species was at the beginning of the progeny, it is necessary to attribute the progenitorial unity to some Omnipotent and Omniscient Being, namely God, whose work is called Creation. This is confirmed by the mechanism, the laws, principles, constitutions and sensations in every living individual.

7. On our earth, only two of the three mentioned above [*Note:* Section 6 listed celestial bodies, elements, and natural bodies] are obvious; i.e., the elements constituting it; and the natural bodies constructed out of the elements, though in a way inexplicable except by creation and by the laws of procreation.

8. Natural objects belong more to the field of the senses than all the others and are obvious to our senses anywhere. Thus, I wonder why the Creator put man, who is thus provided with senses and intellect, on the earth globe, where nothing met his senses but natural objects, constructed by means of such an admirable and amazing mechanism. Surely for no other reason than that the observer of the wonderful work might admire and praise its Maker.

10. The first step in wisdom is to know the things themselves; this notion consists in having a true idea of the objects; objects are distinguished and known by classifying them methodically and giving them appropriate names. Therefore, classification and name-giving will be the foundation of our science.

12. He may call himself a naturalist (a natural historian), who well distinguishes the parts of natural bodies by sight and describes and names all these rightly in agreement with the threefold division. Such a man is a lithologist, a phytologist, or a zoologist.

17. I have shown here a general survey of the system of the natural bodies so that the curious reader with the help of this, as it were, geographical table knows where to direct his journey in these vast kingdoms, for to add more descriptions, space, time, and opportunity lacked.

CAROLUS LINNAEUS
Doctor of Medicine
Given at Leyden, July 23, 1735[1]

Answer the following questions on the basis of your reading of the introduction to *Systema Naturae*. Cite the sections that support your answer for each question.

1. What evidence do you find in Linnaeus's introduction that suggests that he had a biblical worldview?

   See margin for answer.

   _____
   _____
   _____
   _____
   _____

---

[1] Carolus Linnaeus, *Systema Naturae* (1735; Facsimile of the First Edition), trans., M. S. J. Engel-Ledeboer & H. Engel (Nieuwkoop, Netherlands: B. De Graaf, 1964), 18–19

Carl Linnaeus

Question 1 Answer

In Section 1 Linnaeus describes living beings as "God's works." In Section 4 he attributes the work of Creation to God and describes Him as omnipotent and omniscient. In Section 7 he states that natural bodies are constructed "in a way inexplicable except by creation." In Section 8 Linnaeus says that the Creator put the observer, man, on the earth to "admire and praise its Maker."

2. Did Linnaeus believe that organisms could change over time? Explain.

Linnaeus specifically stated in Section 1 that "no new species are produced nowadays."

3. What justification did Linnaeus give for studying classification?

In Section 10 Linnaeus says that the first step of wisdom is "to know the things themselves" and that knowing them consists of "classifying them methodically and giving them appropriate names." Classification is, in his view, "the foundation of our science."

4. What criterion did Linnaeus use to classify "natural bodies" into his system?

In Section 12 he says that the parts of natural bodies are distinguished by sight, that is, on the basis of how they look.

5. What reason does Linnaeus give for publishing *Systema Naturae*?

See margin for answer.

Although not given by Linnaeus as a reason for publishing his book, one of the long-term effects of the widespread acceptance of *Systema Naturae* was the universal adoption of binomial nomenclature, in which each species of organism described by science is given a unique scientific name. The mountain lion (also known as the painter, panther, cougar, catamount, ghost cat, fire cat, puma, ko-icto, katalgar, coowachobee, or klandagi), for example, is *Puma concolor*, regardless of whether a scientist studying one is from Portland or Puerto Vallarta.

6. How does the assigning of scientific names make it easier for scientists to study and communicate about organisms?

Having one scientific name rather than one of many regional names eliminates possible confusion about the identity of an organism being studied or discussed.

The Myxozoa and Modern Classification

Carl Linnaeus died eighty-one years before Charles Darwin published *On the Origin of Species*. In Linnaeus's day most scientists believed in a literal reading of Genesis. They also held to the Aristotelian idea of the fixity of species, the idea that organisms do not change over time (see page 200 in your textbook). Since then, however, most scientists have embraced an evolutionary worldview that rejects both of these ideas. The effects of this change in worldview on the science of taxonomy have been striking. To see some of these effects, take a look at the following abstract written by some modern taxonomists.

Question 5 Answer
In Section 17 Linnaeus says that he has given this general survey "so that the curious reader … knows where to direct his journey"; that is, so that other scientists will use his work to further their own studies.

Article Source

To view Linnaeus's entire article, see the *Myxozoan Classification* link. This link is available as a digital resource.

FROM "GENOMIC INSIGHTS INTO THE EVOLUTIONARY ORIGIN OF MYXOZOA WITHIN CNIDARIA"

"The Myxozoa comprise over 2,000 species of microscopic obligate parasites that use both invertebrate and vertebrate hosts as part of their life cycle. Although the evolutionary origin of myxozoans has been elusive, a close relationship with cnidarians, a group that includes corals, sea anemones, jellyfish, and hydroids, is supported by some phylogenetic studies and the observation that the distinctive myxozoan structure, the polar capsule, is remarkably similar to the stinging structures (nematocysts) in cnidarians. To gain insight into the extreme evolutionary transition from a free-living cnidarian to a microscopic endoparasite, we analyzed genomic and transcriptomic assemblies from two distantly related myxozoan species, *Kudoa iwatai* and *Myxobolus cerebralis*, and compared these to the transcriptome and genome of the less reduced cnidarian parasite, *Polypodium hydriforme*. A phylogenetic analysis, using for the first time to our knowledge, a taxonomic sampling that represents the breadth of myxozoan diversity, including four newly generated myxozoan assemblies, confirms that myxozoans are cnidarians and are a sister taxon to *P. hydriforme*. Estimations of genome size reveal that myxozoans have one of the smallest reported animal genomes. Gene enrichment analyses show depletion of expressed genes in categories related to development, cell differentiation, and cell–cell communication. In addition, a search for candidate genes indicates that myxozoans lack key elements of signaling pathways and transcriptional factors important for multicellular development. Our results suggest that the degeneration of the myxozoan body plan from a free-living cnidarian to a microscopic parasitic cnidarian was accompanied by extreme reduction in genome size and gene content."[2]

Answer the following questions on the basis of your reading of the abstract. Cite examples from the abstract to support your answers.

7. Do the authors of this abstract have a biblical worldview or an evolutionary worldview? Explain.

evolutionary; Phrases such as "evolutionary origin" and "extreme evolutionary transition" indicate that the authors assume that evolution is an established fact.

8. In contrast to the morphology-based taxonomies of previous efforts to classify the myxozoans, what does the abstract state that the most recent revisions are based on?

The abstract makes reference to "phylogenomic analysis," that is, DNA sequencing and analysis. It is assumed that differences in DNA sequences correspond to evolutionary changes in the organisms.

[2] E. Sally Chang, et. al., "Genomic Insights into the Evolutionary Origin of Myxozoa within Cnidaria," Dec. 1 2015 (Washington, DC: National Academy of Sciences).

ANALYSIS

9. Why did Linnaeus classify organisms on the basis of appearance?

In Linnaeus's day there really wasn't any other way to classify organisms. DNA and its role in heredity had not yet been discovered.

10. Why do evolutionists classify organisms on the basis of phylogeny?

Evolutionists start from an assumption that all organisms are descended from a single-celled common ancestor. Their phylogeny-based classification is an attempt to re-create this line of descent.

11. How do these two different methods affect how often classification changes?

The morphology (form) of an organism changes very little, though natural selection may produce new species on occasion. Therefore, morphology-based taxonomies need relatively little modification over time. Phylogeny-based systems change frequently because new genetic data is currently being produced in large quantities, and the interpretations of this data differ as well.

12. Which classification method is more practical? Explain.

See margin for answer.

13. How would Linnaeus have classified myxozoans and cnidarians? Explain.

Linnaeus would have separated the myxozoans from the cnidarians because of their different body forms.

Question 12 Answer

Answers will vary. A morphology-based system is easily accessible to all and is relatively unchanging. Most scientists in the course of their daily work do not need to know which organisms supposedly evolved from which other organisms. Systems that are based on proposed evolutionary connections change as evolutionary theory changes. Scientists do need to communicate with each other frequently, and it is inconvenient to have to learn a new scientific name or classification scheme every few years just because opinion has changed within the taxonomic community regarding an organism's phylogeny.

14. How do modern taxonomists classify myxozoans and cnidarians? (*Hint:* Refer to the abstract that you just read.)

Modern taxonomists believe that myxozoans are a degenerated form of cnidarian, so they classify myxozoans and cnidarians together.

15. Can modern phylogeny-based taxonomic systems ever produce a satisfactory phylogenetic tree that is based on DNA analysis? Explain.

Answers will vary. God created all living things within distinct kinds. Although the organisms within a kind can change, no kind of organism can change into a different kind. Since kinds of organisms have not and cannot evolve from each other, the search for a definitive phylogenetic tree is bound to be a fruitless endeavor.

GOING FURTHER

16. Think back to what you learned about evolution in Chapter 10. Using what you know about evolutionary theory and what you read in the abstract above, why do you think myxozoans are actually a bad example of supposed evolution in action?

Answers will vary. In order for evolution to be true, organisms would have to change gradually into more complex organisms through the accumulation of new genetic information. According to the abstract, modern taxonomists believe myxozoans to be cnidarians that have *lost* a significant amount of genetic information. Such a loss of information would be expected in the creation model in which a world that was formed perfect has since come under sin's curse but is not predicted by the evolutionary model. In fact, it is the opposite of what evolution would predict—rather than complex organisms evolving from simpler ones, myxozoans appear to be an example of simpler organisms devolving from complex ones.

✝ *Evaluate Contrary Thinking*

Myxozoans are but one of many examples of aspects of our fallen world that are predicted by a creationist model far better than by an evolutionary one. Thanks to a loss of genetic information, myxozoans are ill-equipped to deal with any large-scale change in their environment. Many myxozoans are not only host-specific (meaning that they can survive only in one species or perhaps kind of host) but also tissue-specific—they can survive only in one kind of host tissue.

12A LAB

Squeaky Clean

Bacteria Growth and Handwashing

"Make sure you wash your hands with soap and warm water." If you had a dollar for every time you heard these words, you'd probably be rich! But do you know why you need to wash your hands this way?

People haven't always had medical reasons to justify handwashing. Though some cultures, such as the Jews (Matt. 15:1–2), practiced ritual handwashing, it didn't become widespread for medical purposes until the 1800s.

Handwashing saves people's lives. In 1845 Hungarian physician Ignaz Semmelweis pioneered the practice of doctors washing their hands before working with patients after he observed that handwashing greatly reduced the occurrence of dangerous infections in new mothers. Although many doctors rejected his practices, they eventually became standard, and Semmelweis was vindicated years after his death by the formation of the germ theory of disease.

Since the bacteria on your hands are too small to see, in this lab activity you will be growing a colony that will be large enough for you to see without a microscope. To do this, you will use a *petri dish*, a large flat dish with a lid (see Appendix B), containing a gel-like substance called *nutrient agar*. Nutrient agar is a mixture of beef extract, a protein called *peptone*, water, and a gel derived from algae. Nutrient agar provides both the structure and the nutrients that bacteria need to grow into a colony.

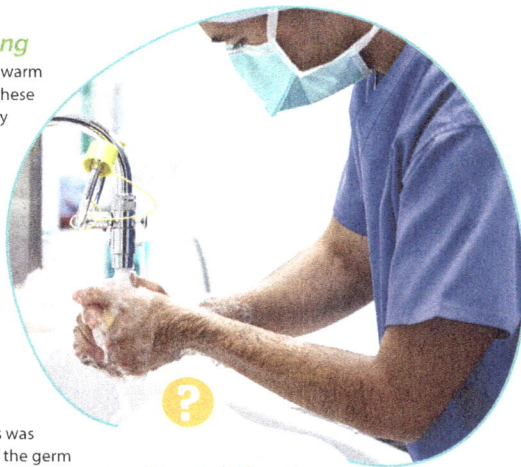

Equipment

microscope
prepared sterile petri dish with cover
ruler
marker
sterile cotton swabs (4)

hand soap
hand sanitizer
plastic bag, large
laboratory apron
goggles

How do different methods of handwashing affect bacteria count?

QUESTIONS

- How dangerous is it when I don't wash my hands?
- What does water do in handwashing?
- What does soap do in handwashing?

- Analyze how the different parts of handwashing affect bacterial count.
- Evaluate the effectiveness of handwashing in preventing disease.

Equipment Notes

Carolina Biological Supply carries a kit that includes the materials necessary to evaluate hand soap effectiveness for a group of thirty students. A smaller kit with similar components is available from Home Science Tools.

You can use either regular or antibacterial hand soap.

You will need to prepare petri dishes ahead of time since the agar needs time to set.

You may decide to do this lab activity in two parts, starting it at the beginning of Chapter 12 and finishing it at least three days later, perhaps as part of a review day. Allow bacteria to incubate in a warm, undisturbed place, such as on the top of a refrigerator. If you have access to an incubator, you won't need to wait the full three days.

Group Size

Groups of three will be ideal for conducting this lab activity, but if you want to use different-sized groups, make sure that all the people whose hands are tested have shaken hands with each other.

Safety Considerations

Having students work with bacterial cultures is potentially dangerous, but as long as you avoid culturing bacteria from the hands of a sick student, there is little risk of culturing and exposing students to a pathogen. Students with cuts or wounds on their hands should protect them with gloves, and students should avoid breathing in bacteria. You could also consider conducting this lab activity as a demonstration.

Make sure that students wash their hands at the end of the activity and work only with sterile equipment. If you don't have a way to sterilize glass petri dishes, use sterile plastic ones that may be disposed of when you are finished.

Preparing the Petri Dishes

You will need to prepare agar for bacterial culturing. If you want to make nutrient agar, you can boil 3.0 g of beef extract, 5.0 g of peptone, and 15.0 g of agar in 1 L of distilled water until dissolved. However, the dehydrated nutrient agar available from many science equipment suppliers is very good, inexpensive, and considerably more convenient since beef extract is not usable after a few years of refrigeration. Dehydrated nutrient agar and nutrient broth are available in single-serving pouches from some suppliers.

1. To prepare the culture dishes, follow the agar preparation instructions, either melting it or mixing it with water. You will mix 1.2 g of powdered nutrient agar with 60 mL of water.

2. Let the agar cool until you can touch it without discomfort (approximately 50 °C).

3. Open the petri dish cover just enough to pour in agar until it covers half the petri dish. Replace the cover and slide the bottom plate around until the entire bottom dish is covered.

4. Let the petri dish stand for one hour or until the agar has set.

You will need one petri dish for each group. To avoid needing to prepare large numbers of petri dishes, consider doing this lab activity in larger groups.

An Environment for Growing Bacteria

Close windows and turn off fans and other blowers. The fewer air currents there are in the room, the less chance there is of contamination.

Incubating Bacteria (Step G)

You may allow as little as forty-eight hours of growth time for bacteria. When incubating petri dishes, be sure that they are upside down with the medium on the upper surface. If the culture is incubated with the lid side up, condensation on the lid will drip down and splash the bacterial culture.

QUARTERS OF YOUR PETRI DISH

1. Nothing applied to agar

2. Dirty hand applied to agar

3. Hand-washed only with water applied to agar

4. Hand-washed with both soap and water applied to agar

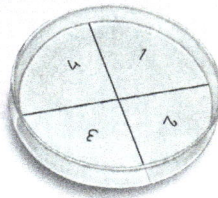

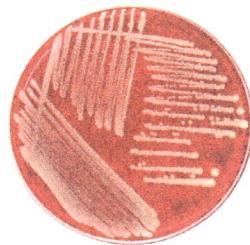

Follow the pattern above when swabbing your petri dish, but keep your swabbing in the appropriate area.

PROCEDURE

Starting Bacterial Cultures

Your teacher will provide you with a prepared, sterilized petri dish containing solidified nutrient agar.

A Turn the petri dish over. Divide the dish into quarters by drawing on the bottom of the petri dish with the ruler and marker. Label the quarters 1, 2, 3, and 4.

B Shake hands with two other people in your group. You will be testing the bacteria count on each person's hands in a different way.

C Swab one of your group member's hands with a sterile cotton swab. Without putting the swab on the benchtop, open the dish just enough to be able to insert the swab. Now swab Area 2 of your petri dish, using a zigzag squiggle. (See margin, below left.) Turn the dish 90° and do it again, still in Area 2. Close your dish as soon as possible.

D Have one person from your group wash his hands with only warm water for 15 seconds. He should dry his hands with a paper towel. Swab his hand with another sterile cotton swab and apply it to Area 3 using the same process as you did in Step C.

E Have one person from your group wash his hands with warm water and hand soap for fifteen seconds. He should dry his hands with a fresh paper towel. Swab his hand with another sterile cotton swab and apply it to Area 4 using the same process as you did in Step C.

1. Why do all of the people whose hands will be tested need to shake hands?

   Shaking hands will allow the bacteria count on everyone's hands to be roughly the same.

2. What does your answer to Question 1 tell you about the cultural practice of shaking hands with people?

   Shaking hands spreads germs! After everyone in the group shakes hands, they should all have similar bacteria counts.

3. Why should you avoid setting your swab on the benchtop?

   Setting the swab on the benchtop could contaminate the bacterial sample.

4. Why should you open the petri dish only partway and close it as quickly as possible?

   Opening the petri dish partway avoids having airborne bacteria settle on the agar.

F Now put some hand sanitizer on a fresh, sterilized cotton swab and swab along the two lines separating the quarters.

G Label a plastic bag with your group's name. Place your petri dish in the plastic bag and seal it. You will not remove the petri dish from the bag for the rest of this lab activity. Your teacher will store this for a few days until the bacterial colonies grow large enough for you to observe.

H Everyone in your group must wash their hands.

5. What is the function of Area 1 if you haven't added anything to the agar?

   Area 1 acts as a control for the experiment.

6. If a bacterial colony grows in Area 1, what would you conclude?

   The growth of a bacterial colony in Area 1 would indicate that either the agar or the petri dish had bacterial contamination.

7. What do you expect to see growing in each quadrant of your petri dish? Explain your reasoning.

   See margin for answer.

8. Why did you swab the lines separating the quadrants with hand sanitizer?

   Hand sanitizer will kill any bacteria and clearly mark off the four quadrants.

Observing Sample Bacteria

I Observe the samples of bacteria provided by your teacher. These are some bacteria that you may see in your petri dish.

9. Which bacteria did you observe?

   Answers will vary depending on the samples that you made available.

10. Describe what you observed and classify the bacteria shape.

    E. coli looks like pink rods. It is a bacillus bacterium. *S. epidermidis* looks like purple dots at about 1000× magnification; it is a coccus bacterium. *B. subtilis* is a purple bacillus bacterium.

When scientists observe bacteria, they need to stain them in a way that makes them easier to observe and helps to identify whether the bacterium's cell walls contain peptidoglycan. If a bacterium's cell wall contains the protein, it looks purple when stained. This kind of bacterium is referred to as *Gram-positive*. If a bacterium's cell wall doesn't contain peptidoglycan, it looks pink when stained, and it is referred to as *Gram-negative*.

11. How does the information about Gram staining relate to what you observed?

    Students may reason that the color of the bacteria resulted from their being stained for visibility under a microscope. They were all Gram-positive except for *E. coli*.

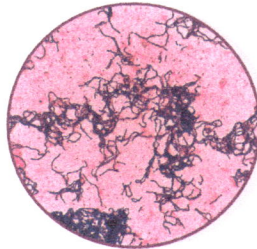

▲
Gram-positive bacteria

Students should expect no or very few bacterial colonies in Area 1. Area 2 should have the most bacterial colonies, followed by Area 3, then Area 4.

✓ **Antibiotics and Bacterial Growth (Question 8)**

The area around an antibiotic agent on an agar plate in which bacterial colonies don't grow is called the *zone of inhibition*. The size of that area is related to the strength of the antibiotic.

✓ **Day 2 Introduction**

The second day of this lab activity begins with the Observing Sample Bacteria section. At this time your students will observe some microscope slides of bacteria, view their bacterial colonies, and make some conclusions about their results.

✓ **Observing Sample Bacteria**

Have students observe any or all of the bacteria that they may see in their colonies: *Escherichia coli, Staphylococcus aureus, Staphylococcus epidermidis*, and *Bacillus subtilis*. You may culture these bacteria yourself and have students observe a wet mount of your culture. You will need to stain your colonies with gram stain if you choose to do this. As an alternative, you may want to obtain preserved slides of these bacteria. If none of these options are available to you, find some good images of each of these bacteria online and allow students to observe them and write a description.

Observing Your Bacteria

After several days of incubation, your teacher will return your petri dish. ***Do not remove the dish from the bag.*** Observe whichever side gives you a better view of the bacterial colonies.

J Count the number of colonies, or clumps of bacteria, in each quadrant and record your data in Table 1.

TABLE 1

Quadrant	Number of Colonies
1	0
2	should be the highest number of the four quadrants
3	should be a number between those of Quadrants 2 and 4
4	should be the lowest number of the three samples

12. What colors and textures did you observe in your bacteria colony? What do these different colors represent?

   Answers will vary, but most bacterial colonies will be white, gray, or light yellow. Textures may vary, and may include smooth, rough, raised, or flat. Different colors represent different species of bacteria.

13. Using the table on the next page, describe the form, elevation, and margins of the bacteria you grew.

   Answers will vary according to the type of bacteria cultured.

FORM	punctiform	circular	filamentous	irregular	rhizoid	spindle
ELEVATION	flat	raised	convex	pulvinate	umbonate	
MARGIN	entire	undulate	lobate	erose	filamentous	curled

ANALYSIS

14. If possible, identify any of your bacterial colonies on the basis of your observations. You may use the internet to help you.

See margin for answer.

_____

_____

K Use a ruler to measure the larger bacterial colonies.

15. Which bacterial culture appeared to grow the fastest? How can you tell?

Answers will vary regarding which is the fastest growing bacterial culture, but the diameter of the colony is an indicator of the speed of bacterial growth.

16. How did your hypothesis in Question 7 compare with what you observed?

Answers will vary according to student hypotheses.

_____

_____

Question 14 Answer

E. coli makes shiny, raised colonies with entire margins. *S. epidermidis* makes small, circular colonies with a convex elevation and an entire margin. *B. subtilis* makes flat colonies with an irregular form and lobed margins with a dry texture.

⚠ *Disposing of Bacterial Cultures*

To dispose of your bacterial cultures, soak agar plates and any tools in a 10% bleach solution overnight. If using plastic petri dishes, place them in a plastic bag before putting in the garbage. Glass petri dishes must be autoclaved before reuse. Make sure that all students wash their hands before leaving the laboratory. For a resource on biological waste disposal, go to the *Biological Waste Disposal* link. This link is available as a digital resource.

Question 19 Answer
Washing hands with warm water reduces the bacteria count by washing the bacteria away. Friction through rubbing also contributes to the removal of bacteria.

17. Did you find any bacterial colonies in Area 1? How does this affect the results that you observed in the other areas of your petri dish?

   Answers will vary. If students had bacterial colonies growing in Area 1, then their petri dish was contaminated, which puts the other results into question. If Area 1 has no bacteria colonies, then the results for the other areas can be trusted.

L Return your petri dish to your teacher for disposal and then everyone in the group must wash their hands.

CONCLUSION

18. On the basis of what you have observed, why do you think it is important to wash hands properly?

   Washing hands properly reduces the bacteria count on our hands, bacteria that could make us sick.

19. What does washing hands with just warm water do?

   See margin for answer.

20. Sometimes soap in dispensers in restrooms becomes contaminated with bacteria. How could you tell whether the soap you were using in this experiment was contaminated?

   There would be more and larger bacterial colonies in the area cultured for washing hands with soap than would normally be expected.

21. Suggest what happens to bacteria during handwashing. How is this different from using hand sanitizer?

   During regular handwashing, bacteria are killed by soap or are washed down the drain. When using hand sanitizer, bacteria are only killed, not washed away; some are more likely to survive on our hands. Additionally, proper handwashing removes some of the oils on the hands that protect many bacterial cells. Most hand sanitizers evaporate before they can penetrate these oils.

Handwashing saves lives. When Ignaz Semmelweis began practicing hand-washing, he watched more of his patients survive to care for the children they had just birthed. The graph below shows the percent of patients who died from puerperal fever at the First Clinic of Vienna Maternity Institution for the years 1841–49.

22. When did Semmelweis notice the biggest decrease in mortality rates in his maternity patients?

May 1847

23. Explain what appears to have caused this reduction in mortality.

Semmelweis began the practice of chlorine handwashing.
Chlorine is an antibacterial agent.

24. Why is it worthwhile to put scientific research into the practice of handwashing?

Handwashing saves people's lives, and we must cherish the lives
of people because they are made in God's image.

Puerperal Fever Monthly Mortality Rates 1841– 49

Loving One's Neighbors through Hygiene

In light of the events surrounding the COVID-19 outbreak of 2019 and its continued effects on daily life in the years following, additional research on personal hygiene may prove especially helpful.

✓ *Taking Bacterial Growth Further*

Guide students to investigate one of the topics that interests them. Research has shown that the friction that occurs when a person lathers up and washes hands has been a surprising factor in reducing bacterial counts.

Make sure that you thoroughly review your students' plans, especially since bacterial studies carry a certain element of risk. You may consider having students prepare agar plates for their extended study.

Have students present the results of their study in a report or presentation to the class so that everyone can benefit from the knowledge they have gained. Students can thus experience one of the last steps of the scientific inquiry process: sharing their results with others.

GOING FURTHER

You can take this experiment further by testing how other factors affect bacterial counts on our hands. Create your own experiment from the following ideas.

» Test other methods of handwashing, including using dirt or ash as is done in some poorer countries.

» Test different kinds of soaps including antibacterial soaps.

» Test how the amount of time lathering affects bacterial count.

» Test whether water temperature affects the results of handwashing.

M You collect lots of bacteria on your hands by touching other people! Physical contact with people is one source of bacteria. For another experiment, try swabbing different places in your home or classroom and culture these samples to find the places in your environment where bacteria make a home.

25. What will you test?

Answers will vary but could include any of the ideas mentioned above.

26. Write out the plan for your experiment.

Answers will vary.

N Report the results from your experiment in a separate report or presentation.

12B LAB

One Slick Solution

Oil-Eating Bacteria

Deep, deep down in the murky depths of the Gulf of Mexico there was a big problem. The Deepwater Horizon oil rig operated by British Petroleum had exploded and sunk on April 20, 2010, killing eleven men and causing a gusher on the ocean floor that leaked oil for eighty-seven days. It was one of the largest oil spills in history. Oil spills are dangerous because oil pollutes the environment and is toxic to living things, especially sea birds, turtles, and dolphins. It took months for a huge team including scientists, volunteers, and robots to cap the leak and clean up miles of ocean and coastline.

Volunteers put out booms to contain the floating oil and cleaned beaches and wildlife. Aircraft pilots skimmed oil off the surface of the ocean. One of their smallest allies may surprise you—bacteria! Scientists injected the genetically engineered bacteria *Alcanivorax borkumensis* near the leak on the sea floor. The bacteria actually ate some of the oil before it floated to the surface.

Eliminating the oil before it rose to the surface was important because it protected wildlife and prevented currents from carrying the oil to beaches in Louisiana, Alabama, and Florida. Using nature to clean up the environment in this way is called *bioremediation*. It's one of the most efficient ways to restore a contaminated environment.

Oil from natural sources regularly seeps into the ocean environment—though not 210 million gallons! Oil is a long molecule made from mostly carbons and hydrogens. Many microbes use oxygen to naturally break down these long molecules into byproducts like carbon dioxide and water that are harmless to living things. It's a process that is similar to cellular respiration, but it uses hydrocarbons instead of glucose.

How can we use our knowledge of bacteria to clean up oil spills?

The US Coast Guard and British Petroleum also used controlled burning of oil on the water's surface to clean up the Deepwater Horizon spill.

LAB 12B OBJECTIVES

- Analyze data of the effects of various chemicals on oil-eating bacteria.
- Evaluate the effectiveness of bioremediation.

✓ Ideas for Prelab Discussion

You may consider beginning this lab activity by showing your students a video of the Deepwater Horizon oil spill, including the cleanup efforts.

You may need to review the concepts of pH and the basics of hydrocarbon chemistry with your students in preparation for this lab activity.

You should also demonstrate how to aerate oil samples using the plastic pipettes. To introduce air into the oil sample, place a pipette into the graduated cylinder and squeeze the bulb. This should create bubbles. Do not release the bulb before removing the pipette from the graduated cylinder.

🧪 Equipment Notes

Oil-eating bacteria are available from science equipment suppliers, allowing students to create a model oil spill and treat it with oil-eating bacteria. For one such kit, see the *Oil-Eating Bacteria Kit* link. This link is available as a digital resource.

Follow your kit's instructions to create enough solution for each group to have 80 mL. Graduated cylinders that are used to contain the oil spill, bacteria, and cultures should have uniform height and width. Possible substitutions for standard graduated cylinders include snap seal graduated cylinders, small plastic cups, or plastic preform soda bottles that look like test tubes. You may need to adjust the required volume of oil and water depending on the containers that you use.

🖥 Learning More about Using Technology to Clean Up Contamination

To learn more about cleaning up oil spills, see the *Cleaning Up Contamination* link. This link is available as a digital resource.

Safety Precautions

Oleophilic bacteria are harmless to people, but students should avoid contacting or ingesting the bacteria. Also, do not have students use mouth pipettes to aerate the graduated cylinders, as this will provide less oxygen and could result in accidental ingestion of bacteria.

Some Items to Review

Make sure that you review pH with students and remind them that a pH of 7 is neutral, a pH less than 7 is acidic, and a pH greater than 7 is basic.

Review carbon chemistry (Student Edition, pages 40–43) with your students to prepare them for thinking about breaking down oil.

You may need to review the basics of writing chemical equations (Student Edition, page 35). Balancing the equation for the chemical reaction in Question 1 is not critical. If you want students to balance the equation, you may need to do a review during your prelab discussion time.

Fertilizer

As an alternative to weighing the fertilizer, pouring it in, and then stirring it until it has dissolved, you may want to buy or make a liquid fertilizer solution for students to add to their graduated cylinders.

Separate Stirring Rods

Be sure to remind students not to cross contaminate their graduated cylinders by mixing more than one graduated cylinder with the same stirring rod.

Pseudomonas, an oleophilic bacteria

Bacteria that process oil are known as *oleophilic bacteria*. In this lab activity you will be working with a mix of different types of oil-eating bacteria to see how they can break down oil into smaller molecules to reduce the amount of floating oil. But to survive, the oil-eating bacteria will need air and some nutrients, such as potassium and nitrogen. During remediation efforts, scientists sometimes add these nutrients to polluted environments in the form of a fertilizer, which works well when cleaning contaminated soil. But this isn't very effective in the open ocean.

Oleophilic bacteria can live in a pH range of 5.5–10 and at a temperature ranging from −2 °C to 60 °C. You'll see how these other factors help bacteria grow so that they can consume more oil.

1. Write a chemical equation, showing the reactants and products, in which oleophilic bacteria metabolize oil. Though petroleum is a mix of hydrocarbons, use octane (C_8H_{18}) in your chemical formula. Bacteria use oxygen to break down octane into water and carbon dioxide.

$$2C_8H_{18} + 25O_2 \longrightarrow 18H_2O + 16CO_2$$

Equipment

laboratory balance (accurate to 0.01 g)
graduated cylinders, 50 mL (4)
graduated cylinder, 10 mL
glass stirring rods (3)
wax pencil
weighing paper
cotton balls
disposable plastic pipettes

oil-eating bacteria suspension, 80 mL
10-10-10 fertilizer
acetic acid (CH_3COOH), 1 M, 5 mL
universal indicator paper or pH meter
milk of magnesia, 8 mL
cooking oil, 60 mL
laboratory apron

nitrile gloves
goggles

QUESTIONS

• What do bacteria need to grow?

• How can we nurture oil-eating bacteria to clean up an oil spill?

PROCEDURE

Setting Up

NUMBERING YOUR GRADUATED CYLINDERS

1. bacterial suspension, oil

2. bacterial suspension, oil, fertilizer

3. bacterial suspension, oil, fertilizer, acetic acid

4. bacterial suspension, oil, fertilizer, milk of magnesia

A Using a wax pencil, label four graduated cylinders from 1 to 4. Add 20 mL of the bacterial suspension to each graduated cylinder.

B Weigh out 0.25 g of fertilizer on three different pieces of weighing paper. Add this fertilizer to Cylinders 2, 3, and 4. Use a separate glass stirring rod to stir the solutions in each graduated cylinder until the fertilizer is dissolved.

2. Predict whether the fertilizer will encourage or inhibit bacterial growth. Explain.

Answers will vary. Fertilizer has the effect of encouraging bacterial growth because it supplies nitrogen and phosphate, both of which encourage bacterial growth.

C Use the 10 mL graduated cylinder to add 5 mL of acetic acid to Cylinder 3 and stir the contents. Determine the pH of the solution using universal indicator paper or a pH meter.

136　Lab 12B

3. What is the pH that you measured? On the basis of your measurement, predict whether the acetic acid will encourage or inhibit bacterial growth. Explain.

The pH should be around 2.0. The acetic acid will inhibit

bacterial growth because it lowers the pH to below the normal

range for oil-eating bacteria.

D Add 8 mL of milk of magnesia to Graduated Cylinder 4 and stir the contents. Determine the pH of the solution using universal indicator paper or a pH meter.

4. What is the pH that you measured? On the basis of your measurement, predict whether the milk of magnesia will encourage or inhibit bacterial growth. Explain.

The milk of magnesia should raise the pH of the mixture to about

10.0. This is at the very high end of the normal pH range of

oleophilic bacteria, so it will act as an inhibitor to bacterial growth.

E Add 15 mL of cooking oil to all four graduated cylinders.

5. What is the purpose of Graduated Cylinder 1?

See margin for answer.

Question 5 Answer
Graduated Cylinder 1 acts as a control for the oil and bacterial suspension.

6. Why does the oil float to the top when you add it to each graduated cylinder?

Oil is less dense than water.

7. How would this property of oil affect how it is cleaned up in a spill?

Because oil floats, spills can be contained with booms, and oil

can be skimmed off or burned on top of the water.

F Measure the volume of the oil layer in each graduated cylinder by subtracting the volume at the bottom of the oil layer from the volume at the top of the oil layer. (Refer to Appendix D if you need help on how to read a graduated cylinder.) Record these values to at least 0.1 mL precision in the *Day 0* column of Table 1.

G Put a cotton ball in the top of each graduated cylinder to prevent contaminants from entering it.

H Place the cylinder in the location specified by your teacher. Wash your hands before leaving the laboratory.

8. Predict which of the graduated cylinders will show the most bacterial growth. Explain your prediction.

Answers will vary. Cylinder 2 should show the most bacterial

growth. We can infer that its pH is in the right range and that it

contains fertilizer.

Measuring Oil Consumption

I After Day 1, read the volumes at the top and bottom of each oil layer to determine the volume of the oil to at least 0.1 mL precision. Record your data in Table 1.

J After measuring the volume of each oil layer, aerate each cylinder. Use a separate pipette for each cylinder. Insert the pipette into the lower liquid layer, then squeeze the pipette bulb, forcing air into the liquid. Do not release the bulb until it has been removed from the cylinder. Make sure that you aerate all graduated cylinders to the same degree.

K Return the cylinder to its safe location. Wash your hands before leaving the laboratory.

L Repeat Steps I–K for the next three days.

9. What would happen if you didn't bubble air through the graduated cylinders?

   The bacteria would deplete the oxygen, and the bacteria would stop consuming the oil and possibly die.

10. Why should you use a separate pipette for each graduated cylinder instead of using only one pipette for all of the graduated cylinders?

    Using the designated pipette prevents cross contamination of the graduated cylinders.

11. Did you observe any changes in the appearance of the liquid in the graduated cylinders? Explain.

    Answers will vary but could include the cloudiness of the water. Some graduated cylinders may also form bubbles.

M Dispose of your graduated cylinders according to your teacher's instructions.

ANALYSIS

N Use Graphing Area A to graph your data from Table 1. Plot the time in days on the *x*-axis and the volume of the oil layers in milliliters on the *y*-axis. Use a different color for each graduated cylinder.

12. Was your prediction in Question 8 accurate? Explain.

    Answers will vary. Cylinder 2, containing water, oil, and fertilizer, should have the most depleted oil layer if the graduated cylinders were aerated uniformly.

⚠ *Disposing Of Bacterial Cultures*

To dispose of your bacterial cultures, either autoclave all materials or soak them in a 10% bleach solution overnight before discarding in the trash can. Make sure that all students wash their hands before leaving the laboratory.

13. Write down any surprising observations and try to explain them.

Surprising observations would include bacterial growth in Cylinder 1, an unchanging oil layer volume in Cylinder 2, or a change in oil layer volume greater than that observed in Cylinder 2.

14. If you had not equally aerated your graduated cylinders, how would your results have been affected?

Unequally aerating the graduated cylinders would have introduced another variable, which would have called all the results into question.

CONCLUSION

15. On the basis of what you have observed, what do you think is necessary for bacteria to be helpful in cleaning up oil spills?

Bacteria need sufficient nutrients and proper environmental conditions (e.g., oxygen levels and pH) to be useful in bioremediation.

Of the oil spilled in the Deepwater Horizon oil spill, 17% was captured from the well and 8% was removed through skimming and burning. Workers also successfully used chemicals to disperse the oil. One of the reasons they did this was to make it easier for bacteria to metabolize the oil. It takes a long time for bacteria to consume this much oil.

16. Suggest one limitation or negative effect of bioremediation for an oil spill using oleophilic bacteria.

See margin for answer.

17. In what ways did your graduated cylinders model a real oil spill? What variables were left out of your model?

See margin for answer.

Question 16 Answer
Answers will vary. Accept any one. *Examples:* dead zones in the ocean where oxygen had been depleted, the long periods of time required to clean up the oil, the fact that fertilizers can't be used effectively in the open ocean

Question 17 Answer
The presence of oil, water, bacteria, and fertilizer from runoff are all present in both the model and a real oil spill. Variables left out of the model include environmental factors such as wave action, currents, and wildlife. A huge variable is the complex nature of petroleum, a mix of many different organic compounds that varies from one region of the world to another. Students may cite other omitted variables.

People have compared the 2010 Deepwater Horizon oil spill to the *Exxon Valdez* oil spill in 1989. Though the Deepwater Horizon event was a larger spill, it contained a lighter oil that had shorter hydrocarbons. The *Exxon Valdez* oil spill happened on the ocean surface and closer to the Alaskan shoreline, resulting in more injured animals. Also, the Gulf of Mexico has more natural seeps, so there were naturally more oleophilic bacteria in the water. Finally, the Deepwater Horizon spill happened in an area of the world where the temperature is much more nurturing to bacteria than where the *Exxon Valdez* spill occurred.

But one of the most significant differences in these two incidents was the massive response of employees and volunteers to clean up the Deepwater Horizon spill, involving 47,000 people. This oil spill was in a much more heavily populated area that was easier to access, resulting in an overwhelming response from the local community.

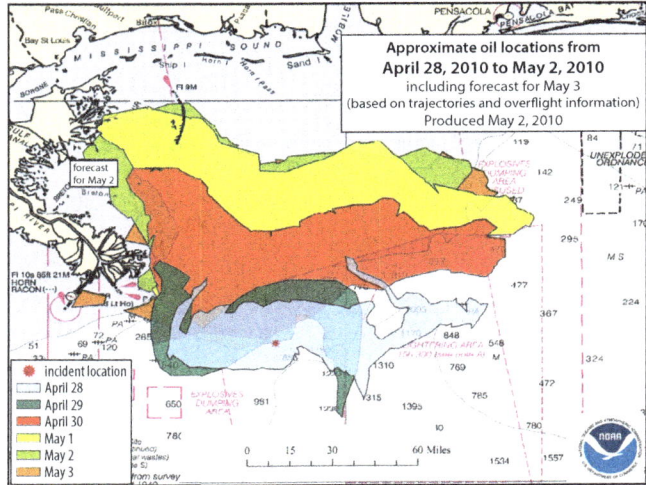

Meteorologists and ocean scientists team up to model the way that environmental factors affect an oil spill in order to predict how the oil will spread.

18. How is the conservation work involved in cleaning up an oil spill glorifying to God?

Cleaning up an oil spill is a clear fulfillment of the Creation Mandate—protecting people from the toxic effects of the accident, wisely using resources, and helping animals.

Oil Spill Trajectory Modeling

To show students a video of how the National Oceanic and Atmospheric Administration (NOAA) predicts the way that an oil spill will spread, see the *Oil Spill Trajectory* link. This link is available as a digital resource.

GOING FURTHER

19. You varied the pH in the graduated cylinders to see how it affected the growth of oleophilic bacteria. What other things could you have varied?

the type of oil used, the temperature at which the graduated

cylinders were stored, and the amount of aeration

20. Choose a variable from your answer to Question 19 and write out a plan to test it. If you'd like to actually carry out this experiment, collect data and share it with others.

Answers will vary. The most difficult variable to test would be

temperature because it would require keeping all the graduated

cylinders at the same temperature for five days.

TABLE 1

GRADUATED CYLINDER	VOLUME OF OIL LEVEL (mL)				
	Day 0	Day 1	Day 2	Day 3	Day 4
1	15.0	15.0	15.0	14.5	14.5
2	15.0	14.5	14.1	13.8	13.2
3	15.0	14.5	14.5	14.5	14.5
4	15.0	14.5	14.2	13.9	13.6

Sample Data

The sample data provided was collected when the lab activity was developed. Students' data should be similar.

GRAPHING AREA A

Volume of Oil Layer over Time

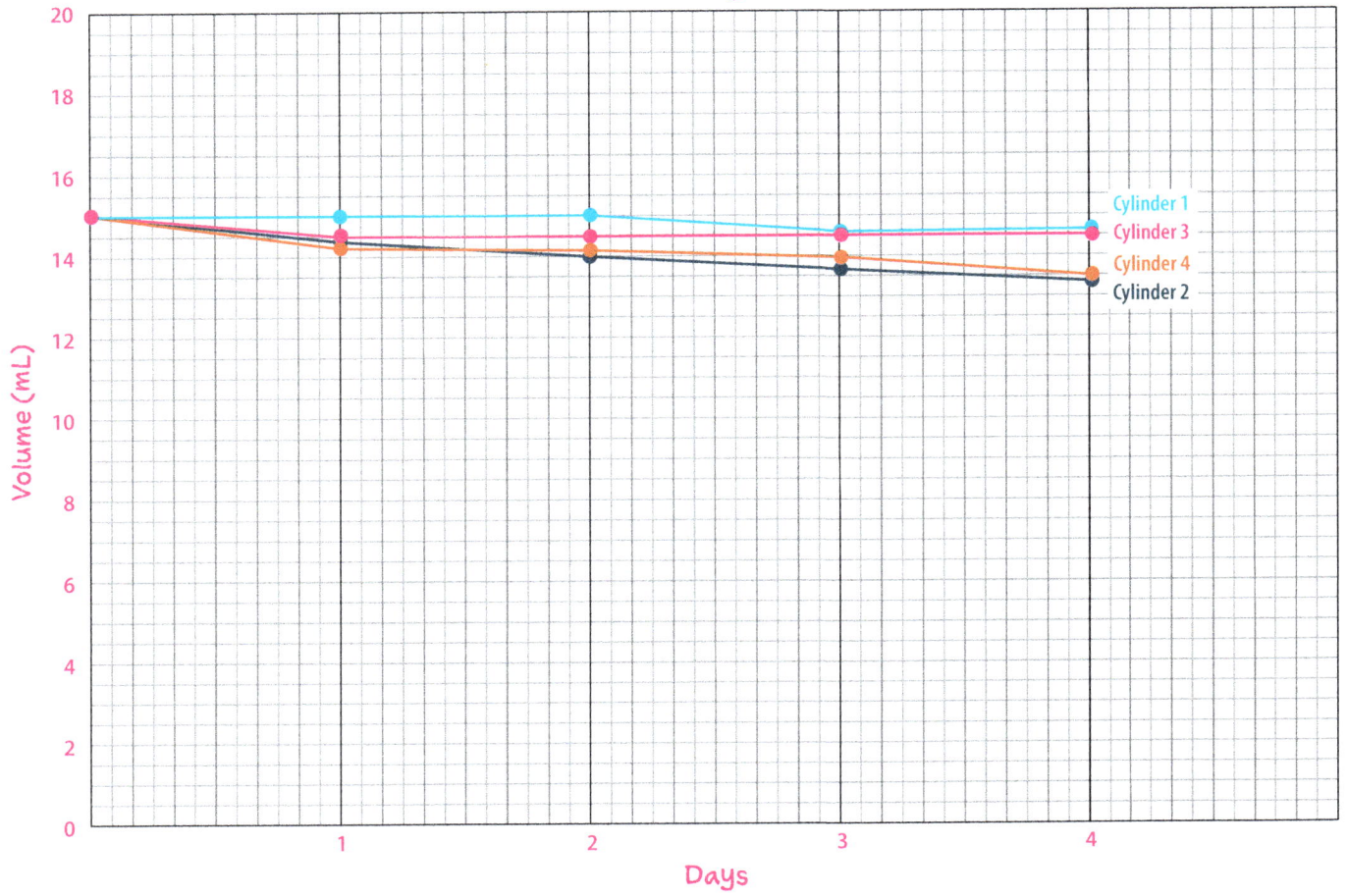

Volume (mL)

Days

Cylinder 1
Cylinder 3
Cylinder 4
Cylinder 2

13A LAB

Wee, Watery World

Exploring the Microscopic World of Protozoans

Hundreds of thousands of idealistic young men fell during the American Civil War. The majority of those who perished were not killed by bullets or shells, but by disease. One of the most common diseases they suffered from was *amoebic dysentery* caused by the parasitic protozoan *Entamoeba histolytica*. The spores of *E. histolytica* survive in contaminated soil and water. Unsuspecting soldiers ingested the spores when they drank from contaminated streams.

Many other protozoans are free-living (nonparasitic) and can be found in water and soil around the globe. They are an important part of aquatic food webs. Protozoans exhibit an astonishing variety of forms and functions. In this lab activity you'll examine some of these amazing—and sometimes deadly—microscopic organisms.

?

What are the distinguishing features and characteristics of protozoans?

Equipment

microscope
reference books or keys for protozoans
pipettes (5)
microscope slides (5)

cover slips (5)
living cultures of amoebas, euglenas, and paramecia
cotton fibers
glycerin or methyl cellulose

toothpick
carmine powder or yeast
pond water

OBSERVING PROTOZOANS

Amoeba

A Obtain a sample from the amoeba culture and prepare a wet mount of it (see Appendix D). Amoebas usually stay close to the bottom of the culture or crawl on some object, but they can be drawn into a pipette by using the following technique.

» Squeeze the air out of a pipette bulb.

» Place the pipette directly above the place where the amoeba should be.

» Release the bulb to suck up one drop.

» Quickly put the entire drop on the slide. (Wait a minute before placing the cover slip on top of the amoeba culture on your slide to allow the amoeba to attach to the slide and begin to move.)

QUESTIONS

- What are protozoans?
- What are some of the different kinds of protozoans?
- How do protozoans move and get food?
- What kinds of protozoans can be found in pond water?

NOTES ON HANDLING LIVE CULTURES

1. Live cultures often contain a few organisms other than the targeted species.

2. Live cultures sometimes go "sour," meaning that the organisms die. If this happens, observe a prepared slide of the desired organism.

3. Keep the cultures covered when not in use.

4. Use a separate pipette for each culture to prevent cross contamination.

5. Make sure that your slides and cover slips are clean and free of any soap residue.

- Compare various protozoans and protozoan phyla.
- Identify examples of protists in pond water.

Equipment Notes

Most science equipment suppliers will ship live specimens so that they arrive just prior to a particular date. Protozoan and algal cultures need to be fresh. Despite careful shipping procedures, though, occasionally a sour culture will arrive. Therefore, you should have on hand preserved slides of all the living organisms that you plan to use.

You might want to reassure students that the amoebas they are viewing are not the dangerous *Entamoeba*.

Pond water containing organisms may be difficult to obtain at the time of year you need it. Unboiled pond water, even though it has been obtained from a cold pond, should produce organisms if it is kept warm for a few days.

One of the problems with culturing pond water is keeping the pond water in adequate light without getting it too warm. If placed in direct sunlight in a window, a container of pond water can easily get so warm that most organisms will be destroyed. Artificial light is probably the best. Try to keep the water temperature between 65 °F and 75 °F.

Got Keys?

If you do not have access to books or keys for protozoans, do an internet search for pond water life or pond water field guides to find online resources.

Start Simple

Probably the easiest culture to keep alive and the organism that students locate most quickly (because of its abundance) is the euglena. However, because of its small size and euglenoid movement, which is almost impossible to stop, it is difficult to see the internal structures. If students become frustrated trying to find the amoeba or slow down the paramecium, suggest that they try observing the euglena and then go back to the amoeba and paramecium.

B Scan the entire slide on low power. Scan first from right to left; then move the slide a little lower and scan left to right. Repeat until you have scanned the entire slide. Anything that moves faster than a snail's pace is *not* an amoeba.

C Be sure that the culture medium does not evaporate. Using the pipettes from the amoeba culture, add more medium to the edge of the cover slip as necessary.

D Observe the movements of the amoeba for a while on both low and high power.

E If you see an amoeba engulfing food or dividing, inform your teacher so that you can share it with the class.

1. Describe the locomotion (movement) of the amoeba that you observed.

   Answers will vary but should include a description of cytoplasm flowing into pseudopods.

2. What is the name for this type of locomotion?

   amoeboid movement

3. Is more than one pseudopod present at one time?

   yes (usually)

4. Besides movement, for what other purpose does an amoeba normally form pseudopods?

   to engulf food particles

F Draw an amoeba in Drawing Area A. Label as many structures as you can.

Euglena

G Prepare a wet mount from the euglena culture. Euglenas will be found throughout the culture medium, so no special capture technique is necessary.

H Scan the slide. Euglenas can move quickly but usually will not leave the microscope field very rapidly. Occasionally it will be necessary to use cover slip pressure or a thicker medium to slow their movement.

I Observe the euglenas on both low and high power.

5. Describe how euglenas move.

   Euglenas move by whipping their flagella and pulling themselves through their medium.

6. Were you able to observe a second kind of movement? If so, describe it.

   Euglenas can also move by means of wormlike contractions and expansions, but this is not always apparent when viewing.

Classification Chaos

You may recall that euglenas were formerly classified as algae. Some sources will still so classify them, but the Catalogue of Life now places euglenas in the phylum Euglenozoa within the kingdom Protozoa. This is another example of the highly fluid nature of modern classification systems (see Lab 11B).

7. Scientists have disagreed over the years about how to classify euglenas, sometimes grouping them with protozoans, and at other times classifying them as algae. What characteristics of euglenas are plantlike? Which are animal-like?

Euglenas have chloroplasts and can photosynthesize, so they are plantlike in that regard. However, like animals, euglenas can locomote and lack cell walls.

J Draw a euglena in Drawing Area B. Label as many structures as you can.

Paramecium

K Prepare a wet mount from the paramecium culture. Paramecia may be found throughout the culture.

L Scan the entire slide. Paramecia move rapidly and will need to be chased across the slide. To slow down or stop the paramecia, your teacher may have you take one of the following actions.

» Have your lab partner use the edge of a paper towel to wick away a small amount of culture medium from beneath the cover slip while you keep the paramecium in your field of view. As the culture is drawn off, the cover slip will exert more pressure on the paramecium. Be careful to not allow the medium to evaporate completely.

» Before placing the cover slip, place a small quantity of cotton fibers on the slide. These will block the paramecia's paths and localize their activity.

» Use a special medium prepared with glycerin or methyl cellulose. Because these media are thicker than the culturing solution, protozoans move more slowly in them.

Be careful not to confuse other protozoans in the culture with paramecia. A paramecium looks like the slipper-shaped illustration seen on page 277 in your textbook and is easy to recognize.

M Observe the paramecium on both low and high power and try to locate all the structures pointed out in your textbook.

N After observing your paramecium for a while, carefully remove the cover slip from your slide and add a few cotton fibers to the medium if you have not done so already; do this only if you are not using the thickened medium.

O Observe the reactions of the paramecia as they encounter the cotton fibers.

8. Describe the movement of the paramecium. Be sure to include the names of the cellular structures involved in its movement.

The paramecium moves by beating its cilia.

Slow Down, Paramecium!

The easiest method of slowing paramecia for this activity is to use cotton. Show the students how much cotton is enough—they are likely to use too much. Five to ten 1 cm long fibers should be adequate.

9. How does the paramecium react when it encounters a cotton fiber?

Answers will vary. When the paramecium hits the cotton, it
moves backward, turns slightly, and goes forward again.

_____

10. How does amoeboid movement compare to paramecium movement?

Amoeboid movement involves a change in body shape to
achieve motion. The paramecium moves through the use of
cilia, but its body shape does not change.

11. Look carefully at a resting or confined paramecium and observe the operation of its contractile vacuole. Describe what you see.

The contractile vacuole gradually expands as it fills with water
and then quickly contracts as it expels water out of the
paramecium.

12. Could a paramecium survive if its contractile vacuole malfunctioned? Explain.

No. The contractile vacuole maintains the proper amount of water
in the paramecium's cytoplasm. Paramecia live in hypotonic
solutions, and water constantly enters from their environment.

_____

P Using a toothpick, place a few grains of carmine powder or yeast on your slide. Observe the path of the powder or yeast as it enters the paramecium and moves within it.

13. What is the function of the carmine powder?

Carmine powder is a dye that allows the movement of food
through the paramecium to be observed.

14. How do paramecia obtain their food?

Food is swept into the gullet by the beating cilia.

_____

15. How is the paramecium's food digested?

After entering the gullet, the food is enclosed in a food vacuole
for digestion.

Q Draw a paramecium in Drawing Area C. Label as many structures as you can.

Do a Dry Run First!

It's a good idea to test this portion of the activity before assigning it to students, as different strains of paramecia perform this task at different rates, sometimes slowly. To save time, omit Question 13 and have students use the Student Edition to answer Questions 14 and 15.

OBSERVING POND WATER

A drop of pond water is often an astonishing hive of activity. The typical pond is home not only to amoebas, euglenas, and paramecia, but a host of other protists as well. These organisms are an important part of a pond's food web.

R Make two wet mounts from the pond water culture provided by your teacher: one from the material near the bottom of the culture and the other from material near the top.

16. Why is it advisable to take samples from both areas?

   Some organisms are free-swimming and can live anywhere in the water column. Others are sessile and live on the bottom or on other surfaces within the culture.

S Use Drawing Area D to draw some of the organisms that you observe in your mounts. Reference the books and keys provided by your teacher to identify each organism, if possible, including the taxonomic groups that each one belongs to (e.g., Protozoa). Label any structures that you are able to identify.

GOING FURTHER

Entamoeba histolytica wasn't identified until 1873, well after the Civil War ended. Later, scientists and doctors noted that most people infected with *E. histolytica* didn't actually get sick. The search for the answer to why that was so is one of the many obscure but interesting stories of science.

T Research information about Émile Brumpt; then answer the following questions.

17. What did Émile Brumpt suggest was the reason why most people did not get sick when infected with *E. histolytica*?

   In 1925 Brumpt proposed that there were two different species of Entamoeba, one pathogenic and the other not. He proposed calling the nonpathogenic form Entamoeba dispar.

18. Why was Brumpt's proposal not initially accepted by the scientific and medical communities?

   Brumpt's proposal was not accepted because there were no apparent morphological differences between the two proposed species.

✓ **Options**

The Going Further section is enrichment material that may be omitted altogether or assigned as homework.

✓ **Brumpt's Proposal Confirmed**

The confirmation and description of a nonpathogenic form of *E. histolytica*— *E. dispar*—were given in a paper by L. S. Diamond and C. G. Clark in 1993.

© 2024 BJU Press. Reproduction prohibited.

Wee, Watery World   147
Wee, Watery World   **147**

How Much Learning Is Enough?

The command from God to have dominion over the earth never set a bar for how much learning is enough. Many scientific discoveries have overturned previously held beliefs that were assumed to be true for centuries. It is likely that some of what we assume to be true today will be considered quaint in years to come. There is still much that we do not know, even about phenomena that we know relatively much about. The task of science is to keep asking questions and to keep seeking answers to those questions. Perhaps one of the students working on this lab activity will make the next big breakthrough in science!

19. How was the question of the existence of a nonpathogenic *Entamoeba* finally resolved?

The differences between *E. histolytica* and *E. dispar* were established by molecular methods in 1993, sixty-eight years after Brumpt first suggested that they were separate species.

20. What lesson can you draw about the nature of science from the story of *Entamoeba*?

Answers will vary. Sometimes, the technology to verify or refute a hypothesis is not available until long after the hypothesis is made.

21. What practical applications might our better understanding of *Entamoeba* have?

Answers will vary. Being able to identify whether the Entamoeba present in a patient are of the pathogenic or nonpathogenic variety might help doctors better decide how to treat their patients.

22. Using what you learned about the study of *E. histolytica*, how do you think the Creation Mandate is further advanced by the continued study of organisms that we already seem to know much about?

Answers will vary. Though we knew much about *E. histolytica* in 1925, our understanding was still incomplete. Further study was necessary to establish the fact that there were two species where previously there was believed to be only one. Similarly, we know much about many other organisms, but our knowledge is incomplete. God's creation is complex enough to keep us continually searching for answers, no matter how much we think we already know. As we search, we discover new information that can help us to better exercise dominion and serve others.

DRAWING AREA A

_____

DRAWING AREA B

DRAWING AREA C

DRAWING AREA D

13B LAB

Fun with Fungi

Observing Fungi

A group of hikers are returning to their vehicle at dusk. Suddenly they see a faint, eerie glow just a few feet in front of them near the ground by the trail. Shining a flashlight toward the glow reveals only a rotting stump covered with shelf fungi. When the light is turned off, the glow vanishes.

These hikers aren't imagining things. The shelf fungi growing on that stump are *Panellus stipticus*, a species of glowing fungi. The glow is very faint, so when the hikers shined their flashlight on the stump, it reflected enough light to their eyes that they could no longer see the faint glow until they had once again readjusted to the dim light.

Fungi may seem irrelevant since we seldom see them unless they have large fruiting structures, but they are very important to life. They are responsible for most of the decomposition of dead organisms—imagine the mess we would have without decomposers! Fungi also include edible mushrooms, yeasts used for baking bread, and molds used for some cheeses, such as blue cheese. Lunch wouldn't be nearly as much fun without fungi!

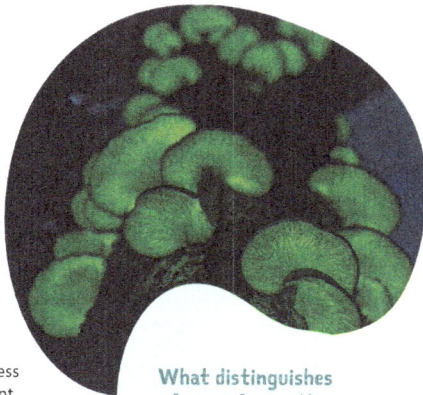

What distinguishes a fungus from other organisms and from other fungi?

Equipment

dissecting microscope or hand lens

microscope

living cultures of *Rhizopus stolonifer* and *Penicillium notatum*

preserved slides of *Rhizopus stolonifer*, w.m.; cup fungus apothecium; and *Coprinus*, c.s.

preserved specimens of puffballs, shelf fungi, and mushrooms

laboratory apron

goggles

QUESTIONS
- What are the characteristics of fungi?
- How do the phyla of fungi differ?
- How are fungi classified?

PROCEDURE

Phylum Zygomycota

The phylum Zygomycota derives its name from the thick-walled sexual structures called *zygosporangia* that are produced when specialized hyphae unite. These fungi are usually sessile, produce spores, and resemble algae in structure. In addition to the sexually produced zygosporangia, zygomycota bear asexually produced spores in sporangia. Examples of this phylum are common and abundant in the soil and air and include molds and yeast. One, *Rhizopus stolonifer*, is a black mold that often grows on bread.

- Identify and describe fungal structures.
- Differentiate among the three phyla of fungi.
- Compare fungi with other organisms.

Prelab Discussion

During this lab activity students are asked to draw the sexual reproductive structures of each of the three phyla of fungi. At the beginning of the activity review the structures that they will be asked to draw. Images will be especially helpful to most students.

Equipment Notes

Try to obtain a living culture of *Rhizopus stolonifer* (black bread mold) either as a laboratory culture or on moldy bread. It might be interesting for students to examine it under a stereomicroscope or a hand lens.

Rhizopus stolonifer is sometimes identified as *Rhizopus nigricans*.

If you have the time and the equipment in your laboratory, you can use the oil immersion setting on your microscope to observe yeast budding. To obtain a yeast culture, put a tablespoon of sugar in a pint jar or 500 mL beaker full of warm water, add about half a packet of active dry yeast, and leave the jar uncovered overnight in a warm place. On the day of the lab activity put a drop of the culture on a slide and then add a drop of dilute methylene blue to the culture. Complete the process by putting a cover slip on the slide and setting up a microscope for the highest magnification available. Have students observe it as soon as possible or project the image from the microscope onto a screen or wall of the classroom. You could also take a picture and project that instead.

Question 1 Answer

Answers will vary. Students should see hyphae, sporangia, and other structures under adequate magnification. Students may note hyphae around the outer edge of the culture and spore-producing structures toward the center.

Preserved Slides of Fungi

Some students find the preserved slides of certain fungi difficult to identify and label. A preserved slide of *Rhizopus*, for example, does not look like the diagram shown in the Student Edition. The hyphae are tangled, broken, and not neatly arranged. Advise students of the differences.

Penicillium notatum

Fungi in the genus *Penicillium* have gone through a number of changes in classification, so the scientific name *Penicillium notatum* is no longer considered valid. However, science equipment suppliers still typically list live cultures of this organism under that name.

A Using either a dissecting microscope or a hand lens, observe cultures of *R. stolonifer*.

1. Describe what you see, including any fungal structures that you can identify.

   See margin for answer.

2. *R. stolonifer* produces hyphae that are clear or white. What causes its dark appearance?

   colored sporangia and spores

B Observe a preserved slide of *R. stolonifer* on high dry power.

C Observe a preserved slide of *R. stolonifer* with zygotes.

D Use Drawing Area A to draw a zygosporangium with two parent hyphae.

Phylum Ascomycota

Fungi in phylum Ascomycota often appear very similar to fungi in phylum Zygomycota but differ in their spore-forming structures. They are named for their microscopic sac-like asci (s. ascus) that hold the sexually produced ascospores.

E Using either a dissecting microscope or a hand lens, observe the laboratory culture of *Penicillium notatum*.

3. Describe what you see, including any fungal structures that you can identify.

   Answers will vary. Students should see white hyphae growing around a bluish-green, spore-covered area.

F Observe a preserved slide of a cup fungus apothecium.

4. Describe the asci that line the apothecium.

   Answers will vary. Students should mention rows of straight asci, each with a single row of ascospores.

G Use Drawing Area B to draw one of the asci that line the apothecium.

Phylum Basidiomycota

As you learned in your textbook, phylum Basidiomycota includes mushrooms, puffballs, and shelf fungi. The phylum's name comes from the sexually produced basidiospores borne on club-shaped cells called *basidia* (s. basidium). On the familiar mushroom, thousands of basidia are attached to the gills under the mushroom's cap, and each basidium has four basidiospores.

H Using either a dissecting microscope or a hand lens, observe preserved and dried specimens of puffballs.

5. Where are the basidia and basidiospores located on puffballs?

   inside the puffball

6. How are the basidiospores released?

When the puffball is disturbed, it breaks open and the spores are released.

I Using either a dissecting microscope or a hand lens, observe preserved specimens of shelf fungi.

7. Where are the basidia and basidiospores located on shelf fungi?

within tiny pores on the underside of the fungi

8. How are spores released from shelf fungi?

They fall out of the pores.

J Using a dissecting microscope or a hand lens, observe preserved specimens or fresh samples of mushrooms.

» Note the cap, gills, stipe, and hyphae of the various species.

» Note the different colors, sizes, and textures of the various species.

K Observe the preserved slide of *Coprinus*. Be sure to observe the cap.

» Find the gills and locate the basidia and basidiospores.

L Use Drawing Area C to draw a section of a gill with basidia and basidiospores.

9. Considering what you've learned in this lab activity, what structure do you think scientists use to classify fungi?

the fungi's sexual structures

10. Why is it important to classify fungi?

Classifying fungi allows us to study them more effectively, possibly leading to improved uses, like lifesaving antibiotics.

11. How do the study and classification of fungi help fulfill the Creation Mandate?

Some fungi can be very useful to humans, providing food and medicines. Other fungi are very toxic. Therefore, it's important to be able to identify fungi both to save human life and to improve human life using products made from fungi.

Wise Dominion— Even over Molds!

Much of the good health we take for granted today can be traced back to Fleming's providential discovery of penicillin. The Creation Mandate extends to every part of God's creation—even to humble molds!

Options

The Going Further section is enrichment material that may be omitted altogether or assigned as homework.

GOING FURTHER

Most of the imperfect fungi are plant parasites, but some are parasites of humans. Although they cause few if any fatalities, they are still very annoying. Millions of dollars are spent each year on antifungal powders, creams, and sprays.

M Two common fungal ailments are athlete's foot and ringworm. Research each of these conditions.

12. Describe athlete's foot and its common symptoms. Explain how infections spread, how they can be prevented, and how they can be cured.

Answers will vary. Athlete's foot is caused by several species of fungi. These fungi cause dry, red, itchy, scaly skin between the toes. The infection is spread either by skin-to-skin contact or by contact with contaminated surfaces. It can be prevented by avoiding contaminated skin or surfaces. It can be killed by applying various treatments containing antifungal compounds.

13. Describe ringworm and its common symptoms. Explain how infections spread, how they can be prevented, and how they can be cured.

Answers will vary. Ringworm is caused by several species of fungi. These fungi cause small rings of hyphae under the skin. It is spread by contact with contaminated soil or infected people or pets. Ringworm can be prevented by avoiding contaminated skin or surfaces. It can be treated by applying antifungal ointments.

DRAWING AREA A

DRAWING AREA B

Fun with Fungi   **155**

DRAWING AREA C

14A LAB

Name That Plant

Identifying Plants

Biologist Nathan Klaus was hiking near Pine Mountain, Georgia, when he came upon an amazing find. Next to the trail were six very special trees—all American chestnuts, a species on the verge of extinction. The tallest tree in this newly discovered stand is the southernmost fruit-bearing American chestnut in the world. There are only a few stands left in the entire country, though over a century ago they made up about a quarter of all the trees in the Appalachian Mountains. American chestnuts were a key species in the Appalachian food web, and Appalachian families depended on the chestnut's decay-resistant wood and abundant yields of edible nuts. But in the early 1900s a fungal disease began to wipe out the population.

How did Nathan Klaus identify these trees among all the different types in the forest? American chestnuts are identified by their flowers, fruit, and leaves. Even the shape of a tree and the color and texture of its bark can help identify it.

Now you get to try your hand at identifying some plants that may be as close as your own backyard! It's one thing to see photographs or drawings of plants but quite another to recognize actual specimens. In this lab activity you will explore plant diversity in a specific location by finding and identifying some specimens.

American chestnut

What kinds of plants live in a particular ecosystem?

Equipment

camera	hand lens
garden gloves	small notebook
garden shears	small plastic bags
field guide	permanent marker

QUESTIONS

- How can I tell what kinds of plants live in an ecosystem?
- How can I find out the names of those plants?
- Why do scientists monitor the plants in an ecosystem?

PROCEDURE

Collecting Specimens

Every scientist needs to record his observations, especially when working in the field. Observations are data, and it is important to write down information about what you observe so that you can remember it later. Sometimes the tiniest pieces of information can lead to discoveries in science. Let's set up your own personal field journal.

A Decide where you are going to observe. Obtain permission from the owner to be on the property, to take pictures, and to collect samples of the plants.

B Write down in your notebook the location you have chosen in which to observe plants. Write down the date, the time of day that you are making observations, and the people in your group. Write down your observations about the weather, wind, and soil conditions, along with the local air temperature.

Name That Plant   157

- Identify plants using a field guide.
- Categorize plants as seedless vascular plants, nonvascular plants, gymnosperms, dicot angiosperms, or monocot angiosperms.
- Classify leaves as one of five types.
- Develop a personal field journal.

Prelab Discussion

In this lab activity students learn how to keep a field journal and to use field guides to identify plants. Consider showing them an example of a good field journal, perhaps of a famous botanist. Also demonstrate how they will choose specimens to document in their journals.

Equipment Notes

A trip through your yard, the schoolyard, or a local park should supply a number of good, fresh specimens. Even in the winter some useful specimens of persistent foliage and conifers can be found. You may choose to have students bring a few specimens from their yards; it is safe to assume that the ecosystems are similar.

You could also make this an indoor activity. In that case, be sure to plan ahead when selecting plants. The unusual leaf venations of some houseplants are very useful. A fern or two would also be helpful.

After obtaining as many fresh plant specimens as possible, try to balance the number of different classifications among them. You will no doubt have more dicots than anything else, but try to vary the types of venation.

Reviewing Biomes

Review the definition of an ecosystem from page 53 in the Student Edition. Also, review the ways that different environmental factors affect life. See pages 56–57 in the Student Edition.

Plant Identification App

To check out an example of a plant identification app, see the *Plant ID App* link. This link is available as a digital resource. Apps designed specifically to identify invasive species (see Question 10) are also available.

#QUICK NOTE

Handling Plants Safely

- Be careful when handling plants, as some may cause dermatitis, as poison ivy can, or have thorns, as blackberries do.

poison ivy

- If you have any significant plant allergies, be sure to tell your teacher.

- In your search for plants do not disturb any animal nests that you encounter in the field.

1. How do weather and climate affect the plants in an ecosystem?

   Certain plants prefer certain amounts of precipitation and certain ranges in temperature. Some cannot survive without specific environmental conditions.

C Take a picture of the place where you are observing plants. Leave a spot in your journal to paste the picture. Write a brief description of the area.

Observations to note in your journal could include the color of the plant's flowers (if you don't collect any), the height of the plant, its location near a body of water, the presence of any insects on the plant, or any discoloration in the plant. Also note the geography of the area, including the presence of rocks or underbrush, or whether the ground is level or sloped.

D Make sure that you put on gloves before handling plants.

E Choose at least ten plants that you want to observe. Take a picture of the place where you collect each sample. Collect samples of each plant and place each in a plastic bag that has been labeled with a number using a permanent marker.

F Write down each specimen's number and identify it as best you can while in the field. Be sure to look at the whole plant, not just at the leaf, to identify each plant. Write the name and bag number in your field journal.

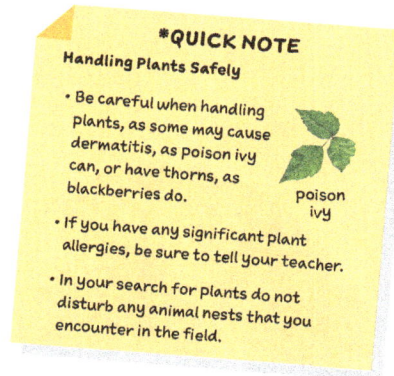

Leave lots of space for additional notes on the specimens that you identify. Use an app, a field guide, or an internet source to guide you.

A page from the field journal that William Clark kept while on the Lewis and Clark expedition

G Make notes in your journal about the specific place where you collected each of your plant samples.

2. What source(s) did you consult to identify your plants?

   Answers will vary. Students may use local web links, apps, and field guides to identify their plants.

These are just preliminary field identifications. Later on you'll be examining these samples again to make sure that you've identified your specimens correctly.

3. Why did you choose the plants that you did?

   Students may mention that a plant was aesthetically pleasing or had flowers or fruit. They may also mention that they could already identify certain plants from previous knowledge.

4. What characteristics of your specimens tell you that they are all plants?

   Students will probably mention that plants have stiffness from cellulose and are green because they contain chlorophyll. They may also mention the presence of a cuticle.

Verifying Specimens

Now take your specimens to a place where you can observe them more closely.

H Look at each of the specimens that you collected. Fill in Table 1 with the information given below for each specimen. You will be graded for accuracy, so be careful!

» common name

» scientific name (at least the genus)

» leaf description (needles, scales, parallel or netted venation, and simple, pinnately compound, bipinnately compound, or palmate shape)

» classification (seedless vascular, nonvascular, gymnosperm, monocot angiosperm, or dicot angiosperm)

I Confirm that your identifications are correct by comparing your specimens to the descriptions in your field guide, app, or other source. Also check a range map to make sure that the plant you've identified is found in your area. If it is not, you may want to reconsider your identification.

J Remove your gloves and wash your hands.

MONOCOT OR DICOT?

If you have trouble knowing whether your plant is a monocot or dicot angiosperm, look at the leaves. If the leaves have parallel venation, the plant is a monocot. If the leaves have netted venation, the plant is a dicot. Also, if your plant has a woody stem, it is a dicot.

dicot leaf

monocot leaf

CONCLUSION

5. After closer examination, you likely changed some of the names that you assigned to your specimens. Explain why.

Answers will vary. Students may have changed answers after looking at range maps or more closely examining their specimens or their field guides.

6. Which category did most of your plant specimens fall into—nonvascular, seedless vascular, gymnosperm, monocot angiosperm, or dicot angiosperm?

See margin for answer.

7. Suggest a reason why there are more plants from this category.

There may be more species in this category in the ecosystem that students are observing.

8. Why is it important to be able to identify and classify plants?

Answers will vary. Students may mention the ability to distinguish poisonous plants from nonpoisonous ones. They may also mention the need to monitor plant populations for controlling invasive species.

GOING FURTHER

9. Suggest one way that scientists can monitor a population of plants like the American chestnut and help them make a comeback.

See margin for answer.

Describing Leaves

Send your students to pages 302–5 in the Student Edition to help them figure out how to describe leaves.

What's the Difference?

Students occasionally have problems distinguishing between a leaflet and a leaf. The key is knowing the difference between a stem and a petiole (see pages 302 and 305 in the Student Edition). If there is doubt, they should look for a stipule, which is found only at the base of a petiole. Thus, everything beyond the stipule is a single leaf. Sessile leaves or leaves without petioles may present problems for some students.

Ferns can occasionally be confused with angiosperms. Any ferns that students have collected should have distinctive fern characteristics (e.g., sori) that can be used to identify them.

Question 6 Answer

Students will probably have collected more dicot angiosperms.

Question 9 Answer

Surveys of tracts of land can be done to monitor plant populations. Existing American chestnuts can be bred to propagate fungus-resistant individuals. Scientists can also work to develop a way to fight the fungus in American chestnuts that are surviving.

✓ Unwelcome Guests

See the following list of invasive plants that students might consider.

- Scotch broom (*Sarothamnus scoparius*)
- Chinese privet (*Ligustrum sinense*)
- kudzu (*Pueraria lobata*)
- common ivy (*Hedera helix*)
- yellow star thistle (*Centaurea solstitialis*)
- milfoil (*Myriophyllum* spp.)

An additional problem that plagues some ecosystems occurs when plants are introduced from another area and begin to take over. These invasive species crowd out the local plant species, which can make the area's biodiversity plummet.

10. Do some research to find an invasive plant species in your area. Write the common name and any recommendations that conservationists and ecologists give for controlling this plant.

   Answers will vary. Students may mention precautions in transferring live plants to other places or uprooting and burning them altogether.

11. Write a description of how you would identify this plant if you were to see it in a local ecosystem.

   For identification purposes, students' descriptions should include something about the plant's leaves, flowers, bark, or shape.

12. Why is it important to monitor plant populations, both those that are struggling and those that are thriving to such an extent that they endanger the health of an ecosystem?

   God's first command to Adam included specific wording about managing the plants on the earth for man's use. (Gen. 1:28–29)

160 Lab 14A

TABLE 1 Identifying Plants

Specimen Number	Common Name	Scientific Name	Leaf Description	Classification
1	wild strawberry	*Fragaria vesca*	netted venation, compound	dicot angiosperm

14B LAB

A Fruitful Lab

Exploring Flowers, Fruits, and Seeds

Have you ever thought about the difference between a fruit and a vegetable? Maybe you're thinking, "Fruit tastes good, and vegetables don't!" What you might not know is that green beans and cucumbers, commonly thought of as vegetables, are actually fruits. Fruits come from the developed ovaries of flowers, while vegetables come from other parts of a plant—leaves, stems, roots, and even flowers.

Flowers are the sexually reproductive parts of angiosperms. There are many types of flowers, but they all have certain characteristics in common. When flowers are pollinated, they produce fruit, which in turn produces seeds. In this lab activity you'll explore the different types of flowers, fruits, and seeds.

What kinds of structures are usually found in different kinds of flowers, fruits, and seeds?

Equipment

scalpel
hand lens
large kitchen knife
single-edged razor blade
fresh flower specimens
fresh fruit specimens
seed specimens

QUESTIONS
- How are flowers similar and different?
- What do all fruits have in common?
- What do the seeds of monocots and dicots look like?

PROCEDURE

Observing Flowers

Not all flowers are variations of the rose, lily, or daisy. Many flowers lack showy petals, and many have very unusual structures, but most of them share the same basic floral parts.

A Using your scalpel, carefully dissect a fresh flower to see the various internal structures. The best method of dissection is to slice through the ovary area from the top down. Examine the structures with a hand lens if necessary.

B Using Drawing Area A, draw a longitudinal section of your flower. Label all the parts of a flower that you learned about in your textbook.

If a flower has sepals, petals, and at least one stamen and one carpel, it is called a *complete flower.* Flowers lacking any of these structures are called *incomplete flowers.*

1. Is your flower complete or incomplete? Explain.

   Answers will vary. Students should indicate the presence or absence of sepals, petals, stamens, and carpels.

LAB 14B OBJECTIVES

- Identify and describe various structures of flowers.
- Identify and describe various types of fruits.
- Compare various kinds of flowers and fruits.
- Identify the parts of a seed.

Equipment Notes

To give your students the best experience possible, be sure to examine the teacher notes on the following pages next to the procedures for the three sections of this lab activity for recommendations and tips on flowers, fruits, and seeds to dissect.

Consider bringing paper plates and clean kitchen utensils for the dissections. Eating the fruits when you are finished adds interest and entertainment. Do not allow students to eat fruit that has been touched with laboratory equipment, such as a scalpel!

Options

If you can do only part of this activity, the most important and productive part is the flower dissection. The other sections can be done as a demonstration.

Virtual Dissections

If you would like students to do this lab activity at home or have practice before doing the flower dissection, consider researching virtual flower dissection for some helpful information.

Drawing Flowers

If you would prefer, you may have students take pictures of the flowers instead of drawing them. They can post pictures in Drawing Area A.

Reviewing the Differences between Monocots and Dicots

For review, consider helping your students make a T-Chart where they can contrast the leaves, stems, roots, seeds, and flowers of monocots and dicots.

Getting Flowers for Observation

If you can obtain a good supply of different flowers, this section can be extremely profitable. The flowers dissected should be of the simple-flower type. Since one gladiolus stalk supplies several flowers, samples obtained from a local florist work well. Roses are also good but are considerably more expensive. Mums, daisies, and carnations are poor choices because they are complex and the flower parts are tiny despite large blooms. When obtaining flowers, look for other flowers that can be dissected. Lilies, tulips, daffodils, azaleas, and snapdragons are good choices that are readily available. Be sure to get flowers with showy parts so that students can easily see the male and female reproductive structures.

If you choose to have students dissect multiple flowers, have them answer Questions 1–6 for each flower.

🔬 *Getting Fruits for Observation*

Make sure that the fruits your students observe include at least one pome, drupe, true berry, and modified berry; also include a legume. Besides the usual apple, tomato, orange, pear, banana, grape, cantaloupe, cucumber, squash, plum, peach, and watermelon, try to obtain unusual fruits. Snow peas, okra, corn, kiwifruit, star fruit, tomatillos, pomegranates, avocados, papayas, figs, dates, strawberries, green peppers, cranberries, raspberries, and pineapples are all interesting and often available in a form usable for this activity. Some other fruits common on grocery shelves include peanuts (in the shell), unpitted olives, pickles (whole), and star anise (dried; found in the international food section). Have fun with this!

You may decide to assign a different fruit to each lab group, or you may choose to dissect the fruit ahead of time.

🔄 *Classifying Fruits*

Your students may find the table on page 317 in the Student Edition to be helpful as they categorize the fruits they dissect.

✅ *Drawing Fruits*

If you would prefer, have students take pictures of the fruits and paste them in Drawing Area B.

---

2. Is your flower male, female, or both? Explain.

See margin for answer.

_____

_____

_____

_____

Holly trees have imperfect flowers, and each tree bears only male or female flowers. For holly trees to produce berries, there needs to be a male and female tree near one another. These kinds of plants are described as *dioecious*.

Flowers that have either male or female structures are called *imperfect flowers*. Flowers with both male and female structures are *perfect flowers*.

3. Is your flower perfect or imperfect? Explain.

If a student answered Question 2 as male or female, then the flower is imperfect. If he answered "both" to Question 2, then the flower is perfect.

Some flowers are actually several flowers held in one receptacle. These kinds of flowers are called *composite flowers*.

4. Is your flower composite or not composite? Explain.

If the flower is composed of many small flowers, it is composite. *Examples:* daisies, chrysanthemums, dandelions, clover

In Chapter 14 you have learned about the differences between the leaves, stems, roots, and seeds of monocots and dicots. But how can you know whether a plant is monocot or dicot by looking at its flowers? Monocots have petals or sepals in groups of three, while dicots have petals or sepals in groups of four or five.

Trilliums are considered monocots because their sprouts have one cotyledon. You could know that trilliums are monocots by looking at their flowers—they have three petals.

5. Is your flower from a monocot or a dicot plant? Explain.

If petals or sepals are in groups of three, the flower is a monocot. If petals or sepals are in groups of four or five, the flower is a dicot.

_____

| superior ovary | half-inferior ovary | inferior ovary |

Now look at the ovary of your flower (if it has one). If the ovary is above the sepals, it is a *superior ovary*. If it is halfway past the base of the sepals, it is a *half-inferior ovary*. If it is below the sepals, it is an *inferior ovary*. If your flower doesn't have an ovary, skip Question 6.

6. Categorize your flower's ovary as superior, half-inferior, or inferior. Explain your classification.

If the ovary is above the base of the sepals, it is superior. If it is halfway past the base of the sepals, it is half-inferior. If it is below the sepals, it is inferior.

Observing Fruits

A fruit is a ripened ovary. Although many fruits are like the familiar apple and orange, many are quite different. Fruits have been classified into groups on the basis of the way their structures develop.

C Using a scalpel (or large kitchen knife), dissect the fruits that your teacher gives you.

You are working with a few different kinds of fruit. These include a pome, a drupe, a true berry, a modified berry, and a legume.

D Using Drawing Area B, draw a longitudinal section of your fruit.

E Use the Key to Common Fruit Types on pages 166–67 to help identify your fruit types. Record your classifications in Table 1.

7. Describe some of the variation that you observed in your fruits.

   Answers will vary. Students may describe the location of seeds and the skins of the different fruits.

8. Suggest reasons for the differences in fruit, especially as it relates to the differences in flowers.

   Students should recognize that the position of the ovary and the number and position of oviducts influence the structure of fruit.

Observing Seeds

The three basic parts of a seed are the plant embryo, stored food, and seed coat. The diversity of these structures found in different plants, however, is almost as wide as the diversity found in flowers and fruits.

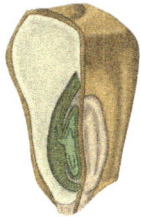

monocot corn seed   dicot bean seed

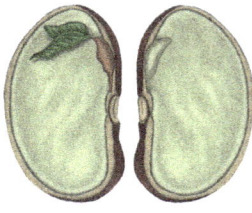

F Using a hand lens, observe the exterior of your seeds. Draw your seeds in Drawing Area C, being careful to identify each one by its fruit.

G On your drawing, label the following parts of your seeds.

  » hilum—the point where the seed is attached to the ovary

  » seed coat—the hard, protective covering of the seed

H Using a scalpel (or large kitchen knife), dissect the seeds by cutting them in half. It may take some effort to open some seeds.

I Draw your dissected seeds in Drawing Area D, being careful to identify each one by its fruit.

9. Describe some of the variation that you observed in your seeds.

   Answers will vary. Students may notice the difference between monocots and dicots in the differing number of cotyledons.

GOING FURTHER

10. Observe some of the other flower, fruit, and seed specimens. What did you observe in these other specimens that you didn't see in your original observations?

    Answers will vary.

11. Why is it important for people to study the different kinds of flowers, fruits, and seeds?

    We need to know what types of flowers, fruits, and seeds are edible. We also need to know how they are best pollinated so that we can cultivate them to feed people.

Getting Seeds for Dissections

Try the various seed dissections before you ask students to do them. Give special instructions when needed. Tools like nutcrackers and even saws may be needed for some seeds. Dried beans can be soaked in water overnight to make them easier to open.

Drawing Seeds

If you would prefer, have students take pictures of the seeds and paste them in Drawing Areas C and D.

You've learned about some of the different types of flowers, fruits, and seeds in this lab activity.

J Do some research on the different types of vegetables.

12. Name some of the different categories of vegetables and give examples of each.

Students may mention allium, which includes leeks, garlic, shallots, and onions. The stems and bulbs of these vegetables are eaten. Brassica includes vegetables whose flowers and greens are eaten, including Brussels sprouts, cabbages, kale, and broccoli. Legumes include vegetables with edible seeds such as peas, lentils, and beans. Roots and tubers include potatoes, Jerusalem artichokes, and certain spices. Herbs and spices are usually obtained from leaves or seeds.

Key to Common Fruit Types

1. (a) Single ovary that may have one or more chambers for ovules, usually without other floral parts . . . . go to 3

   (b) Collection of ovaries, usually with other floral parts . . . . go to 2

COMPOUND FRUITS

2. (a) Several separate ovaries of a single flower that ripen individually, usually on an enlarged receptacle . . . . aggregate fruit

   (b) Several ovaries from separate flowers that ripen fused together, usually on an enlarged receptacle . . . . multiple fruit

3. (a) Fruit dry at maturity . . . . go to 4

   (b) Fruit fleshy at maturity . . . . go to 10

DRY SIMPLE FRUITS

4. (a) Fruit open when ripe . . . . go to 5

   (b) Fruit closed when ripe . . . . go to 7

5. (a) Ovary wall thin, single-chambered ovary with many seeds; opens along one or two sides when ripe . . . . go to 6

   (b) Ovary wall thin, multiple-chambered ovary, each chamber with many seeds; opens when ripe . . . . capsule

6. (a) Opens along one side . . . . follicle

   (b) Opens along two sides . . . . pod

7. (a) Fruit with thin wing formed by ovary wall . . . . samara

   (b) Fruit without wing . . . . go to 8

8. (a) Thick, hard, woody ovary wall enclosing a single seed . . . . nut

   (b) Thin ovary wall . . . . go to 9

A fruit with a winged seed like a propeller blade is called a *samara*.

9. (a) Ovary wall fastened to a single seed .... grain

   (b) Ovary wall separated from a single seed .... achene

FLESHY SIMPLE FRUITS

10. (a) Fleshy portion develops from receptacle enlargement;
        ovary forms leathery core with seeds .... pome

    (b) Ovary fleshy .... go to 11

11. (a) Ovary two-layered: outer layer fleshy and inner
        layer forming hard woody stone or pit, usually
        enclosing one seed .... drupe

    (b) Entire ovary fleshy .... go to 12

12. (a) Thin-skinned fruit with divided ovary, usually with
        each section containing seeds .... true berry

    (b) Thick, tough-skinned, with divided ovary, usually
        with each section containing seeds .... modified berry

TABLE 1 Identifying Fruit Types

Name of Fruit	Type of Fruit
raspberry, blackberry	aggregate fruit
pineapple, fig	multiple fruit
milkweed	follicle
bean, pea, peanut	pod
maple, elm	samara
acorn, chestnut	nut
corn, wheat	grain
sunflower, dandelion	achene
apple, pear	pome
peach, olive, nectarine, plum	drupe
cranberry, tomato, blueberry	true berry
cantaloupe, cucumber, orange, lemon	modified berry

DRAWING AREA A

DRAWING AREA B

DRAWING AREA C

DRAWING AREA D

15A LAB

Plant Processes

Investigating Plant Hormones and Ripening

Chances are, you've eaten a banana or two in your life. It's equally likely that you don't live anywhere near where those bananas were grown. Bananas grow in the tropics, where year-round temperatures are high and rainfall is abundant. You probably also know from experience that ripe bananas are soft and bruise easily, so shipping them when ripe is probably not a good idea. So how is it that you can find a ripe banana from Costa Rica in a supermarket in Cedar Rapids on just about any given day of the year?

To peel away the mystery of bananas, let's start by looking at how a banana changes during the ripening process.

Why can some fruits be harvested while unripe but later sold in ripened form?

Equipment

hot water bath
laboratory spatula
petri dishes (4)
disposable plastic pipette
test tubes (2)

graduated cylinder, 10 mL
unripe banana purée
iodine solution
Benedict's solution
ripe banana purée

nitrile gloves
goggles
laboratory apron

PROCEDURE

Testing an Unripe Banana

Iodine can be used to test for the presence of starches, which turn a dark blue color when exposed to iodine. Benedict's solution is used to test for the presence of simple sugars. It turns green or yellow in low sugar concentrations; orange or red results indicate higher sugar concentrations.

A Obtain a small sample of unripe banana purée from your teacher.

B Scoop a small portion of the sample into a petri dish. Use a pipette to apply a few drops of iodine solution to this portion.

1. Describe what you observe when you apply iodine to the unripe banana.

The banana and iodine turn dark blue.

2. What does this result indicate about the presence of starch in an unripe banana?

Starch is present.

C Now place 5 mL of the remaining unripe banana purée into a test tube.

D Place 5 mL of Benedict's solution into the test tube. Shake the tube well and place it in the hot water bath provided by your teacher.

QUESTIONS

- What is the difference between an unripe banana and a ripe one?
- How do bananas ripen?
- What role does ethylene play in the processing of bananas prior to sale?

- Explain the difference in the starch and sugar contents of ripe and unripe bananas.
- Explain the role of ethylene gas in the ripening of bananas.
- Explain how the Yang Cycle can be interpreted as evidence for a creationist worldview.

✓ Introduction to Lab 15A

Ethylene is briefly mentioned as a plant hormone on page 330 in the Student Edition. This lab activity will explore the process of ripening and the role of ethylene in that process.

✓ Prelab Discussion

You might want to have on hand a very green and a very ripe banana and have students discuss the differences between them.

🧪 Equipment Notes

Any small dishes can be substituted for petri dishes. Students may rinse, dry, and reuse dishes in between purée samples.

Any small scooping tool, such as a measuring spoon or wooden tongue depressor, can be substituted for a laboratory spatula.

You can make the unripe and ripe banana purées before class. Place the bananas in a blender with an approximately equal amount of water and blend until the bananas have been completely liquefied. For the students to see a discernible difference, you should buy the greenest and the brownest bananas possible; you might need to purchase some bananas in advance and let them ripen, then purchase green bananas as close to lab activity time as you can. When making the two purées, take care that they do not contaminate each other.

✓ Temperature Setting

The hot water bath should be set at 99 °C, or just below boiling.

E Allow the tube to sit in the bath for ten minutes.

3. Describe what you observe in the tube at the end of the ten minutes.

See margin for answer.

4. What does this result indicate about the concentration of simple sugars in an unripe banana?

The concentration of simple sugars in an unripe banana is low.

Recall from Lab 7B that starch can be converted to simple sugars.

5. Using the results from testing an unripe banana for starch and simple sugars, coupled with what you learned from Lab 7B, make a prediction about the levels of starch and simple sugars that you would expect to find in a ripe banana.

The starch content of a ripe banana should be low, and the

simple sugar content should be high.

Testing a Ripe Banana

F Repeat Steps A through E of the previous section using a sample of ripened banana purée.

6. Describe what you observe when you apply iodine to the ripened banana purée.

The iodine solution remains brown.

7. What does this result indicate about the presence of starch in a ripe banana when compared to that in an unripe one?

The amount of starch in a ripe banana is much less than that in

an unripe one.

8. Describe what you observe in the tube of ripened banana purée at the conclusion of the Benedict's solution test.

The solution appears dark red or even brown.

9. What does this result indicate about the presence of simple sugars in a ripe banana when compared to that in an unripe one?

A ripe banana has more simple sugars than an unripe one does.

10. In Lab 7B the conversion of starch to glucose was due to fermentation by yeast. Using your results from that lab activity and your personal knowledge of ripening bananas, what do you think suggests that bananas ripening is *not* due to yeast fermentation?

Yeast fermentation produces a characteristic smell along with

carbon dioxide gas. Ripening bananas do not exhibit either of

these characteristics.

11. Summarize what you have learned so far about the difference in starch and sugar content between unripe and ripe bananas.

Ripe bananas contain less starch and more sugar than unripe bananas. The increase in sugar content cannot be due to alcoholic fermentation because no characteristic gas or smell is produced during the ripening process.

Thinking about Ripening

Let's get back to the question of how your supermarket can offer ripe bananas for sale. Obviously, they can't be shipped while ripe since they would get damaged during the trip. You'll need to do a little sleuthing to examine this issue!

G Research ripening fruit and climacteric fruit.

12. What is a climacteric fruit?

a fruit that continues ripening after harvest

13. Are bananas climacteric or non-climacteric?

climacteric

14. What significance does the answer to Question 13 have for the harvest and shipment of bananas?

Since bananas are climacteric, they can be harvested and shipped while still green.

H Now do an internet search on bananas. Look for information on how bananas are harvested, shipped, and prepared for market.

15. Write a summary of what you learned.

Answers will vary. Bananas are harvested while green, shipped in refrigerated containers, and exposed to ethylene gas in ripening sheds prior to distribution.

I Finally, do an internet search on degreening.

16. What is degreening?

Degreening is the use of ethylene gas to enhance the color of citrus fruits.

17. What do degreening and the processing of bananas have in common?

Both are techniques used to enhance the appearance of fruit prior to sale. Both use ethylene gas.

18. What is the major difference between degreening and the processing of bananas? (*Hint:* Recall the difference between climacteric and non-climacteric fruits.)

Citrus fruits are non-climacteric. Degreening doesn't ripen the fruit; it only changes the color of the fruit.

Yang Cycle

For more information on the Yang Cycle, see the *Yang Cycle* link. This link is available as a digital resource.

Irreducible Complexity

The Yang Cycle is an example of irreducible complexity. All the parts of the cycle have to be present in order for the cycle to work. This is a significant problem for evolutionists but no problem at all for an all-wise Creator.

GOING FURTHER

Techniques for accelerating the ripening of fruit have been known since ancient times, but the biological mechanisms that control ripening have only recently been discovered. Ethylene's effects on plant growth were first described in 1901, and it wasn't until 1934 that ethylene was determined to be a plant hormone. The process by which plants produce ethylene wasn't completely described until the 1980s by Dr. Shang Fa Yang at the University of California at Davis. The ethylene production process in plants is called the *Yang Cycle* in his honor. Similar to cellular respiration or photosynthesis, the Yang Cycle is a complex series of chemical reactions involving intermediate products and requiring specialized enzymes. Regulator genes turn the cycle on or off, so that ethylene is produced only when the plant needs it. One popular science magazine says that how such a complex chemical pathway could have arisen by mutation and natural selection continues to puzzle evolutionary scientists.

19. What features of the Yang Cycle pose problems for the evolutionary model?

Answers will vary. Each of the genes and enzymes needed for the Yang Cycle is an information-rich molecule whose creation by purely random means is statistically unlikely. The cycle can also not evolve stepwise since it requires all of its component parts in order to function.

20. How do those same features strengthen the position of the creation model?

Answers will vary. The Yang Cycle, like all biological cycles, shows strong evidence of design. The evolutionary model cannot account for apparent design in nature, but the creation model predicts, and the Bible confirms, that creation would bear witness of its having been created.

15B LAB

Too Salty?

Experimentation and the Flood

When irrigation water seeps through cropland, dissolved mineral salts in the water are left behind, a process known as *salinization*. It is more difficult for crops to grow when the soil has an abnormally high salt content, and that problem can have a significant impact on a culture. For example, as salt built up over centuries in the soil of ancient Mesopotamia, entire civilizations collapsed.

Receding floodwaters from Noah's Flood likely left salt water on the land. After drying, the land would have had a higher-than-normal salt content. How did plants grow and reestablish ecosystems after the Flood?

?

If seawater from Noah's Flood was salty, how did plants survive the Flood?

Equipment

containers to soak seeds (2)	spray bottle	paper towels
containers to hold seeds during germination (3)	paper sack	seawater
	plant seeds	distilled water

PROCEDURE

Background

This question was partially addressed by the paper "Seed Germination, Sea Water, and Plant Survival in the Great Flood," by George F. Howe in *Creation Research Society Quarterly*, Vol. 5, No. 3 (December 1968), pp. 105–12.

Howe's paper went through a *peer review* process before it was published, a standard practice in scientific research. When an article is peer reviewed, it is examined by other scientists working in the same field before it is published. They read the article and make suggestions for improvement. The peer review process raises the quality of the article.

QUESTIONS

- How does salt water affect seed germination?
- How can experimentation aid our thinking about biblical issues like the Flood?
- What is peer review?

Too Salty? 175

LAB 15B OBJECTIVES

- Describe the parts of a scientific paper.
- Evaluate the effects of salt water on seed germination.
- Write a lab report in the style of a scientific paper.

✔ Prelab Discussion

The buildup of salt in soil is a problem similar to the one studied by Howe. Students who live in agricultural areas, especially where heavy irrigation is practiced, may be familiar with soil salinization. Plants don't grow well in soil with a high salt content. It reduces the soil's ability to grow food, which, in turn, discourages population growth.

🧪 Equipment Notes

Most seeds from the grocery store or garden store will work fine. Bean seeds are a good choice.

Actual seawater would be ideal. If you don't live near the ocean, add enough distilled water to 35 g of salt to produce 1 L of solution. Sea salt is an ideal choice if you need to make your own salt water; the salt mix used to set up a saltwater aquarium would also be an excellent choice. A pet store that sells saltwater fish will usually have this in stock. This option is more expensive but will closely replicate actual seawater.

💻 Soil Salinity

For more information on soil salinity and plant growth, see the *Soil Salinity* link. This link is available as a digital resource.

The paper by Howe follows a common style of scientific writing. Notice the following sections of the paper.

TITLE

The title conveys the content of the paper in a concise way.

ABSTRACT

A well-written abstract allows a scientist to read the abstract and decide whether to read the rest of the paper. It also allows the scientist to come away with the major details of the paper and not have to read the paper in depth. An abstract summarizes the entire paper in one or two paragraphs and is often limited to 200–300 words. It usually is the last part of the paper to be written, even though it is found next to the title. An abstract will typically answer the following questions.

» What was the purpose of the study?

» What was the experimental design of the study?

» What were the major findings of the study?

» What were the conclusions of the study?

INTRODUCTION

The introduction starts broad and then focuses the reader on the work reported in the paper. A good introduction will answer the following questions.

» What was studied?

» Why was this topic important?

» What did we know about this topic before this study?

» What new things will we learn because of this study?

MATERIALS AND METHODS

The materials and methods section will sometimes be referred to as simply "methods." The purpose of this section is to explain how the study was done. It will contain details about control groups, experimental groups, experimental techniques, procedures, and data analysis methods.

RESULTS

The results section reports data collected. This section will often have data tables and graphs.

DISCUSSION

The discussion section involves the interpretation of the results. The original research question or hypothesis is compared to the results, which are then compared to previous research. A few sentences are usually dedicated to suggestions for extending the research.

REFERENCES

The references section lists the papers and books referred to by the article.

The format for scientific writing is very efficient at presenting information. Because this format is fairly standard across all scientific disciplines and because scientists are accustomed to using and reading papers written in that style, it is relatively easy for the reader to move around in the paper depending on his level of interest.

Howe's Research

A Read the above-referenced paper by Howe. Your teacher will give you directions about how to access it.

1. What scientific principle explains why we expect that salt water will affect seed germination? Explain.

   The principle of osmosis leads to this expectation. The salt water will cause cells to shrink by losing water to the environment. Salt water has the same effect on plants as a drought does.

2. Why did Howe thoroughly rinse the soaked seeds before planting them?

   We already know that salt affects germination in soils. Rinsing the seeds washes away the salt on the seeds. This is consistent with the purpose of the lab activity. We are testing how the soaking in the salt water affects germination. We want to prevent any extra salt interactions with the soil.

3. This exercise is a model for the Flood and post-Flood environments. What is the counterpart in the Flood narrative for the rinsing of seeds before planting?

   The seeds would have been deposited during the Flood. After the Flood, rainwater, which is fresh water, naturally rinsed the seeds and the soil.

4. Why did Howe collect fruits with seeds and soak them in the different water mixtures rather than just soaking the seeds alone in the different water mixtures?

   This procedure is closer to the events in nature. It is more likely that fruits and seedpods, not just seeds, would have been buried during the Flood. The fruit structure would also have acted as a shield to protect the seeds from the seawater.

Reading a Scientific Paper

It is important for students to read an actual scientific paper. It's a different style of writing from what they are accustomed to in English and literature classes.

Accessing Howe's Paper

To access Howe's paper, search for it by title at the Creation Research Society's website. Download and print the article for your students or give directions to them to find the article and download it at home.

One advantage of this paper is its clear biblical worldview. Evolutionists often claim that creationists don't do scientific research. This paper is evidence to the contrary. Evolutionists also claim that believing in the Creation record is a science-stopper. This paper shows that believing the Bible leads scientists to ask questions that can be tested scientifically.

The paper points out that Charles Darwin did some good experimental science, despite the fact that Christians sometimes demonize him. Nonbelievers *can* do good science. To view Darwin's paper, see the *On the Action of Seawater* link. This link is available as a digital resource.

Notes on Howe's Experiment

Tap Water versus Distilled Water

Howe used tap water in his experiment. In our day it would make more sense to use distilled water because of the existence of so many water softeners in the country. Soft water has some salt dissolved in it.

Proof of Concept

The study by Howe should be considered to be a *proof of concept* study. A more thorough study would involve populations of seeds on the order of hundreds instead of ten for each group.

Seed Notes

Number of Seeds

A weakness of Howe's study is the small number of seeds he used—only ten seeds for each of the five species he studied. Since students are studying only one species, they will increase the number of seeds per group. This will give better results.

Soaking the Seeds

Howe soaked his seeds for 140 days in order to model Noah's Flood. If students are going to model the Flood, you will need to start soaking the seeds during the summer. However, soaking the seeds for several days to several weeks is still a worthwhile and practical experiment. It would be a simulation of seawater flooding crop fields. You will need to choose how long to soak the seeds and inform students about regular observations of the seeds.

Howe had to regularly monitor and change the water in his experiment; students will probably have to do the same.

Sprouting the Seeds

You could also plant the seeds in small pots with potting soil, but the seeds will germinate more quickly in the moistened paper towels, and students will find the germination process easier to observe.

Howe lists four groups of seeds that he experimented on:

» fruits stored in paper sacks

» fruits soaked in seawater

» fruits soaked in seawater mixed with tap water

» fruits soaked in tap water

5. Explain why the fruits stored in paper sacks were the control groups.

They are the fruits closest to the natural order of things. They are not stored in water of any kind.

6. Explain why fruits soaked in tap water was a necessary group for the experiment.

If the fruit soaked in salt water showed a dramatic change in seed germination, the result might have been caused by the salt or by the water. Including tap water shows the effect of water *alone* on seed germination, adding confidence to the conclusions about the seeds soaked in seawater.

7. Why did Howe's study include a group of fruits soaked in seawater mixed with tap water?

Including this group allowed Howe to make a beginning estimate of how the concentration of salt water affects seed germination. It allowed him to consider a mathematical relationship between salt concentration and seed germination.

8. What conclusion did Howe draw from his experiments?

Howe determined that the plants he tested could still germinate after being in seawater for 140 days.

9. List two suggestions for future work that Howe suggested on the basis of his experiment.

Answers will vary. Howe suggested research into the incidence of resistance to prolonged soaking in flowering plants, the factors that prevent germination in certain seeds after soaking, and the quantification of soaking resistance in olive plants.

Seed Germination Experiment

B Set up two containers and fill one with fresh water and one with seawater.

C Place 20–25 seeds in each container and in a paper sack. Allow the seeds in the containers to soak for the time prescribed by your teacher. You will also be instructed regarding any needed maintenance of the seeds while they are soaking.

D Research the effects of salt on plant growth for background material to use in the introduction of your lab report.

E After the seeds have soaked for the specified time, remove them from the water and wash them thoroughly with distilled water.

F Place paper towels two layers thick in a container. Place the seeds from the salt water on top of the paper towels and place another two layers of paper towels over the seeds. Spray the paper towels with distilled water until they are soaked.

G Repeat Step F for the seeds from the fresh water and from the paper bag.

H Examine the seeds regularly over a couple of weeks.

I Write a lab report of your findings using the style of a scientific paper. Use Howe's paper as an example to follow. The lab report should contain the sections outlined on page 176.

10. Write a paragraph about how this study demonstrates ways in which a biblical worldview affects science.

Answers will vary. Students should mention that because Bible-believing scientists take the Bible literally, it leads them to ask questions that other scientists would not think to ask, such as how long seeds could last in seawater.

✓ Grading the Formal Lab Report

A rubric for assessing students' formal lab reports is provided in Appendix G. The rubric is also available as a digital resource..

✝ Worldview Guides Investigation

Since they hold to an evolutionary worldview, many scientists will not even think to ask questions related to how the Flood may have affected Earth and its processes. On the other hand, creationists like Howe have done much research supporting the plausibility of the biblical Flood scenario. Worldview *does* affect how one does science!

16A LAB

The Immortals Next Door

Investigating Hydras

What if there were a way to stay ever young, fit, and healthy—a real-life Fountain of Youth? Some scientists think that a certain animal, one that might not seem an obvious candidate, may hold the key to unlocking the secret of eternal youth: the lowly hydra. Hydras are common in freshwater lakes and streams in temperate zones around the world, but you may have never noticed them because of their very small size. They can sting like jellyfish, but their venom is too weak to be a danger to humans. Let's take a look at this tiny representative of the phylum Cnidaria.

How do hydras move, feed, and react to their environment?

Equipment

microscope
hand lens
petri dish
disposable plastic pipettes (3)
probe

concavity slide
cover slip
preserved slides of a budding hydra and hydra cross section
spring water

live cultures of hydras and brine shrimp or *Daphnia*
dilute acetic acid

PROCEDURE

Observing Preserved Hydras

A Observe a preserved budding hydra on low power.

B Make an outline drawing in Drawing Area A. The entire hydra will not fit into the microscope field, so you will need to move the slide several times as you draw.

C Label the tentacles, mouth, basal disk, and buds.

D Using high power, draw a cross-section view of a portion of the hydra body wall in Drawing Area B. This drawing should include internal structures and not just be an outline.

E Label as many of the cellular structures of the hydra as you can (see page 358 in your textbook).

QUESTIONS

- How do hydras move and feed?
- Do hydras respond to changes in their environment?
- Why are hydras the focus of much current research?

LAB 16A OBJECTIVES

- Describe the structure of a hydra.
- Describe the responses of living hydras to various stimuli.
- Explain the role of hydras in current medical research.

Suggested Review

Students may find it helpful to review the information about hydras on page 358 in the Student Edition.

Equipment Notes

Portions of the lab activity that call for observation using a hand lens can be accomplished using a dissecting microscope if one is available.

A toothpick may be used in place of the probe.

If tap water is used for the hydra, the water must be treated with chemicals used in freshwater aquariums. This will remove excess chlorine and other chemicals that harm the hydra. Boiled pond water or water from a freshwater aquarium may also be used.

The hydra is the most perishable of the living organisms needed for the Chapter 16 lab activities. Planarians should keep well. Order both organisms at the same time to save on shipping costs.

The hydra is a bit finicky. The wrong water, wrong temperature, or dissolved materials in the water can render even a healthy hydra unresponsive. To get the best possible results, supply the hydra with quiet, stable conditions before the students begin. But be prepared to warn students in the event that the hydras do not cooperate and advise that they should do the best they can to answer the questions.

Set up a disposal station for "used" hydras.

Transferring a Hydra

Quickly flush the hydra from the pipette into the petri dish. If it becomes attached to the inside of the pipette, the repeated flushing necessary to extract it will likely damage the hydra.

Downtime

The living hydra will require rest time between experiments. Suggest to students that they do their microscope work during these intervals.

Observing Hydra Responses

F Fill a clean petri dish with spring water.

G Using a pipette, draw in some water from the hydra culture provided by your teacher.

H Select a hydra and use the pipette to flush it, dislodging the specimen.

I Draw the dislodged hydra into your pipette and then flush it into your petri dish.

J Use a hand lens to observe your living hydra.

1. Describe any forms of locomotion or movement that you observe.

   Movement of tentacles is usually seen. Somersaulting occasionally occurs. Floating may also be seen.

2. On the basis of what you have read about the cnidarian nervous system in your textbook, make a prediction about what kinds of responses you will be able to observe in your hydra.

   Because hydras have only a nerve net and no true brain, their responses to stimuli will be limited.

K Very gently swirl the water around the hydra.

3. What is the hydra's reaction to the movement of water?

   Answers will vary. The typical reaction is a body contraction.

L After the hydra has recovered from the above experiment (which may take several minutes), touch your probe as gently as possible to its base.

4. What is the hydra's reaction to the probe at its base?

   Answers will vary. Often a body contraction can be seen.

M After the hydra has recovered from the probe in Step L, touch your probe as gently as possible to one of its tentacles.

5. What is its reaction to the probe at its tentacles?

   Answers will vary. Usually a body contraction can be seen.

N After the hydra has recovered from the probe in Step M, arrange your probe so that the hydra may touch the probe of its own power.

6. What is the hydra's reaction to the probe when it touches it of its own power?

   Answers will vary. Usually a movement of tentacles toward the probe can be seen.

Observing a Feeding Hydra

O Using a pipette, place a few living *Daphnia* or brine shrimp in your petri dish near the hydra. Be careful not to add the food so rapidly that you disturb the hydra. Do not try to force-feed the hydra but keep the food within tentacle reach.

P Using a hand lens, watch the actions of your hydra carefully. Especially note any activity in the mouth region.

7. Describe the feeding process that you observe.

   Tentacles touch the food and then move it toward the mouth, which opens to receive the food.

8. In what structure of the hydra does extracellular digestion take place?

   the gastrovascular cavity

9. What happens to any substances that cannot be digested?

   They are egested through the mouth.

Observing a Hydra's Response to Acid

Q Carefully remove your hydra and set it in a large drop of spring water placed on a concavity slide.

R Place a cover slip over the slide and observe the hydra (or sections of it) with a microscope.

S After the hydra has recovered from the transfer and you have been able to focus on the cells of its tentacles, place a small drop of the dilute acetic acid on the edge of the cover slip. Watch the hydra carefully. For the best results, one partner should put the acetic acid on the slide while the other partner observes the reaction through the microscope.

10. What did you observe as the acid reached the hydra?

    The hydra coiled up and released its nematocysts in response to the acid.

CONCLUSION

11. How specialized to different stimuli are the responses of the hydra? In other words, is the hydra able to respond differently to various kinds of stimuli? Explain.

    Answers will vary. The hydra appears to have limited reactions, including muscular contractions or coiling and releasing nematocysts. It can also respond to the presence of food items by going after them.

Feeding a Hydra

Students frequently become frustrated while trying to get a hydra to eat. You cannot force-feed a hydra. It must take the food on its own initiative. Simply keep the food within tentacle distance without disturbing the hydra. Sometimes energetic food will disturb it; damaging but not completely killing the food with the probe will sometimes stimulate a hydra to eat.

✓ Ideas for Further Exploration

To further explore cnidarians, try the following.

- Have some preserved hard corals and jellyfish on hand for comparison with hydras.

- Do an internet search for moon jellyfish. You should find several videos of these common jellyfish. Have students compare the differences between hydras and moon jellyfish in form (polyp versus medusa) and manner of locomotion.

- Do an internet search for *Myxobolus cerebralis*, the causative agent for whirling disease, an important disease factor in the management of commercial fish hatcheries. As we saw in Lab 11B, myxozoans are now considered extremely reduced cnidarians. Note how myxozoans are both similar to and different from free-living cnidarians.

✝ Worldview and Death

A secular worldview assumes that death is a purely natural phenomenon and therefore a purely natural cure is possible for it, perhaps even likely. A biblical worldview sees death as part of God's curse following mankind's fall. Because mankind willingly disobeyed God's command, death has both a physical and a spiritual component. We may be able to dabble with the physical aspect of death (though never overcoming it), but we are completely incapable of remedying the spiritual side of death. Only the work of Christ can undo the Curse, and that work is received only by faith. It is beyond the realm of science.

12. Why are the reactions of the hydra limited?

The hydra's nerve net provides only the capability of responding to stimuli rather than "thinking" with a brain to respond to stimuli.

To remove your hydra, flush the slide with spring water while holding it over a separate culture container for "used" hydras. Do not return the hydra to its original culture.

GOING FURTHER

In a 1998 article in the journal *Experimental Gerontology* researcher Daniel Martinez claimed that hydras are biologically immortal. Subsequent research seems to back that claim. This phenomenon is known as *non-senescence*, meaning that the mortality rate, or rate at which individuals within a population die, does not increase with age as is typical for most other organisms. Scientists are still investigating what makes this possible.

Researching the hydra is just one of the many efforts currently underway to try to radically lengthen the human life span. There's even a name for this movement: *immortalism*.

13. According to a biblical worldview, why does physical death exist within God's creation?

According to the Bible, death entered the creation when Adam first sinned in the Garden of Eden.

14. John 3:16 clearly states that eternal life is possible. According to this verse, how is eternal life obtained?

Eternal life is obtained by believing in the only Son of God.

15. In view of the truth of John 3:16, do you think that science can ever solve the problem of physical death? Explain.

Answers will vary. The Bible plainly teaches that death is a result of our fallen condition and is a part of the Curse. Since all humans are sinners, all are under the penalty of death. The redeemed who have placed their faith in Christ are no longer spiritually dead, but their bodies still experience physical death. In 1 Corinthians 15:26 we learn that death is the last enemy that Christ will overcome.

16. Can research aimed at extending human life expectancy be compatible with a biblical worldview? Explain.

See margin for answer.

17. Why should a Christian not be afraid of death? (See 1 Corinthians 15:55.)

The Bible promises the Christian that physical death is not the end for the believer. In fact, the apostle Paul states that, for the Christian, physical death is gain. Though they may physically die, those who by faith have trusted in Jesus will experience eternal life with Him—the cure for the sting of eternal separation caused by Adam's fall.

Question 16 Answer
Answers will vary. The aging process is often accompanied by many problems, such as increasingly poor health, memory loss, diminishing vitality, reduced mobility, and other ailments. Research into longevity may shed new light on how to treat these conditions, and advancing treatment options would certainly be a way of exercising dominion and serving people. But the Bible is clear that only Christ can overcome physical death.

DRAWING AREA A

DRAWING AREA B

16B LAB

Fish Tank Fiend!

Observing Planarians

Al the Amateur Aquarist came home after a hard day at work, grabbed a cold soda from the refrigerator, headed for the living room, and plopped into his comfortable recliner. From there he cast his gaze onto his most prized aquarium—a 300 gallon hi-tech, heavily aquascaped beauty housing exotic plants and fish from the Amazon basin. Al had spent many months and a lot of money getting this tank just right. For Al there was no better way to unwind than to watch the dazzling shoal of cardinal tetras, the stately promenade of angelfish, the gentle swaying of the plants in the outflow of the filter, and—wait a minute. What were those little brown things slithering along the inside of the front glass panel? Al got up to get a closer look. Could they be? Oh no, they were—planarians! Argh!!

For Al and other aquarium keepers, planarians are unsightly pests that probably hitched a ride into their tanks on bundles of plants from the local tropical fish store. But for certain scientists, planarians are a subject of intense research. Not only are planarians an important part of aquatic ecosystems, scavenging the bottom for detritus, but planarians may hold the key to a much sought-after advance in medical treatment. Let's begin by observing a live planarian.

?

How do planarians interact with their environment?

Equipment

petri dish
disposable plastic pipettes (several)
hand lens
forceps
probe
spring water

culture of living planarians
Epsom salt
index card or small piece of poster board
raw beef liver, boiled egg yolk, or flaked or pelleted fish food

QUESTIONS
- How do planarians behave?
- Why do people study planarians?

PROCEDURE

Observing Planarian Movement

First let's look at how planarians move.

A Fill a petri dish with spring water until the bottom is just covered.

B Obtain a planarian from the culture provided by your teacher.

C Agitate the water around the planarian by making small currents with a pipette until the planarian floats around in the water, and then quickly draw it into the pipette.

- Describe the motion of planarians.
- Describe the responses of planarians to various stimuli.
- Describe the structure and special characteristics of planarians.
- Compare planarians with previously studied organisms.

Suggested Review

You may want to have students review the material on worms on pages 361–64 in the Student Edition.

Students should be familiar with anatomical locator terms such as ventral, dorsal, and lateral. See Appendix C.

Equipment Notes

A dissecting microscope can be used in lieu of a hand lens if necessary. A toothpick may be used in place of a probe.

You might want to encourage students to take their planarians home after the exercise is completed and keep them as pets (provided their parents do not object). They can be fed about once a week with a small piece of liver, a bit of boiled egg yolk, or flaked or pelleted fish food. Excess uneaten food should be removed, and the water will need to be changed when it becomes cloudy following each feeding. Water taken from a spring, well, or aquarium is best, but tap water that has been left standing for several days can be used. Additional information can be found online or in books on culturing live fish foods. Encourage students to consider performing regeneration experiments on their planarians at home. Leftover planarians can be fed to aquarium fish but sparingly to ensure that the worms are eaten and not allowed to start up an unwanted infestation.

Why View Live Worms?

Observing a live planarian is a valuable exercise. Students seeing something that they have only read or heard about helps make concrete for them what a creature really is.

Where Are the Cilia?

Though not described in the Student Edition, students should be able to deduce the location of the cilia by observing the gliding motion of the planarian.

D Expel the planarian into the petri dish that you have prepared before it has a chance to attach itself to the inside of the pipette.

E Give the planarian five minutes to acclimate to its dish before continuing.

F Using a hand lens, study the two ways planarians move: ciliary movement and muscular movement.

1. By observing the movement of your planarian within its dish, determine where you think its cilia are located.

   on the ventral surface

2. After observing the planarian moving by using its cilia, describe this type of movement.

   It appears to glide along.

G To see muscular movement, place a few grains of Epsom salt into the water right next to the planarian. Watch carefully as the Epsom salt begins to dissolve near it.

H After you have observed the movement that results, use forceps to remove and discard the grains of salt, and then add some fresh spring water to dilute the salt.

3. Describe muscular movement in comparison to the ciliary movement that you described above.

   The planarian twists its body, frequently raising itself from the surface on which it rests. It does not move as smoothly with muscular movements as it does with ciliary movements.

Observing Planarian Responses

Now that you've observed how planarians move, let's look at how they respond to stimuli in their environment. You should conduct each of the following tests several times to be sure that your planarian is exhibiting an actual response instead of simply making a random movement.

I With a clean probe, very lightly touch one side of the planarian.

4. Describe the planarian's response.

   Answers will vary. Usually, the planarian turns away from the touched side.

J Lightly touch one of the planarian's auricles (the side points of its head).

5. Describe the planarian's response.

   Answers will vary. Usually, it moves away from the touched side.

K Lightly touch the planarian's posterior end.

6. Describe the planarian's response.

   Answers will vary. Usually, it moves forward.

7. What do the planarian's responses all have in common?

   The planarian always moves away from the stimulus.

Now observe your planarian's response to current.

L Fill a clean pipette with water and slowly force the water out to produce a current. Direct the current to one side of the planarian in the petri dish. (Have ready several pipettes full of water.)

8. Describe the planarian's response.

   Answers will vary. The planarian should turn to move toward the current.

M Direct the current toward the side of the planarian's head.

9. Describe the planarian's response.

   Answers will vary. The planarian should turn to move toward the current.

N Direct the current toward the posterior end of the planarian from directly behind it.

10. Describe the planarian's response.

    Answers will vary. It should turn to move toward the current.

O Next, observe your planarian's response to light. Cover half of the petri dish with an index card, small piece of poster board, or similar, so that half the dish is shaded. Make sure that your planarian is in the lighted half of the dish.

11. Describe the planarian's response to light.

    Answers will vary. The planarian should move into the shaded portion of the dish.

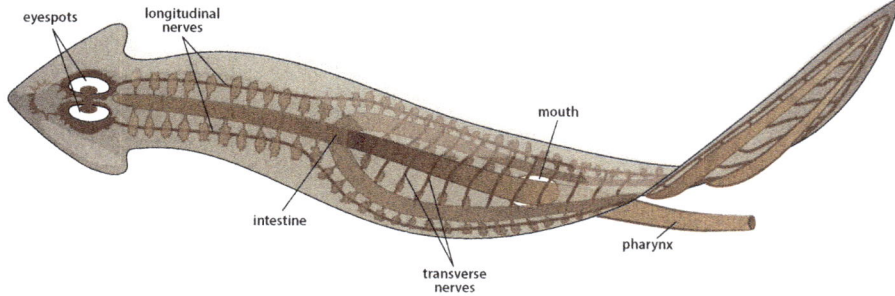

eyespots | longitudinal nerves | mouth | intestine | pharynx | transverse nerves

Observing Planarian Feeding

Refer to the sketch above to answer Questions 12–15.

12. Describe the mouth's location on the body.

   The mouth is a hole near the middle of the ventral surface.

13. Describe the appearance of the pharynx.

   It looks like a tube.

14. Describe the area within the planarian where digestion begins. Be sure to use the proper name for this structure.

   Digestion begins in the intestine, a branched cavity inside the worm that has a single opening.

15. After the structure described above completes preliminary digestion, what must happen to any matter that cannot be digested? Explain.

   Since the planarian digestive tract has only a single opening, any matter that cannot be digested must be expelled through the pharynx.

P Place a small piece of food near your planarian and observe its behavior.

16. Describe the planarian's response to food.

   Answers will vary. The planarian should respond by beginning to feed on the food item. It extrudes its pharynx to suck in particles of food.

17. How does the planarian's range of responses (how specific its responses are) compare to that of the hydra?

   Answers will vary. They are similar, but students should observe that the planarian's responses are more specialized.

GOING FURTHER

There are a variety of ways that our flustered aquarist friend Al could get rid of his planarian pests, but one method he should probably *not* consider is catching the worms, chopping them up, and feeding them back to his fish. They might enjoy the treat, but the problem with this method is that every piece of uneaten worm has the potential to grow into an entire new worm. In fact, in one study planarians grew back from pieces as small as *1/279th* of the original worm! This amazing ability to regrow from pieces is called *regeneration*, and although other animals can regenerate to one degree or another, planarians are particularly adept at it.

✓ Going Further: Parasitic Flatworms

If you have the time and inclination, you could have students examine preserved slides of beef tapeworms and human liver flukes. We suggest preserved slides of beef tapeworm (*Taenia saginata*) scolex, proglottid, and bladders in meat as well as preserved slides of human liver fluke (*Clonorchis sinensis*). If the slides of the beef tapeworm are not easily available, slides of other species of tapeworm may be used for the scolex and proglottid, but for the bladder stage a photograph of beef containing bladders is best.

Have students look for differences and similarities between these parasitic worms and planarians.

✓ Time Management Options

The material in the Going Further section may be omitted in the interest of time, or it may be assigned as homework.

✓ Current Research

The focus in planarian research is on understanding the genetic and cellular mechanisms by which planarian pluripotent cells, called *neoblasts*, differentiate into the various types of cells that planarians need to regenerate different kinds of tissue. By understanding such mechanisms, researchers hope that one day it might be possible to use human stem cells more effectively for various medical treatments. Keep this in mind when helping students consider which websites of the ones they find are most relevant in answering Questions 18–22. You may also remind them of the brief description of cell differentiation on page 169 in the Student Edition.

✝ When Does Dead Mean Dead?

Stem cell research is just one of the ethical questions that have arisen as a result of advances in medicine and technology. Questions that used to be easy to answer (e.g., "When is a person actually dead?") are now more difficult. Fallen humans can be tempted to "play God." A thoroughly biblical worldview is needed to properly ascertain whether new discoveries and techniques honor God and serve mankind or usurp God's authority and serve self. This question is further explored in the ethics feature in Chapter 25.

Q Do an internet search for planarian regeneration. After looking at a few sources, answer the following questions.

18. What specific aspects of regeneration in planarians do scientists find particularly intriguing?

Answers will vary. Scientists are particularly intrigued by the pluripotency (ability to differentiate into any type of cell) of planarian neoblasts (a type of stem cell), the rapidity with which planarians regenerate, and the abundance of neoblasts throughout a planarian's body.

19. What is the current focus of research on planarian regeneration?

Answers will vary. Much current research is focused on the genetic and biochemical control mechanisms involved in directing the differentiation of cells.

20. Why is an understanding of the mechanism of planarian regeneration of interest to human medical research?

Answers will vary. It is believed that an understanding of the mechanism of regeneration in planarian neoblasts will advance the treatment of disease using human stem cells.

21. Why does the subject of regeneration of human tissue pose ethical dilemmas?

Answers will vary. Stem cell research uses either embryonic stem cells or adult stem cells. Embryonic stem cells have a much greater capacity for differentiation than adult stem cells, but the use of embryonic stem cells, harvested from human embryos, poses an ethical question since life begins at conception.

22. Is stem cell research compatible with a biblical worldview?

Answers will vary. Stem cell research has great potential for medical breakthroughs, but the use of embryonic stem cells violates the biblical view of the sanctity of life. However, the use of adult stem cells, which can be safely harvested from adults, is compatible with a biblical worldview and is showing great promise.

17A LAB

Take a Crack at Crayfish

Dissecting a Crayfish

Crustaceans are a diverse group of animals. Crayfish are typical crustaceans that look like miniature lobsters. As you dissect one today, keep in mind that the structures you observe can be found in many other crustaceans.

What are the structures in a typical crustacean?

Equipment

dissection pan	plastic bag	nitrile gloves
dissection kit	fine-tip permanent marker	goggles
camera	preserved crayfish	
culture dish	glue, rubber cement, or tape	
flower head pins (8)	laboratory apron	

PROCEDURE

The External Anatomy of the Crayfish

A Place your crayfish in a dissection pan and carefully observe it. You will notice that its body is made of multiple segments, each bearing a pair of appendages. Count the pairs of appendages on your crayfish. As you read the names and descriptions of the body structures in the key on page 193, locate them on your specimen.

B Write the number associated with each structure shown on page 192 with the structure's name on the key.

> **QUESTIONS**
> - What arthropod characteristics can be seen in a crayfish?
> - What appendages does a crayfish have?
> - How are a crayfish's internal organs specialized for its lifestyle?

Take a Crack at Crayfish 191

✔ Prelab Discussion

You might mention the concept of *umwelt*, the way that an animal perceives its surroundings. We often mistakenly assume that an animal perceives the world just as we do, but an animal such as a crayfish probably relies more heavily on some senses and less on sight than we do. Ask students to consider the animal's *umwelt* as they work through this lab activity.

🧪 Equipment Notes

Medium-sized crayfish roughly 4 in. from head to tail work well for this exercise. Exceptionally large specimens are not worth their extra cost, while tiny crayfish are almost useless for dissection purposes.

Be sure to order injected specimens. The very small difference in cost is more than outweighed by the enhanced student learning experience.

Flower head pins are commonly available at fabric stores or in the craft section of the local department store. They are commonly used for quilting. You may choose to have students label their own pins or create sets yourself prior to lab day. Simply write the numbers 1–8 on the pins with a fine-tip permanent marker.

If you need more time to complete this activity, the dissected crayfish can be kept for up to several days by covering them with a wet paper towel. Check the specimens daily to make sure that the towels stay moist. When it is time to dispose of specimens, those preserved with nontoxic solutions may usually be wrapped in a paper towel and discarded in the trash. Check your local regulations regarding the disposal of biological waste.

What Am I Looking At?

Students can have difficulty comparing what's in their dissection tray with what's shown in a stylized textbook drawing, and unskilled efforts at cutting and slicing can further complicate matters. A dissected crayfish mounted in acrylic can help students locate key anatomical structures in their own specimens. Check with your science equipment supplier for availability.

EXTERNAL ANATOMY of a CRAYFISH

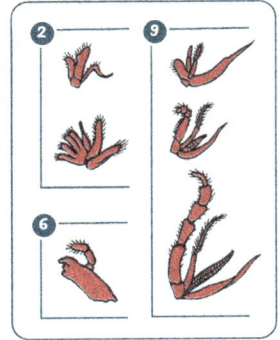

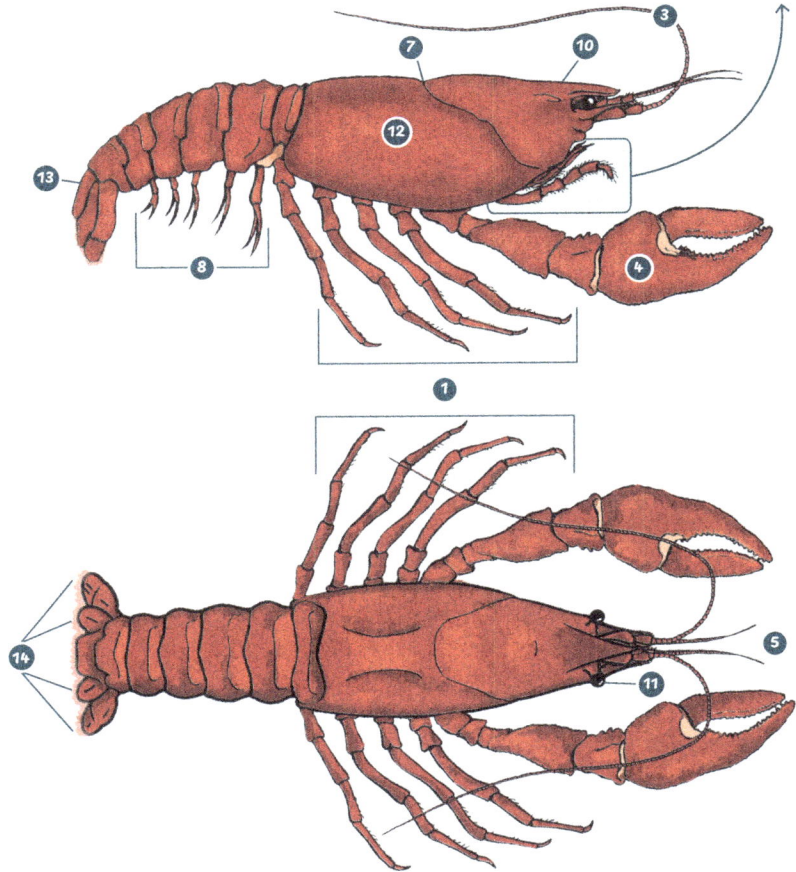

Key to the External Anatomy of a Crayfish

___5___ **1.** antennules—These most anterior appendages are used in the senses of balance, taste, and touch.

___3___ **2.** antennae—Longer than the nearby antennules, these are used in the senses of taste and touch.

___6___ **3.** mandible—Also called true jaws and located just posterior to the antennae, these small, hard coverings of the mouth pulverize food.

___2___ **4.** maxillae (two pairs)—Just posterior to the mandibles, the maxillae assist in chewing.

___9___ **5.** maxillipeds (three pairs)—Just posterior to the maxillae, the highly branched maxillipeds are used to hold food in place.

___11___ **6.** compound eyes—These are technically not appendages.

___12___ **7.** carapace—This single section of exoskeleton covers the cephalothorax.

___10___ **8.** rostrum—This extension of the carapace forms a horny beak between the eyes.

___7___ **9.** cervical groove—This depression marks the division between the head and the thorax.

___4___ **10.** chelipeds—These large claws are used for protection and capturing food.

___1___ **11.** walking legs (four pairs)—Notice that the legs are different from one another.

___8___ **12.** swimmerets (five pairs)—These abdominal appendages move water over the crayfish's gills and function in reproduction.

___14___ **13.** uropods—These paired appendages grow from the posterior segment of the abdomen.

___13___ **14.** telson—This structure also grows from the posterior segment.

Dissection Quiz

After students have had sufficient time to study their dissected specimens, consider giving them a quiz on the crayfish's anatomical structures. Have them remove the pins from their specimens, then call out five structures (numbered 1–5) for the students to pin with the appropriate number.

Question 15 Answer
13 (or 14 if a student counts the appendage-less segment preceding the segment bearing the antennules)

Question 18 Answer
The front two pairs often have pincers; the back two pairs frequently do not.

Question 21 Answer
Answers will vary. Check to see that the student has correctly identified the sex of the crayfish.

C Examine the cephalothorax of your preserved crayfish.

15. How many segments does a crayfish's cephalothorax have?

See margin for answer.

16. Locate and examine the mouth parts. Human mouthparts move vertically (up and down). In what direction do the mouthparts of a crayfish move?

horizontally

17. How do the walking legs differ from one another?

They get longer and thinner toward the posterior.

18. Which of the walking legs, if any, have pincers?

See margin for answer.

D Examine the abdomen of the crayfish.

19. How many segments does the abdomen have?

6

20. Locate the anal opening on the ventral side of the crayfish. On which segment is it located?

on the sixth and most posterior segment of the abdomen

In males, the most anterior swimmerets are enlarged and point anteriorly. In the female, the anterior swimmerets are greatly reduced in size.

21. What is the sex of your crayfish?

See margin for answer.

22. The telson and the uropods form a powerful tail fan. How would the crayfish use these?

The crayfish bends its abdomen under itself, using the tail fan to swim backward.

The Internal Anatomy of the Crayfish

Carefully read the following directions for each body part before starting your dissection. The following procedures must be done in order. You will be expected to know the functions and locations of the organs that have been italicized.

THE BODY CAVITY

Place your animal in the dissection pan with its dorsal side up.

E Carefully insert the point of the scissors under the dorsal surface of the carapace at the posterior edge of the cephalothorax. Cut anteriorly along the midline of the body to the rostrum.

F Reposition the scissors just behind the eyes and make a transverse cut.

G Carefully remove the two pieces of the carapace without disturbing the structures underneath.

194 Lab 17A

© 2024 BJU Press. Reproduction prohibited.

THE RESPIRATORY STRUCTURES

H Remove a few *gills* and place them in a culture dish of water.

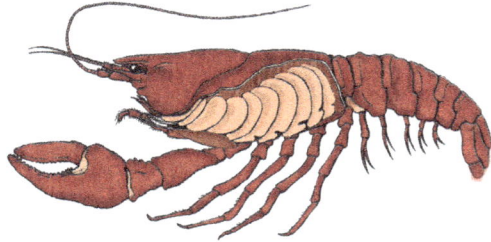

The gills of a crayfish are located just under the carapace.

23. Describe their function.

   Answers will vary. Students should mention that the gills
   exchange carbon dioxide for oxygen with the surrounding
   water.

I Carefully remove the rest of the gills.

24. When a crayfish spends time on land, it holds water inside its carapace. Why would it do this?

   Since a crayfish's gills are able to exchange gases only when
   wet, when on dry land the carapace carries water to enable the
   crayfish to breathe.

THE CIRCULATORY STRUCTURES

J For easier handling of the animal, remove the legs attached to the thorax.

K Carefully separate the dorsal tissues in the thorax and locate the *mid-dorsal heart*.

L Locate the main blood vessels attached to the heart.

The crayfish, like most arthropods, has an open circulatory system.

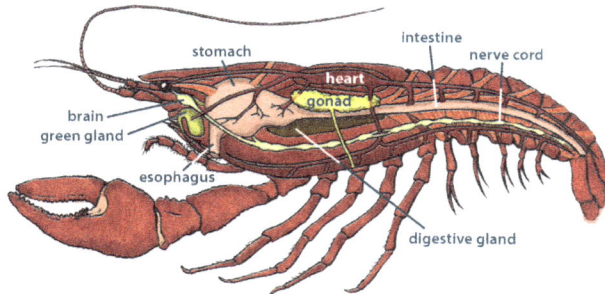

25. How does an open circulatory system work?

The blood leaves the blood vessels and bathes the organs.

26. What force does an open circulatory system rely on that a closed circulatory system does not?

gravity

27. Why would being immobilized on its back be lethal to a crayfish?

The blood would collect around the heart and not flow over the organs.

THE REPRODUCTIVE AND DIGESTIVE STRUCTURES

M The two light-colored masses extending along each side of the body cavity beyond the cervical groove are the *digestive glands*. Look between the digestive glands to find the reproductive structures.

» If your crayfish is a male, look for a small pair of white *testes* and *coiled ducts*.

» If your crayfish is a female, look for a large mass of dark-colored eggs inside the *ovaries*.

28. What is the function of the digestive glands?

The digestive glands produce digestive enzymes.

N To expose the *intestine*, insert the point of the scissors underneath the dorsal side of the exoskeleton covering the abdomen. Cut posteriorly to the final segment.

» Open the abdominal exoskeleton along the cut. The intestine appears as a tube on the dorsal side of the abdominal muscles.

» Do not confuse the intestine with the dark-colored dorsal blood vessel.

O Trace the intestine anteriorly to the portion of the cephalothorax where the intestine joins the large, thin-walled *stomach*.

P Number your pins one to eight using a permanent marker. Insert the following pins in these organs:

» Pin 1—heart

» Pin 2—primary reproductive structure

» Pin 3—either of the digestive glands

» Pin 4—stomach

» Pin 5—intestine

Q Take a picture of your crayfish with the pins and paste it in Photo Area A.

R Remove most of the crayfish's internal organs.

» Just behind the eyes, cut the bands of muscles leading to the stomach.

» Pull the stomach posterior and cut the short esophagus located just below the stomach.

S Carefully lift out the organs all in one piece. Notice the mesenteries (connective tissue), which keep the organs all together.

THE EXCRETORY STRUCTURES

T Clean out the remaining tissue in the head to expose the *green glands* just posterior to and below the antennules. These soft, small, and only slightly green organs filter nitrogenous wastes from the crayfish's body, similar to how kidneys function in a vertebrate.

U Being careful, look for the small, sac-like bladder, which is connected to the green glands.

THE NERVOUS SYSTEM

V At the front of the head cavity, between the eyes, note the *brain*, a tiny mass of white tissue.

W Trace the nerves that go from the brain to the antennae and the eyes.

X Trace the *nerve cord* from the brain to the abdomen by cutting the hard tissue on the floor of the thorax with a scalpel.

» Spread the abdomen apart and pull out the large muscles.

» The nerve cord should now be exposed on the ventral side of the abdomen.

Y Insert the following pins in these organs:

» Pin 6—either of the green glands

» Pin 7—brain

» Pin 8—nerve cord

Z Take a picture of your crayfish with the pins and paste it in Photo Area B.

29. What are the swollen portions of the nerve cord called?

ganglia

30. Why is it an advantage for the crayfish to have its nerve cord on the ventral side rather than on the dorsal side, as it is in humans?

Answers will vary. It is closer to the appendages, and it is better

protected, especially during molting.

PHOTO AREA A

PHOTO AREA B

17B LAB

Cricket Caper

Inquiring into House Crickets

The humble house cricket has had a checkered relationship with humans. Some people have found them inspiring, keeping them as pets or elevating them to respectable status in children's stories such as George Selden's *The Cricket in Times Square*. Other people know them as household pests, worthy only of a scornful swipe of a broom. Somewhere along the way someone discovered that crickets were an easily cultured form of live food for more desirable pets such as birds and reptiles. Even people themselves eat crickets in certain parts of the world—in Thailand they are deep-fried and considered a tasty snack. Don't scoff—they're actually a good source of protein!

But some scientists, like Dr. Rajat Mittal at Johns Hopkins University, are interested in crickets for reasons other than their literary merit or culinary qualities. Mittal and his students are practitioners of "bioinspired engineering." They use high-speed photography to study the jumping motion of spider crickets in an attempt to understand how the crickets' spindly legs power their amazing leaps—the equivalent of a human jumping the length of a football field. They hope to apply what they learn to the manufacture of tiny jumping robots that can navigate uneven ground. It is your task in this lab activity to think of an aspect of cricket behavior that you can investigate.

What can I learn from studying cricket behavior?

Equipment

crickets

PROCEDURE

Plan and Write Scientific Questions

First, you'll need a hypothesis to test. There are many aspects of cricket behavior that you could investigate and different methods that you could use to test those behaviors. To help get you started, think about three broad categories of behaviors.

» How do crickets respond to their environment?

» How do crickets interact with other crickets?

» How do crickets interact with other organisms?

Brainstorm with your group to come up with a list of possible questions to investigate. Consider which ones interest you the most. You'll need to take into account whether you have the necessary equipment on hand (or can easily obtain it) and enough time to adequately test your possible hypotheses. Narrow down your list of questions to the one you will investigate.

1. What question will you investigate regarding cricket behavior?

Answers will vary. *Example:* Do crickets respond to light?

QUESTIONS

- What are some cricket behaviors that can be investigated?

- What kinds of experiments can be designed to test those behaviors?

- Develop a testable hypothesis regarding cricket behavior.

- Design and successfully carry out an experiment to test a hypothesis.

- Report findings in the form of a scientific paper.

Cricket Caper 199

Cricket Caper 199

Scientific inquiry requires testable statements related to the question being investigated. These statements are called *hypotheses*. Think about a testable hypothesis that you can use to help you answer your question about cricket behavior. Have your teacher approve your hypothesis.

2. What is the hypothesis that you will test?

Answers will vary. *Example:* If placed in a tank with one dark end and one lit end, crickets will tend to move toward the lit end.

Designing and Conducting Scientific Investigations

Next, you will need to design an experiment to test your hypothesis. Get approval for your experimental design from your teacher before beginning your experiment.

3. Briefly describe your experimental setup.

Answers will vary.

4. What will be your control group?

Answers will vary.

5. What will be your experimental group?

Answers will vary.

6. What data will you be collecting?

Answers will vary.

7. How will you analyze your data?

Answers will vary.

Set up and run your experiment. Make sure to run several trials to solidify support for your conclusions.

Developing Models

Once you have collected your data, you will need to analyze it by using mathematical tools, such as finding and comparing averages or creating graphs.

Scientific Argumentation

In Lab 15B you learned about scientific writing. Recall that a scientific paper has distinct sections, including an abstract, an introduction, a discussion of materials and methods, the results, a discussion of the meaning of the results (including suggestions for further research), and a list of references. Your teacher will instruct you on whether to report your findings in a formal lab report (a type of scientific paper), an oral report, or a slide presentation. As part of your report or presentation, describe how the behavior you observed gives evidence of God's wisdom in providing for the cricket's needs.

✚ *Even Small Things Matter*

Although seemingly insignificant, the cricket is nonetheless a part of God's creation, and He has provided for the cricket just as He has done for all His creatures. Encourage students to think about what aspects of cricket behavior God has "programmed" into His creature that enable it to fulfill its basic needs for food, shelter, and propagation.

17B LAB *Teacher Guide*

Cricket Caper

An inquiry lab activity allows students a degree of freedom, but the teacher must guide the process to achieve the desired educational outcomes. As teachers start using inquiry lab activities, they often struggle with this process of guiding students. The best method is to think through the goals that you want students to achieve and then use questions to guide them to reach those goals. You want to balance students' freedom to investigate the question in the manner they choose with the need to meet your educational goals.

This activity is about animal behavior. Students will model what actual scientists must regularly do; that is, they will decide on a behavior to test, identify a question to ask about that behavior, and design a test that will help answer the question.

Inquiry lab activities provide you with a degree of flexibility in deciding how the activity should be carried out. Students will need to collaborate to design a workable test; you may choose to have students do this within their group, with each group deciding on its own question and test, or you can have a class discussion resulting in a single question and test. If the latter, allow students to generate the ideas. Prompt them with questions to guide the discussion in directions beneficial to the activity. You may choose to have students do part of this activity as homework. For example, students may do their research as homework, use class time to collaborate on a test design, conduct their tests at home, and finally present their findings in class.

MATERIALS

Equipment needs for this activity will be determined by what aspect of cricket behavior students decide to test. At a minimum, students will need suitable containers for housing their crickets for up to several days; these might be anything from a small cardboard box to a small terrarium. Naturally, crickets need food and water. Crickets can be fed chopped bits of starchy vegetables (e.g., carrots, potatoes, squash), dry fish food flakes, or dry cat food. Water should be placed in a shallow dish with a few pebbles in it to prevent drowning. Remember that the kinds and amounts of food and water provided might figure into the questions that students choose to ask about cricket behavior. For example, students that underfeed their crickets may discover that hungry crickets will readily resort to cannibalism!

PROCEDURE

Planning/Writing Scientific Questions

A great many possible cricket behaviors can be tested within the context of this lab activity. Consider the following ideas:

- movement or chirping in response to changes in temperature
- responses to the presence of predators
- responses to the presence of other crickets of the same or opposite sex (Adult female crickets are easily identified by the presence of a long ovipositor.)
- responses to varying intensity, duration, or color of light
- food preferences
- preference for different substrates
- response to high- or low-frequency sound
- sensitivity to infrared or ultraviolet light

Once students have settled on a question to investigate, they will need to draft a testable hypothesis. Each of the behaviors listed above lends itself to several possible hypotheses that can be tested. Make sure that students' hypotheses are testable before they begin designing their tests.

Designing Scientific Investigations

After students have formulated testable hypotheses, they need to write specific procedures to test their hypotheses. Make sure that you review their procedures prior to their starting to collect data. Help students think through how they will organize their procedures and data.

Conducting Scientific Investigations

Once the groups have good procedures, allow them to collect data.

Developing Models

Students will analyze their data to determine whether identified trends support their hypotheses.

Scientific Argumentation

Students should state a claim about their procedures and analysis. As always, claims should be supported with specific evidence from their data. Consider having students present their findings in the form of an oral report, slide presentation, or formal lab report (see Appendix F). A grading rubric can be found in Appendix G. This rubric is also available as a digital resource.

18A LAB

Something Fishy Going On

Observing Bony Fish

In the sandy depths of Lake Malawi in East Africa there lies a fish—literally. The locals call it *kalingono*—the sleeper. Only the fish isn't sleeping. It's hunting. As it lies on its side on the sand, the splotchy brown and white sleeper looks remarkably like a rotting corpse. When a smaller fish comes near to investigate—wham! The sleeper lunges and swallows the unsuspecting prey whole.

With over 30,000 described species so far, as many or more than all other vertebrates combined, you'd expect the many varieties of fishes in the world to exhibit a wide range of behaviors. Some, like the sleeper's, are truly bizarre. In this lab activity you'll take a closer look at the external form and behavior of a fish of your choice.

Nimbochromis livingstonii, the sleeper

How do fish behave?

Equipment
live fish to observe
notepad and pencil

PROCEDURE

First, you'll need a live fish to observe. It can be your own fish, a neighbor's fish, a fish at a pet store, or a fish in a public aquarium. Whatever fish you choose, it is important that the fish is in a setting that makes it feel comfortable so that it will behave as naturally as possible. A fish in a harshly lit, overcrowded aquarium with no hiding places will be stressed and not look or act its best.

Observing any kind of wildlife takes time. That means that you, the observer, need to give some thought to your own comfort as well. You'll need to observe your fish for an extended period of time without a lot of motion. Once you have chosen a fish to observe and have gotten yourself comfortably positioned, give the fish about five minutes to become acclimated to your presence. Some fish held in captivity quickly learn to associate humans with food; they need time to figure out that you're not there to feed them and go back about their business. Other fish are shy and retiring and need a few minutes to be sure that you're not a danger to them.

When your fish has become acclimated to you and you are ready to start recording observations, try to be slow and deliberate with your motions—don't make sudden movements. If the fish is startled and makes a dash for cover, give it a few minutes to relax again before continuing.

QUESTIONS
- What are the key external features of a fish?
- How does a fish move and function in its environment?
- How does a fish respond to its environment?

LAB 18A OBJECTIVES

- Describe the external features of a fish.
- Describe the behavior of a fish, especially with regard to its interactions with its environment and with other fish.
- Discuss the ethics of fishkeeping as a hobby from a biblical worldview.

Ideas for Prelab Discussion

There are as many described species of fish as all other vertebrates put together, yet because they are aquatic, they are difficult to observe in their natural habitats. Happily, aquarium keeping is one of the world's most popular hobbies, meaning that many fish can be easily viewed far from their remote and inaccessible home waters. Students may need some coaching on what sort of fish to observe. A fish that is extremely active will be difficult to scrutinize closely. On the other hand, a fish that is sedentary may not display any interesting behaviors while it is being observed. Some fish are nocturnal and will be active only at night. Ideally, a student should observe a fish in a setting that allows it to interact both with others of its own kind and with members of other species.

Choosing to observe fish at a pet store or public aquarium presents special challenges since doing so may involve high foot-traffic areas with the potential to keep fish being viewed in a state of perpetual stress. It is usually possible, though, to find a time when there are fewer observers or a display that doesn't attract as much attention. With a little effort, students may be able to find an aquarium in a quieter location, such as in a university biology department, an office display, or a restaurant. Larger cities frequently have fishkeeping clubs; their members are often happy to open their fish rooms to inquiring young minds.

Equipment Notes

This lab activity is written for the study of bony fishes, so students should limit their choices of fish to bony fish. No sharks or other cartilaginous fishes should be included.

Alternative Methodology

You may have students summarize their findings from this activity in the form of a report or other presentation. They may use the questions to guide them as to what should be included in their reports.

Question 1 Answer
Answers will vary. A few fish do not have common names.

✔ **Fin Note**

There is considerable variation in the form and function of fish fins. The answers in this Teacher Lab Manual are generalized for the kinds of fish your students are most likely to encounter.

Form and Features

1. What is the common name for your fish?

   See margin for answer.

2. What is the scientific name of your fish?

   Answers will vary.

A Observe your fish from several angles with an eye for its overall shape and form.

3. Describe the body shape as viewed from the side.

   Answers will vary.

4. Describe the body shape as viewed from the front.

   Answers will vary.

5. In what ways is the fish's body form ideally suited for its aquatic environment?

   Answers will vary. Most fish body forms show some form of streamlining for ease of movement in water.

B Study the fish's fins closely and indicate their number, location, and shape in the appropriate columns of Table 1 (see page 401 in your textbook for help). In the *Rays* column, indicate whether each fin has soft rays, spiny rays, both, or none. In the *Function* column, include whether each fin is used primarily for steering, stability, or propulsion.

The highly modified fins of a lionfish

6. Is the mouth of your fish superior (upturned), terminal (pointing directly ahead), or inferior (downturned)?

   Answers will vary.

7. What does the position of your fish's mouth suggest about its feeding strategy?

   Answers will vary. Fish with superior mouths tend to be ambush predators or to take food items from the water's surface. Fish with inferior mouths tend to be bottom feeders. Fish with terminal mouths employ a variety of feeding strategies.

C Notice the size and position of your fish's eyes.

8. What does the size of your fish's eyes suggest to you regarding the importance of vision for fish?

   Most fish have large eyes relative to their overall size. This suggests that vision is an important sense for fish.

9. What does the position of your fish's eyes tell you about its field of vision?

   Most fish have eyes placed on the sides of their heads, giving them a very wide field of view with only a small blind spot directly to the rear.

D Try to locate your fish's lateral line (see page 402 in your textbook).

10. What is this line made of, and what is its function?

    The lateral line is a series of sensory pores. It detects vibrations in the water.

11. How does the arrangement of the lateral line enhance its role as a sensory organ?

    Answers will vary. By being arranged in a line on each side of the body, the sensory pores can tell from which direction a vibration is coming.

E Watch your fish's breathing carefully.

12. What are the two primary external structures involved in fish breathing?

the mouth and operculum

13. Describe how these two structures work together as the fish breathes.

The fish draws water in through its mouth while its operculum is closed. Then the fish closes its mouth and opens its operculum, allowing water to flow over the gills.

14. Describe any other interesting external features your fish might have.

Answers will vary. Fish may have barbels (whiskers), visible teeth, tubercules, or other features. Students may mention "nostrils" (nares) located on the fish's snout.

15. Describe your fish's coloration.

Answers will vary.

F In Drawing Area A, make a sketch of your fish. Label the fins and any structures that you were able to identify in Questions 6–14.

Behavior

G Now take some time to carefully observe how your fish interacts with other fish and with its surroundings.

16. Where in the water column does your fish spend most of its time (upper, mid, lower)?

Answers will vary.

17. Is your fish active, swimming unceasingly about its enclosure? Or does it limit its movement to one particular area? Describe your fish's general activity.

Answers will vary.

18. Does your fish prefer open water, or does it prefer a sheltered area? How does your fish utilize the space available to it?

Answers will vary.

19. How are your answers to Questions 17 and 18 related?

Fish that are active will generally prefer open areas. Fish that prefer shelter usually keep to the vicinity of that shelter.

20. How does your fish propel itself forward?

Answers will vary. Most fish use their caudal fins for propulsion, though some propel themselves with their pectoral fins.

21. Can your fish move backward? If so, how?

Some fish can paddle backward with their pectoral fins.

22. Compare your fish with other fish in the tank. Does there appear to be a relationship between the activity of your fish described in Questions 17 and 18 and the location of its pectoral and pelvic fins? Explain.

See margin for answer.

23. Does your fish spend its time in the company of other fish of the same species (schooling), or does it keep to itself?

Answers will vary.

✓ *Sketching Option*

As an alternative to sketching, allow students to attach a photo of their fish in the drawing area.

Question 22 Answer

Active-swimming fish tend to have pectoral fins lower on their body and pelvic fins located farther back. Fish that keep stationary in the water column tend to have pectoral fins mounted higher along the margin of their opercula (for more precise movement) and pelvic fins directly beneath their pectoral fins (for more precise steering).

Head down, fins flared. This tiny fish is telling its tankmates, "Back off!"

© 2024 BJU Press. Reproduction prohibited.

24. Does your fish exhibit any aggression toward others of its kind? If so, describe its actions.

   See margin for answer.

   _____

   _____

   _____

25. Does your fish show signs of aggression to other species of fish in the tank? If so, describe its actions.

   Answers will vary. Aggressive behaviors may be similar to those in Question 24.

   _____

   _____

26. Does your fish show any signs of territoriality (having a territory)? If so, describe its actions.

   Answers will vary. Many fish will defend a territory that includes a shelter, food source, or breeding site. Aggressive behaviors may be similar to those in Question 24.

   _____

27. Describe any other behaviors that you observe during your viewing session.

   Answers will vary. Some other behaviors that might be observed include feeding, flashing (dashing of the operculum against the substrate to dislodge gill irritants), yawning, scraping or foraging in or on the substrate, and gravel sifting or moving.

204   Lab 18A

GOING FURTHER

Fishkeeping is big business. Some sources have estimated that roughly 10% of all American households have an aquarium, and the hobby is popular in Europe and Asia as well. Several billion dollars are spent each year on aquariums, fish, and supplies. Large quantities of aquarium fish are bred on farms, but many are also harvested from the wild. This is particularly true of saltwater reef fishes. Do an internet search for "ethics and tropical fish." After reading some articles, answer the following questions.

28. What are some of the ethical issues pertaining to the harvesting and keeping of tropical fish?

Answers will vary. Answers may include overharvesting of wild fish, poisoning of reefs through the use of cyanide in harvesting reef fish, the high mortality rate of wild-caught fish, the number of fish that die because of inadequate care, and the amount of energy used to maintain aquariums.

29. Can the hobby of keeping tropical fish be reconciled with a biblical worldview? Explain.

See margin for answer.

Tropical fish farms provide an alternative to wild-caught fish.

✔ For Further Exploration: Setting Up a Breeding Tank

One particularly fascinating behavior that most students will probably not have a chance to see during this exercise is the breeding behavior of the fish they are observing. One group of freshwater tropical fish, the cichlid family, is well known for its courtship behavior, spawning site preparation, and unusually high degree of parental care given to its young. Several smaller species of cichlid are relatively easy to keep and breed in aquariums as small as 15 to 20 gallons. The author can particularly recommend the kribensis (*Pelvicachromis pulcher*) for areas with soft water and the lyretail cichlid (*Neolamprologus brichardi*) or masked julie (*Julidochromis transcriptus*) for regions with hard water. Tropical fish retailers will usually offer in-store credit in exchange for any offspring successfully reared. More information on the care and breeding of cichlids can be found online or at your local library.

Question 29 Answer

Answers will vary. Dominion includes the wise use of all Earth's resources, and enjoying the aesthetic appeal of tropical fish is certainly a valid use of that particular resource. But just as for any resource, the use of tropical fish must be done in a manner that maintains a healthy environment at the present time and ensures the availability of the resource for future use.

✔ Search Tip

An internet search on ethics and tropical fish may direct students to the websites of some high-profile environmentalist organizations, such as PETA and Sea Shepherd. Although such groups make some valid points regarding conservation issues, it is important to keep in mind that they do so with a blatantly antibiblical, preservationist worldview, so discernment is in order. Students may need to scroll a bit to find resources that offer a more balanced assessment of the issues.

✝ Formulate a Christian Understanding

Thinking from a Christian perspective about something as mundane as having an aquarium might be something new for your students. Yet keeping an aquarium is first of all an exercise in aesthetics, and creating and enjoying beauty is part of what makes mankind distinct from the rest of God's created order; it is a mark of God's image in us. And tropical fish are, in a basic sense, simply another of Earth's resources. In that sense they are something that can be used and managed wisely in keeping with the Creation Mandate.

TABLE 1 Fish Fins

NAME	NUMBER	LOCATION	SHAPE	RAYS	FUNCTION
Pectoral Fin	2	varies by species; usually directly posterior to the head	varies by species	usually soft; may have a spine	steering or propulsion
Pelvic Fin	usually 2	varies by species; usually below and behind the pectoral fins	varies by species	usually soft; may have a spine	stability
Dorsal Fin	usually 1 or 2	along the back	varies by species	varies by species	stability
Anal Fin	usually 1	ventral, posterior to the anus	varies by species	varies by species	stability
Adipose Fin	1 or none	dorsal, posterior to the dorsal fin	usually small and slightly rounded	none	stability
Caudal Fin	1	posterior of the fish; the "tail"	varies by species	varies by species	propulsion

DRAWING AREA A

18B LAB

Reptile Repasts

Inquiring into Reptile Methods of Locating Prey

Lizards eat a variety of foods, and their methods of finding food depend on the species. In this lab activity you will design an experiment to test a particular species of lizard's method of hunting.

How do lizards detect their prey?

?

Equipment
terrarium
lizard

PROCEDURE

Plan and Write Scientific Questions

QUESTIONS
- How do lizards find food?
- Do lizards rely on some senses more than others when hunting?

A Think about how your lizard might use its senses to find food. Brainstorm with your group to come up with a list of possible questions to investigate. Consider which areas of investigation interest you the most. You'll need to consider whether you have the necessary equipment on hand (or can easily obtain it) and enough time to adequately test your possible hypotheses. Narrow down your list of questions to the one that you will investigate.

1. Which of your lizard's senses will you be testing?

   See margin for answer.

2. Write a hypothesis about how your lizard uses the sense you've selected to find its prey.

   See margin for answer.

Designing and Conducting Scientific Investigations

You now need to design an experiment to test your hypothesis. Get approval for your experimental design from your teacher before beginning your experiment.

3. Briefly describe your experimental setup.

   Answers will vary.

- Develop a testable hypothesis regarding a lizard's ability to detect food.
- Design an experiment to test a hypothesis.
- Report findings in the form of a scientific paper.

Question 1 Answer
Answers will vary. Students should concentrate on the lizard's use of its Jacobson's organ, its ability to see the visual outline of its prey, its ability to see its prey's movement, or its ability to hear its prey. While lizards have other senses, those senses play little role in finding prey.

Question 2 Answer
Answers will vary. Students' hypotheses should suggest that, when hunting, a lizard uses its Jacobson's organ, the visual outline of its prey, its prey's movement, or its hearing.

4. What will be your control group?

Answers will vary.

5. What will be your experimental group?

Answers will vary.

6. What data will you be collecting?

Answers will vary.

7. How will you analyze your data?

Answers will vary.

Set up and run your experiment. Make sure to run several trials to solidify support for your conclusions.

Developing Models

Once you have collected your data, you will need to analyze it by using mathematical tools, such as finding and comparing averages or creating graphs.

Scientific Argumentation

In Lab 15B you learned about scientific writing. Recall that a scientific paper has distinct sections, including an abstract, an introduction, a discussion of materials and methods, the results, a discussion of the meaning of the results (including suggestions for further research), and a list of references. Your teacher will instruct you on whether to report your findings in a formal lab report (a type of scientific paper), an oral report, or a slide presentation. As part of your report or presentation, describe how the behavior you observed gives evidence of God's wisdom in providing for the lizard's needs.

18B LAB *Teacher Guide*

Reptile Repasts

An inquiry lab activity allows students a degree of freedom, but the teacher must guide the process to achieve the desired educational outcomes. As teachers start using inquiry lab activities, they often struggle with this process of guiding students. The best method is to think through the goals that you want students to achieve and then use questions to guide them to reach those goals. You want to balance students' freedom to investigate the question in the manner they choose with the need to meet your educational goals.

This activity is about animal behavior. Students will model what actual scientists must regularly do; that is, they will decide on a behavior to test, identify a question to ask about that behavior, and design a test that will help answer the question.

Inquiry lab activities provide you with a degree of flexibility in deciding how the activity should be carried out. Students will need to collaborate to design a workable test; you may choose to have students do this within their group, with each group deciding on its own question and test, or you can have a class discussion resulting in a single question and test. If the latter, allow students to generate the ideas. Prompt them with questions to guide the discussion in directions beneficial to the activity. You may choose to have students do part of this activity as homework. For example, students may do their research as homework, use class time to collaborate on a test design, conduct their tests at home, and finally present their findings in class.

MATERIALS

Equipment needs for this activity will be determined by what aspect of lizard behavior students decide to test. At a minimum, students will need suitable containers for housing their lizards for up to several days; a terrarium is best for this purpose. Choose one that is large enough to give the lizard room enough to move around as it hunts. Lizard food preferences vary by species, though most small carnivorous lizards sold as pets will happily take crickets and other small insects. Water should be placed in a shallow dish.

Just about any type of carnivorous lizard will work for this lab activity, though smaller ones are easier to handle and are not a threat to students' safety. Geckoes work well. These can be bought at many pet stores, but you should give thought to what you plan to do with your specimens after the lab exercise is over. You could make them a part of your classroom, but you might want to read up on the proper care of pet lizards before taking this step. Some pet stores will buy back lizards; check with the store before purchasing one. Another possibility is to borrow a lizard. Many children have pet lizards that they might be willing to lend to you for a day.

You might also consider finding several species of carnivorous lizards and having different teams of students work with different species.

If you don't have access to reptiles, don't worry. This activity can easily be modified to examine the hunting behaviors of ectothermic vertebrates other than lizards.

18B LAB *Teacher Guide*

PROCEDURE

Planning/Writing Scientific Questions

If students have difficulty coming up with research questions on their own, suggest one of the following:

- Can a lizard locate prey by smell alone? (Design a setup that allows a lizard to smell a prey item without being able to see it.)

- Can a lizard detect prey that is motionless? (Design an experiment that presents a lizard with an immobilized prey item and compare the number of strikes on motionless prey versus prey that is moving.)

- Can a lizard hear its prey? (Design an experiment in which a lizard is exposed to recordings of crickets chirping and see whether hunting or stalking behavior is elicited.)

- Will a lizard strike at a prey item in the absence of smell or sound from the prey? (Design an experiment to test whether a lizard will strike at a facsimile of a prey item, and, if so, to see whether there is a difference in the number of strikes on the facsimile versus real prey.)

Once students have settled on a question to investigate, they will need to draft testable hypotheses. Each of the questions listed above lends itself to several possible hypotheses that can be tested. Make sure that students' hypotheses are testable before they begin designing their tests.

Designing Scientific Investigations

After students have formulated testable hypotheses, they need to write specific procedures to test their hypotheses. Make sure that you review their procedures prior to their starting to collect data. Help students think through how they will organize their procedures and data.

Conducting Scientific Investigations

Once the groups have good procedures, allow them to collect data.

Developing Models

Students will analyze their data to determine whether identified trends support their hypotheses.

Scientific Argumentation

Students should state a claim about their procedures and analyses. As always, claims should be supported with specific evidence from their data. Consider having students present their findings in the form of an oral report, slide presentation, or formal lab report (see Appendix F).

19A LAB

Our Fine, Feathered Friends

Creating a Birding Log

Many migratory birds travel thousands of kilometers each year on their annual migrations. That's a lot of frequent flier miles! It would be impossible for any ornithologist—a scientist who studies birds—to keep a constant watch on such well-traveled birds. But happily, ornithologists have a large, dedicated, and energetic group of unpaid assistants who are only too eager to provide them with valuable data on bird numbers and movements. They're *birders*.

Amateur birdwatching, or birding, is a popular pastime for many people and an invaluable tool for scientists as well. Many birders report their bird sightings to birding organizations. Scientists use these reports to estimate the number of different bird species and their population sizes. They also provide valuable data on the ranges and migration paths of those species.

How large and diverse is the bird population in your area?

Equipment
birding app or field guide
camera
notebook

PROCEDURE

Your goal is to identify ten or more species of birds that live in your area.

A During the time period indicated by your teacher, look for birds in the wild. This can be in a city park, in a rural field, or anywhere else you see birds.

B When you find a bird, use a birding app or field guide to identify the bird. If you have a camera, take a picture of the bird. Record its common and scientific names, its habitat (e.g., field, stream bank, urban neighborhood), and its location in Table 1. (If available, include the street address or location name. At the very least, record the name of the county and state. If you have access to a GPS device, record the GPS coordinates of the location.)

C For each species of bird that you identify, record the number of individuals of that species that you see during your project in the *Number* column of Table 1.

D If you took pictures of your identified species, attach them to a separate sheet of paper. Add a label identifying each one by its common and scientific names.

QUESTIONS
- How many birds live in your area?
- How many species of birds live in your area?
- How many individuals of each species of bird live in your area?

LAB 19A OBJECTIVES

- Estimate the number of birds in an area.
- Estimate the diversity of birds in an area.
- Estimate the relative abundance of each species of bird in a particular area.

Prelab Discussion

You may want to talk with students about organized amateur bird counts such as the Christmas bird counts coordinated by the Audubon Society.

Equipment Notes

Several free birding apps are available on-line. One of the most comprehensive is the Audubon Bird Guide app. If your students do not have smart devices, they can complete this lab activity with a good field guide and a notebook.

Students may use either the table provided in the activity or a field notebook at your discretion. If students use a field notebook, have them copy the Table 1 headers into their notebooks to keep their data organized.

It is unlikely that students will be able to find ten different species of birds in a single class period, and since the object is to estimate the population and variety of birds in your area, it would be better to have them work individually or in groups on their own time. You might consider assigning or beginning the activity one week and having them present their findings during the next lab period.

Picture Options

If your students have good cameras, showing pictures of the species they identify can be a very rewarding experience. If printed pictures are not an option, you could have students create and give an A/V presentation of their own pictures to the class.

Number of Species

You may need to adjust the number of required species on the basis of local conditions.

Species Ranges

Almost all birding apps and field guides have range maps, and these can be very helpful when trying to identify a bird that resembles two species whose ranges do not overlap. But students should realize that since birds are flying animals, they sometimes can be found outside their normal ranges.

Simpson's Diversity Index

If you did not assign Lab 4A or if your students are not strong mathematically, you should review the basic principles of Simpson's Diversity Index (see page 29 of Lab 4A in the Student Lab Manual).

Question 8 Answer
Answers will vary but should be a decimal number between 0 and 1. (The values for D and the answer for Question 8 should add up to 1.)

---

1. How many species of birds did you identify?

    Answers will vary.

2. What was your most numerous species?

    Answers will vary.

3. What was your least numerous species?

    Answers will vary.

4. Were any of the species that you identified outside of their normal range? If so, which ones?

    Answers will vary.

E A bird outside its species' normal range is called a *vagrant*. Do some research on bird vagrancy.

5. Write a paragraph summarizing some of the causes of vagrancy in birds.

    Answers will vary. Students should mention that birds may get lost during migration, possibly because of a poor navigation sense, and that other birds get blown off course by storms. A few birds, primarily waterbirds, have a tendency to wander.

F Compare the species that you identified with those your classmates identified.

6. Which species was identified by the most people?

    Answers will vary.

G For each of those species for which two or more individuals were observed by the class as a whole, add the total number of individuals and record the common name and the number—the value for n—in the first two columns of Table 2.

7. Which species had the highest total number of individuals counted by your class?

    Answers will vary.

H Create a bar graph in Graphing Area A using the results that you obtained in Step G.

ANALYSIS

Do you remember Simpson's Index from Lab 4A? It uses the formula

$$D = \frac{\sum n(n-1)}{N(N-1)},$$

where D is the index value, the Greek letter sigma (Σ) means "sum of," n is the number of individuals of any one species in the sample, and N is the total number of organisms in the sample.

I Calculate the value of Simpson's Index for the data that you recorded in Table 2.

(1) Beginning with the first species listed in your sample, record the value for $n-1$ and $n(n-1)$ in Table 2. Repeat this step for each species in your sample.

(2) Add up all the values for n; this sum is N. Record the value for N in Table 2.

(3) Add up all the values for $n(n-1)$ calculated in Step (1). This is $\Sigma n(n-1)$. Record this value in Table 2.

(4) Multiply the total number of individuals of all species in your sample (N) times that number minus one ($N-1$). This is $N(N-1)$. Record this value in Table 2.

(5) Divide the value for $\Sigma n(n-1)$ calculated in Step (3) by the calculated value for $N(N-1)$ from Step (4) to obtain Simpson's Index (D) for your sample. Record this value in Table 2.

(6) Simpson's Diversity Index, the actual number that you need to assess the diversity of your sample, is slightly different from Simpson's Index. To obtain the value of Simpson's Diversity Index for your sample, subtract the value that you calculated in Step (5) from 1 (i.e., $1-D$).

8. What is the value of Simpson's Diversity Index for your sample?

    See margin for answer.

You may remember from Lab 4A that a low value for Simpson's Diversity Index indicates a low biodiversity.

9. What does the Simpson's Diversity Index value that you calculated for your sample tell you about the avian biodiversity of your sample?

See margin for answer.

GOING FURTHER

A study published in Royal Society Open Science in 2016[1] considered populations of English wrens using data collected by volunteer birders. The scientists who analyzed that data found that more wrens lived in southern Britain than in northern Britain and that the wrens in the north tended to be larger than their cousins in the south.

10. How was the work of volunteers valuable to the scientists involved with this study?

The volunteers were able to gather more data than the scientists could have on their own.

11. The scientists who conducted this study concluded that the size difference was a result of adaptation to climate differences. How does this line up with a biblical worldview?

This research fits with a biblical worldview. God created populations of animals with great genetic diversity that allows them to adapt to different environments.

---

[1] Morrison, Catriona A., Robinson, Robert A., Pearce-Higgins, James W., "Winter Wren Populations Show Adaptation to Local Climate," Royal Open Science, Vol. 3, No. 6 (June 2016).

Question 9 Answer
Answers will vary. A low value indicates a low biodiversity, while increasingly higher values indicate greater biodiversity.

Winter Wrens

To view the Royal Society Open Science article, see the *Winter Wren Populations* link. This link is available as a digital resource.

Sample Data

A sample entry is shown in Table 1.

Table 2 should be filled out in the same manner as Table 1 of Lab 4A (p. 30).

TABLE 1 Observed Bird Species

	COMMON NAME	SCIENTIFIC NAME	HABITAT	LOCATION	NUMBER
1	varied thrush	*Ixoreus naevius*	old growth redwood	Bull Creek Flats, CA	1
2					
3					
4					
5					
6					
7					
8					
9					
10					

TABLE 2 Bird Count Data

COMMON NAME	NUMBER (n)	$n - 1$	$n(n - 1)$
	$N =$ _____	$N(N - 1) =$ _____	$\Sigma n(n - 1) =$ _____ $D =$ _____

GRAPHING AREA A

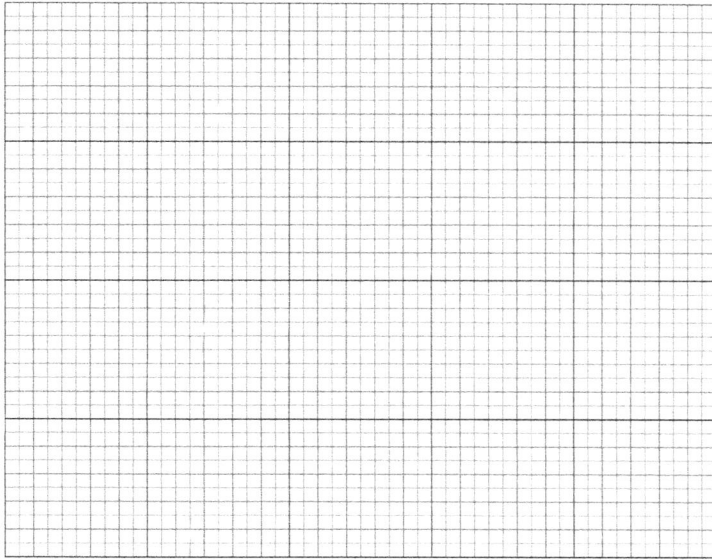

19B LAB

Warming Up to Research

Doing Preliminary Research

Scientists are curious people—they wonder about things that other people might not ever think about. Do dogs dream in color? Can an owl survive on a vegetarian diet? Do bears get grumpy if they are awakened in the middle of hibernation? These are the kinds of questions that motivate scientists to keep acquiring ever more knowledge about how God's creation works.

But how does a scientist know for certain that a question has never been answered before, perhaps not even investigated? Funding for scientific research is often limited, and the people who finance research are probably not interested in funding studies whose questions have already been asked and answered. And what if a question *has* been asked and investigated but not answered? How can a scientist know what kinds of investigative work have already been done? After all, there's no need to start at square one if a scientist can build on someone else's foundation.

The answer to these questions is research. Preliminary research is one of the early steps in the process of scientific inquiry. Good preliminary research helps scientists figure out where to focus their efforts and how to proceed with an investigation. In this activity you'll have a chance to practice this very important scientific skill.

?
How do I find research that has already been done on a scientific question?

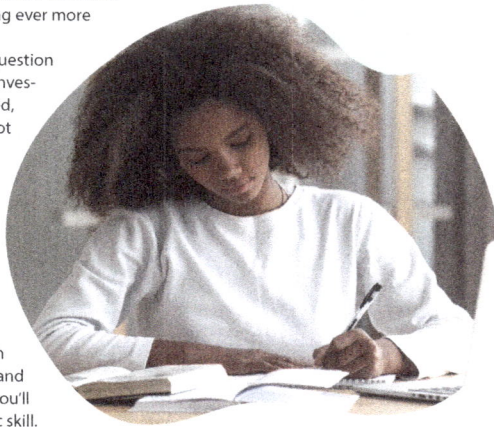

Equipment
computer with internet connection

QUESTION
- How do scientists narrow down the focus of their research?

PROCEDURE

A Think of an endothermic animal that interests you—be specific. Then think of some questions that you could ask about that animal. Do a preliminary internet search on current research to get an idea of what some scientists are presently investigating. Narrow down your list of questions to the one that interests you the most.

1. What animal will you be researching?

   Answers will vary. *Example:* bottlenose dolphin

2. What question will you research?

   Answers will vary. *Example:* How are bottlenose dolphin populations estimated?

- Conduct preliminary research on a particular scientific question.
- Write a testable hypothesis on the basis of preliminary research.

✓ Doing Scientific Research

The goal of this activity is to expose students to the rigor of doing academic research. This necessary science skill may seem daunting to some students, so adjust your instruction and task requirements accordingly. Advanced students may benefit from focusing on a single article and doing a critical assessment of it. Other students may benefit from simply doing targeted internet searches, a skill that is useful for any classroom discipline. Consider walking students through the research process by doing an example in class.

Remind students of the importance of peer-reviewed literature in scientific research. Although the aim of this activity is to expose students to scientific writing of the type found in peer-reviewed journals, such writing may tax the abilities of some students; if so, you may choose to allow them to cite popular science literature as well. Increase or decrease the required number of cited articles according to the abilities of your students.

B Now do an internet search for information related to your specific question.

3. Choose three articles, then cite each one and write a brief summary of the article's topic and findings.

    a. Article: _____

    Summary:

    b. Article: _____

    Summary:

    c. Article: _____

    Summary:

4. Propose a hypothesis on the basis of your research that, if tested, could expand what is already known about the subject under investigation.

    Answers will vary. _____

    _____

5. On the basis of what you have learned from your research, suggest another avenue of investigation for any future scientist who may be interested in the subject.

    Answers will vary. _____

    _____
    _____
    _____

20A LAB

Chill Out!

Inquiring Into the Skin's Ability to Maintain Homeostasis

Have you ever been caught outside when the temperature suddenly fell? Perhaps you've had the unhappy experience of forgetting to bring a coat to a late-autumn football game. As the chill sets in, your body's heat conservation mechanisms activate in an attempt to maintain homeostasis. Your skin is the first line of defense. The blood vessels that supply your skin constrict, reducing the flow of blood to your skin's surface. Your fingers and toes begin to feel numb, and your skin is cold to the touch.

These early effects of cold on the human body, if tended to, are not permanent of course. Wrapping oneself in a thick blanket and enjoying a cup of hot cocoa will quickly put things right. The constricted blood vessels reopen, and soon your fingers and toes feel normal again. It doesn't happen instantly though. It takes a bit of time. But does every part of your body require the same amount of time to recover its normal temperature after being chilled? That's the question you'll explore in this lab activity.

Does chilled skin on different parts of the body recover temperature at the same rate?

LAB 20A OBJECTIVES

- Formulate a hypothesis regarding the ability of different areas of the skin to recover temperature.
- Design an experiment to test a hypothesis.
- Present findings in a scientific paper or other form of presentation.

Equipment

to be determined by your investigation

QUESTION

- Are all parts of the skin equally effective at regulating temperature?

PROCEDURE

Plan and Write Scientific Questions

A Think about how different parts of your body respond after being chilled, then brainstorm with your group to come up with a list of possible questions to investigate. Consider which ones interest you the most. You'll need to take into account whether you have the necessary equipment on hand (or can easily obtain it) and enough time to adequately test your possible hypotheses. Narrow down your list of questions to the one that you will investigate.

1. Which areas of skin will you be comparing?

   Answers will vary.

2. Write a hypothesis about how one area of skin's temperature recovery will compare with another's.

   Answers will vary. Refer to the Teacher Guide on page 218a for additional information.

Designing and Conducting Scientific Investigations

Next, design an experiment to test your hypothesis. Get approval for your experimental design from your teacher before beginning your experiment.

3. Briefly describe your experimental setup.

   Answers will vary. Refer to the Materials section of the Teacher Guide for additional information.

4. What will be your control group?

   Answers will vary.

5. What will be your experimental group?

   Answers will vary.

6. What data will you be collecting?

   Answers will vary.

7. How will you analyze your data?

   Answers will vary.

Set up and run your experiment. Make sure to run several trials to solidify support for your conclusions.

Developing Models

Once you have collected your data, you will need to analyze it by using mathematical tools, such as finding and comparing averages or creating graphs.

Scientific Argumentation

In Lab 15B you learned about scientific writing. Recall that a scientific paper has distinct sections, including an abstract, an introduction, a discussion of materials and methods, the results, a discussion of the meaning of the results (including suggestions for further research), and a list of references. Your teacher will instruct you on whether to report your findings in a formal lab report (a type of scientific paper), an oral report, or a slide presentation.

20A LAB *Teacher Guide*

Chill Out!

Your students learned about the skin's role in regulating temperature on pages 462–63 in the Student Edition. But what the textbook doesn't tell them is whether all parts of the skin are equally effective at regulating temperature.

Lab 20A is an inquiry activity. An inquiry lab activity allows students a degree of freedom, but the teacher must guide the process to achieve the desired educational outcomes. As teachers start using inquiry lab activities, they often struggle with this process of guiding students. The best method is to think through the goals that you want students to achieve and then use questions to guide them to reach those goals. You want to balance students' freedom to investigate the question in the manner they choose with the need to meet your educational goals.

This activity is about the skin's ability to maintain homeostasis. Students will model what actual scientists must regularly do; that is, they will identify a question to ask about the skin's role in temperature regulation and then design a test that will help answer the question.

Inquiry lab activities provide you with a degree of flexibility in deciding how the activity should be carried out. Students will need to collaborate to design a workable test; you may choose to have students do this within their group, with each group deciding on its own question and test, or you can have a class discussion resulting in a single question and test. If the latter, allow students to generate the ideas. Prompt them with questions to guide the discussion in directions beneficial to the activity. You may choose to have students do part of this activity as homework. For example, students may do their research as homework, use class time to collaborate on a test design, conduct their tests at home, and finally present their findings in class.

To get ideas for how this lab activity might be done, do an internet search for skin temperature lab activities.

MATERIALS

Equipment needs for this activity will be determined by the specific hypotheses that students decide to test. Since skin temperature recovery is being tested, students will need to consider several things. First, they will need a reliably repeatable way to chill skin, such as using a cold pack or ice bath. Human skin can take a surprisingly long time to recover its basal temperature after chilling. This fact has been accounted for in the sample procedure provided below—instead of waiting for a complete recovery, we have measured how much recovery takes place during a two-minute interval. Naturally, students will discover this on their own if they try to wait until the skin has fully recovered after its first chill, but that will use up a lot of class time, so you may want to warn them of this in advance.

Second, they will need to be able to measure skin temperature. The best way to do this is by using a digital sensor, such as the one made for use with the Labdisc interface. A thermometer may also be used, in which case students should be cautioned against breakage. Thought needs to be given to the subject of equilibration, that is, how long it takes for a thermometer or probe to reach equilibrium with the substance being measured—skin, in this instance. It may take a thermometer or probe 30–60 seconds to reach equilibrium with chilled skin, during which time students may be surprised to see the skin's temperature continue to "drop." Step C in our sample procedure below has this phenomenon in view.

Lastly, they will need a way to monitor time; probeware will typically have this function built in. Other items that may be useful include first aid tape for securing a sensor or thermometer to the skin, towels for blotting—not wiping, which adds heat due to friction—excess water from the skin if an ice bath is used, and rubbing alcohol or soap and water to remove skin oils, which will help the tape adhere better.

PROCEDURE

Planning/Writing Scientific Questions

For this particular inquiry lab activity the question has already been given: Does chilled skin on different parts of the body recover temperature at the same rate? The question readily lends itself to one of two related hypotheses: (1) one area of skin will recover temperature faster or slower compared to a different area of skin or (2) two areas of skin will recover temperature at the same rate. Make sure that students' hypotheses are testable before students begin designing their experiments.

Designing Scientific Investigations

After students have formulated testable hypotheses, they need to write specific procedures to test them. Make sure that you review their procedures prior to their starting to collect data. Help students think through how they will organize their procedures and data.

Conducting Scientific Investigations

Once the groups have good procedures, allow them to collect data.

20A LAB *Teacher Guide*

Developing Models

Students will analyze their data to determine whether identified trends support their hypotheses. If they are using probeware, such as a Labdisc module, make sure that they follow the manufacturer's instructions for analyzing data or exporting data to an external device.

Scientific Argumentation

Students should state a claim about their procedures and analyses. As always, claims should be supported with specific evidence from their data. This activity provides an excellent opportunity for students to practice writing a formal lab report (see Appendix F), but you may also give them the option to present their findings in the form of an oral report or slide presentation.

SAMPLE PROCEDURE

Equipment

probeware interface with
   external temperature sensor
cold pack
rubbing alcohol
first aid tape
towel

A Wipe a 1 cm square patch of skin on the back of the subject's hand with rubbing alcohol, then use the cold pack to chill the patch of skin for thirty seconds.

B Remove the cold pack and allow the skin to recover for two minutes.

C Again chill the skin for thirty seconds, then use first aid tape to secure the temperature sensor to the subject's skin and begin recording data. In Table 1, record the lowest temperature reading obtained during the sampling interval as the *Initial Temperature* of the skin.

D Record the *Final Temperature* of the skin after two minutes.

E Record the difference between the final and initial temperatures as the *Recovered Temperature*.

F Repeat the test for two additional trials.

G Repeat Steps A–F for the inside of the subject's wrist.

TABLE 1

TRIAL NUMBER	INITIAL TEMPERATURE (°C)	FINAL TEMPERATURE (°C)	RECOVERED TEMPERATURE (°C)
Back of Hand			
1	26.3	27.8	1.5
2	26.0	27.8	1.8
3	26.2	27.5	1.3
Inside of Wrist			
1	24.4	25.0	0.6
2	23.8	24.6	0.8
3	24.0	24.8	0.8

20B LAB

Are You Aware?

A Case Study in Advocacy

?

How should a
Christian view health
awareness issues?

Equipment

LYMPHATIC SYSTEM DISORDERS

As you learned in Chapter 20, lymph is a fluid that fills the spaces in between cells. It is regularly collected via the lymphatic system and returned to the bloodstream. But sometimes this collection system doesn't work the way it's supposed to. In such instances lymph accumulates in one or more parts of the lymphatic system, causing edema (swelling). This condition is known as *lymphedema*. As the disease progresses, the affected part of the body becomes greatly enlarged, with dark and thickened skin. In severe cases limbs become deformed. The affected person may no longer be able to wear normal-fitting clothes or shoes. Lymphedema may hinder the person's ability at work or in enjoyment of favorite activities and hobbies.

No one knows what the exact cause of lymphedema is. It can be inherited, in which case it is known as primary lymphedema. Noninherited, or secondary, lymphedema, is most often seen in cancer patients, especially those who have had radiation therapy or lymph nodes removed as part of their treatment. It is especially common in breast cancer patients.

Lymphedema is not the only lymphatic system disorder, nor is it the most serious. Diseases of the lymphatic system range from swelling of the lymph nodes in response to an infection to serious cancers such as Hodgkin's lymphoma. Do an internet search on lymphatic system disorders. Choose one disorder to research, and then answer the following questions.

QUESTIONS

- What is lymphedema?
- What is advocacy?
- How can Christians decide where help is needed most?

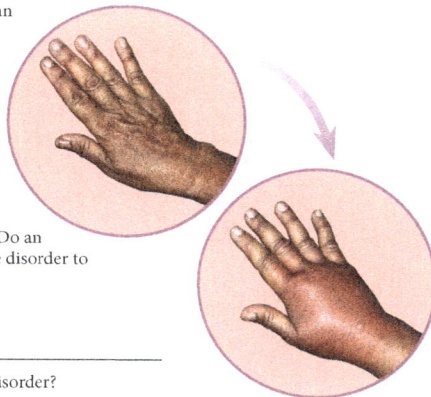

1. What lymphatic system disorder did you research?

   See margin for answer.

2. What part of the lymphatic system is affected by this disorder?

   Answers will vary.

3. What are the symptoms of this disorder?

   Answers will vary.

LAB 20B OBJECTIVES

- Give a biblical justification for advancing the study and treatment of diseases.
- Identify some of the characteristics and limitations of health awareness campaigns.
- Prioritize the giving of time and money to charitable causes on the basis of a biblical worldview.

✓ ### Introduction to Lab 20B

This lab activity touches on the potentially sensitive subject of advocacy and health awareness. It is quite likely that one or more of your students has had or knows someone who has had a disease that is the focus of a particular advocacy group, or perhaps they have volunteered for such a group. It is important to emphasize that this activity is not about whether such advocacy is good or bad but to get students to think critically about the topic from a Christian perspective.

Question 1 Answer
Answers will vary. In addition to those mentioned in the Lab Manual, examples include Castleman's disease, lymphadenitis, and lymphocytosis.

4. What treatments are used for this disorder, if any?

Answers will vary.

5. What are some current areas of research with regard to this disorder?

Answers will vary.

✝ ✓ *Advocating for Others*

This portion of the exercise is well suited to a roundtable discussion format. The goal here is to get students to recognize that physical needs versus spiritual needs is not an either-or proposition. Jesus demonstrated love and compassion both by preaching the good news of redemption and by healing the sick and infirm. Christians should therefore also be concerned with meeting people's physical as well as spiritual needs. However, since a body is temporal while the spirit is eternal, ministering to physical needs should be accompanied by addressing spiritual needs.

Question 6 Answer

Obeying the Creation Mandate includes striving to exercise dominion over diseases and disorders. One of the signs that Jesus was the Messiah was His healing ministry. Pushing back against consequences of the Fall, like diseases, is part of the Christian mission. The second great commandment to love our neighbor as ourselves means that treating diseases and searching for cures also demonstrates godly concern and compassion for those who have been afflicted.

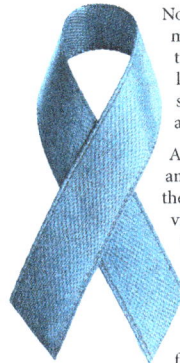

A blue ribbon is the symbol for lymphedema awareness.

A LOOK AT ADVOCACY

No one knows exactly how many people are affected by lymphedema, but many experts believe that the disease is underreported. Some estimates put the number of cases at as many as 250 million worldwide. Despite its prevalence, there's a good chance that you've never heard of this disease. That shouldn't be too surprising. After all, there is a huge number of diseases and disorders out there!

And therein lies a problem. There are thousands of different kinds of diseases and disorders, and hundreds of millions of people are affected by them, but the resources for research into finding cures are not unlimited. How can the victims of these diseases and the people who treat them obtain the resources they need to combat the disease? One way to do this is through advocacy.

Advocacy is a concerted effort to promote support for a particular cause. It's likely that you've seen a pink ribbon sticker on a car or that your favorite sports team has a day during their season when they wear pink as part of their uniform. These symbols are part of the effort to raise awareness of and financial support for breast cancer research and treatment. This effort has recruited many thousands of volunteers and raised many millions of dollars. Yet the issue of advocacy raises some interesting questions. Think about the following questions and be prepared to share your responses with your classmates.

6. Why is it worthwhile to spend money on finding cures for diseases?

See margin for answer.

7. As of 2015 there were 212 health awareness days, weeks, or months recognized by the United States Federal Government, such as National Folic Acid Awareness Week. Do you think that this is an effective strategy for raising health awareness? Explain.

Answers will vary.

8. Why do you think Congress has decreed so many health awareness dates?

   Answers will vary.

9. What might be some of the weaknesses in an awareness campaign for raising funds and support for treating disease?

   Awareness campaigns do not allocate donor resources according to areas of greatest need. The campaigns that are the best organized and most vocal raise the most support, without regard to whether the health issue at stake is one of critical importance.

10. Read Mark 1:29–39. What does this passage suggest about how Christians should balance health awareness and other Christian activity?

   See margin for answer.

11. What is the best way for Christians to wisely use their money to promote health-based charity?

   Christians should carefully research any charitable organization to make sure that the funds it receives are being used in a wise and God-honoring manner. Christians should also prayerfully seek God's guidance about which organizations to support.

GOING FURTHER

In 2016 the US Senate officially recognized World Lymphedema Day, which is observed in early March of each year. With so many health awareness days being set aside each year, there are probably several that you have never heard of before. Do an internet search on national health observances and find a health awareness day that interests you but with which you are unfamiliar. Find out what health issue the day targets and learn a little about it. Discover what organizations promote the day and what sorts of activities they organize in order to increase public awareness. Summarize your findings in a paragraph or two and be prepared to share your findings with your classmates.

Question 10 Answer
The first priority of the church is to fulfill Christ's Great Commission, which is to go and make disciples of all nations. Curing diseases is good, but mankind's greatest need is to hear about God's cure for sin. Without redemption in Christ, a healthy body is of no eternal benefit. The best humanitarian aid also helps people spiritually by linking meeting their physical needs with their spiritual ones.

21A LAB

Dry Bones

Exploring the Skeletal System

People who have experienced one know that a broken bone is no fun. Although broken bones are not necessarily uncommon, the fact that they do not occur more frequently is a testament to God's marvelous design. It takes a lot of force in the right direction to break a bone. The bones, along with the ligaments, joints, and cartilage, provide the support that the body needs. Today you will have an opportunity to study the skeleton at both the microscopic and the macroscopic levels.

How do the parts of the skeletal system work together to support the human body?

Equipment

microscope
preserved slides of dry ground bone, c.s.

model of a human skeleton
skeleton diagrams

PROCEDURE

Microstructures

A Observe the preserved slide of dry ground bone, c.s., and draw a diagram of a complete Haversian system in Drawing Area A. Label your diagram. Include the Haversian canal, lamellae, lacunae, canals, and any other structures observed in the preserved specimen.

Macrostructures

B Memorize the bone names and markings indicated on page 478 of your textbook.

C Locate and label these bones on the models of the skeletons on page 225 of this lab activity.

D Memorize the types of joints on page 479 of your textbook. Fill in the missing information in Table 1.

1. How does the structure of the bones show God's care for His creatures?

The bones give structure to the body. The microstructure ensures that the bones remain strong.

QUESTIONS

- What is the microstructure of a typical bone?
- What are the bones of the skeletal system?
- What are the sesamoid bones?

LAB 21A OBJECTIVES

- Identify the microstructures and macrostructures of the skeletal system.
- Name the various bones and joints of the human body.

Introduction to Lab 21A

The main reason for this lab activity is to give students experience in using anatomical terms properly. These terms are often used by students but are not always understood. Through the activities of labeling diagrams, completing charts, and researching interesting structures, students should learn some of the material that is presented in this unit without having to work at memorizing it.

Options

If time becomes a factor, consider assigning the sesamoid bone section in Going Further (next page) as extra credit.

You may choose to adjust the number of required memorized bone names according to the needs of your students.

Equipment Notes

Dry ground bone is bone that was dried before it was mounted for microscopic viewing. Thus, although the Haversian canal is present, the Haversian blood vessel is not. The lacunae are there but not the osteocytes. Tell students that they are looking at bone that was dead and dried before it was made into a slide. This information should give them a clue as to what structures they will and will not see in a dry ground bone slide.

Quality plastic, life-sized skeletons are as effective in high-school classrooms as real ones. They usually cost less than natural ones and are considerably more durable, making them practical for student use.

Wall charts of the bones of the human body may be used instead of a skeleton. You may be able to obtain these from your physician. Charts such as these are often given away by pharmaceutical sales representatives or surgical equipment distributors.

Divine Engineering (Question 1)

The human skeleton is another awe-inspiring example of God's supreme engineering talents. When we consider how much use we make of this multifunction marvel along with how much wear and tear that usage entails, it is astonishing that it can work as well as it does for so long a time.

✓ Identifying Macrostructures

Students will not be able to copy from the Student Edition all the information needed to fill out the Joints of the Human Body table (p. 226) in this lab activity. Tell them that they will need to think carefully in order to come up with some of the answers.

GOING FURTHER

Do an internet search on sesamoid bones, and then answer the following questions.

2. What are sesamoid bones?

   Sesamoid bones are bones that are not present at birth but develop later in the tendons near the joints.

3. How do sesamoid bones develop?

   Usually friction and pressure on a joint cause sesamoid bones to develop.

4. List some examples of sesamoid bones.

   the kneecap (patella), extra bones in the joints of the fingers and toes

DRAWING AREA A

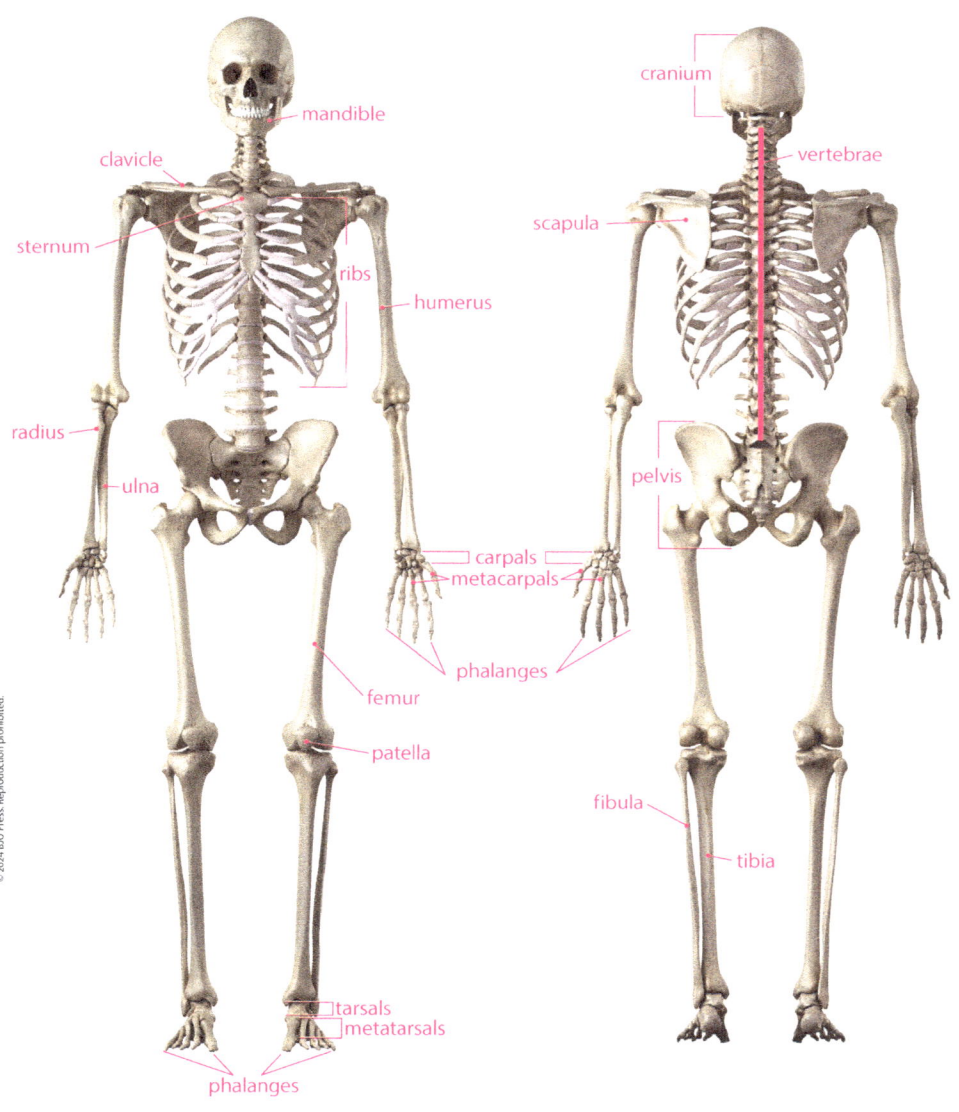

mandible

clavicle

sternum

ribs

humerus

radius

ulna

carpals

metacarpals

phalanges

femur

patella

tarsals

metatarsals

phalanges

cranium

vertebrae

scapula

pelvis

fibula

tibia

TABLE 1 Joints of the Human Body

CATEGORY OF JOINT	TYPE OF JOINT	DESCRIPTION OF MOVEMENT	EXAMPLES
Movable	ball and socket	rotation and free movement in all directions	shoulder, hip
	hinge	bend in one direction	elbow, knee, phalanges
	pivot	rotation	the two vertebrae at the base of the skull
	gliding	limited lateral and vertical movement	wrists, ankles
Slightly Movable	slightly movable	bending, twisting, and slight compression	between vertebrae
Immovable	suture	no movement (because two bones have grown together)	skull sutures

21B LAB

I'm So Tired!

Investigating Muscular Function

We rely on muscles to allow us to move, but our muscles can take only so much. Sooner or later, a muscle used too much can move no more and must rest. This is called *muscle fatigue*. Although the mechanisms of muscle fatigue are not completely understood, it is known that people such as athletes, who use their muscles often, are able to use their muscles longer before getting muscle fatigue.

In this lab activity you will be experimenting with muscle fatigue and some of the factors that can improve or hurt muscle performance.

What factors affect muscular function?

Equipment

plastic jug, 1 gal, filled with water
stopwatch
hand gripper
dishpan
cold water, approximately 10 °C

PROCEDURE

Muscle Fatigue in the Dominant and Nondominant Arms

A Extend your dominant arm parallel to the ground while holding the gallon jug of water.

B Have your lab partner time how long you can keep your arm in its parallel position.

C As soon as your arm drops out of its parallel position, have your lab partner note the time and record it in Table 1.

D Repeat Steps A through C with your nondominant arm.

E Repeat Steps A through D an additional four times.

F Using a graphing calculator or spreadsheet, calculate and record the average and standard deviation of your data for each arm in Table 1.

G Graph your results from Step F on Number Line 1. Since the time that a person can hold up a jug of water can vary quite a bit, you will create your own scale for the number line.

1. Considering the data in your number line, can you say that one arm was stronger than the other? Explain.

Answers will vary. The data in the sample graph is inconclusive since the standard deviations overlap.

QUESTIONS

- What is muscle fatigue?
- How does muscle fatigue affect muscular function?
- How does temperature affect muscular function?

LAB 21B OBJECTIVES

- Explain muscle fatigue.
- Explain how muscle fatigue affects muscular function.
- Describe how temperature affects muscular function.

Equipment Notes

Any type of commonly available hand grippers will work for this lab activity.

Standard Deviation

You might need to explain how to graph on a number line the means and standard deviations of two data sets. The standard deviation of a set of data can be obtained quite easily with a spreadsheet or a graphing calculator. Since students may not be familiar with standard deviation, you may have to instruct them in how to use a spreadsheet or their calculators to determine it.

Answers will vary. Most students will probably find that their dominant arm is stronger.

2. If one of your arms was stronger than the other, was your dominant or non-dominant arm stronger?

   See margin for answer.

3. Suggest a possible reason why you observed the results that you reported in Questions 1 and 2.

   Answers will vary. Students who reported that their dominant arm was stronger may suggest that they use their dominant arm more than their nondominant arm, resulting in stronger muscles in their dominant arm.

Muscle Fatigue and Muscle Strength

H Count the number of times that you can squeeze a hand gripper in 30 seconds and record the number in Table 2.

I Using the same arm, repeat the experiment an additional nine times with as little time as possible between each trial.

J Graph the results with a scatterplot in Graphing Area A. Create a line of best fit for your data.

4. Did your ability to squeeze the hand gripper change over time? If so, how did it change?

   Answers will vary. Almost all students will find that their ability to squeeze the hand gripper will decrease over time until it eventually levels off.

5. Do some research on exercising. How can a person develop the ability to work a muscle for a longer period of time without experiencing muscle fatigue?

   Answers will vary. Consistently exercising a muscle results in it becoming stronger, allowing it to work longer before becoming fatigued.

Temperature and Muscle Fatigue

K After your hand has rested, squeeze the hand gripper as rapidly as possible for 30 seconds and record the number of squeezes in the *Warm Arm* column of Table 3.

L Repeat Step K an additional four times.

M After your hand has rested, submerge your forearm in cold water for 1 minute. Then, using the same arm, squeeze the hand gripper as rapidly as possible for 30 seconds and record the number of squeezes in the *Cold Arm* column of Table 3.

N Repeat Step M an additional four times.

O Find the average for both the warm arm and the cold arm. Record these averages in Table 3.

Cold Water

To avoid injury, the cold water should not be much colder than 10 °C.

P Graph the results using a bar graph in Graphing Area B.

6. Why do you think you were instructed to chill your forearm instead of your hand?

   The muscles that control the fingers, and therefore grip strength,

   are located in the forearm, not in the hand.

7. On the basis of your data and your graph, do you think that there is any meaningful difference between the time it took for muscle fatigue to set in for a warm muscle compared with a cold muscle?

   Answers will vary. Most students are not likely to notice a

   significant difference.

8. On the basis of your data, suggest a relationship between muscle temperature and fatigue.

   Answers will vary. Students will typically observe little or no

   difference in the rate at which fatigue sets in for a warm versus a

   cold arm, suggesting that muscle temperature is not a significant

   factor under the conditions being tested.

9. Stretching is important before exercising. Do some research on why it is important to perform a warmup routine before stretching. Write two or three sentences on what you find.

   Answers will vary. Stretching before warming up can lead to

   torn muscles. Warming up increases blood flow and makes your

   muscles more flexible.

10. What are some fields of work that require knowledge of muscle function and muscle fatigue?

    Examples: coaching, athletics, athletic training, physical therapy,

    law enforcement, emergency response, the military

11. Choose one of the fields from your answer to Question 10 and write a paragraph on how a person in that field can use knowledge about muscle function and muscle fatigue to show love to other people.

    Answers will vary.

Sample Data

The data shown on this page and graphed on the next page is typical. Students' results may be different.

TABLE 1 Muscle Fatigue in the Dominant and Nondominant Arms

TRIAL #	DOMINANT ARM (time in seconds)	NONDOMINANT ARM (time in seconds)
1	23	9
2	24	22
3	11	5
4	25	19
5	13	5
Average	19	12
Standard Deviation	5.9	7.2

NUMBER LINE 1

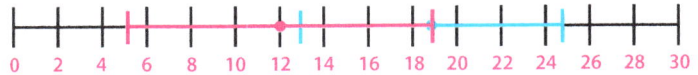

TABLE 2 Muscle Fatigue and Muscle Strength

TRIAL #	NUMBER OF SQUEEZES IN 30 s
1	49
2	31
3	26
4	22
5	20
6	19
7	21
8	20
9	24
10	23

TABLE 3 Temperature and Muscle Fatigue

TRIAL #	WARM ARM (number of squeezes)	COLD ARM (number of squeezes)
1	65	62
2	56	57
3	56	53
4	57	49
5	56	56
Average	58	55

Gripper Squeezes vs. Trial Number

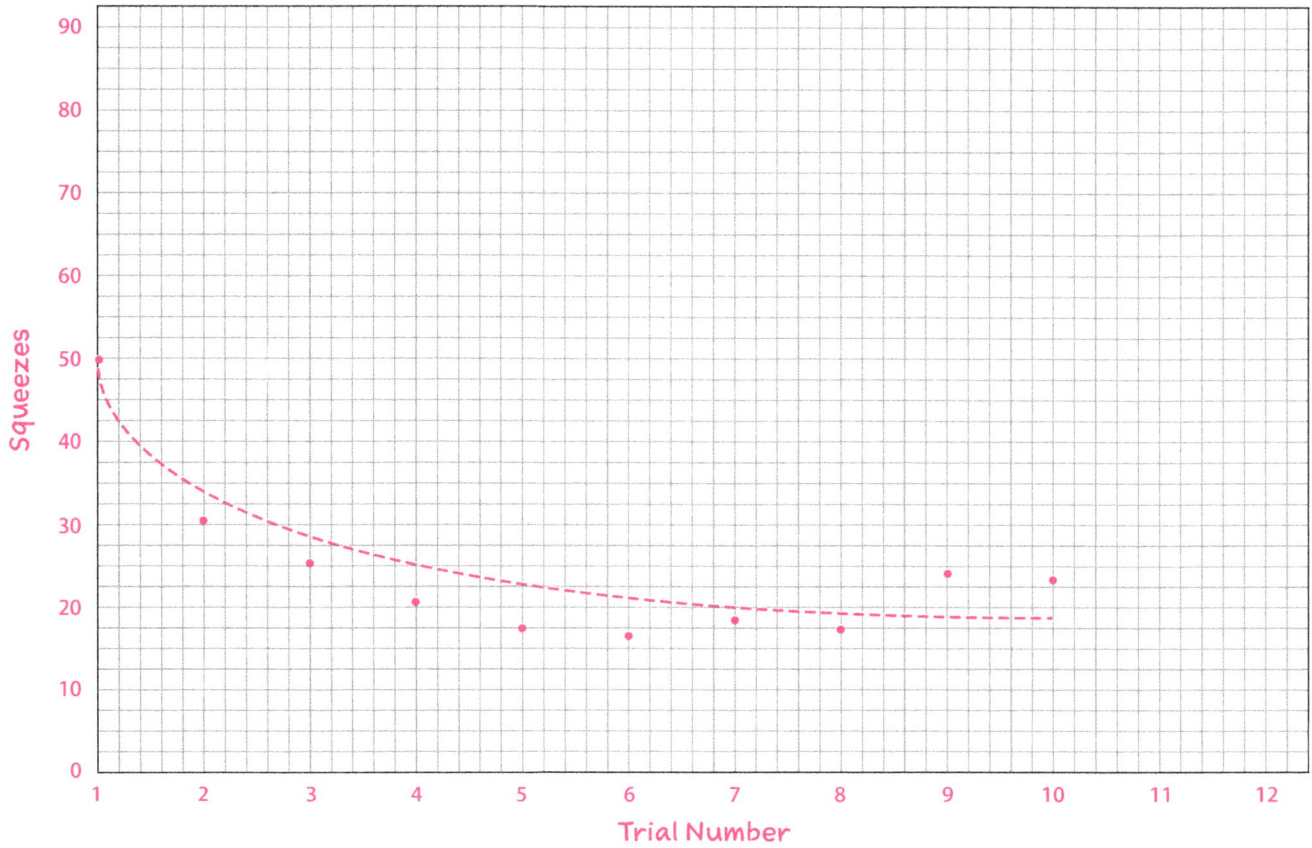

Squeezes

Trial Number

Gripper Squeezes vs. Temperature

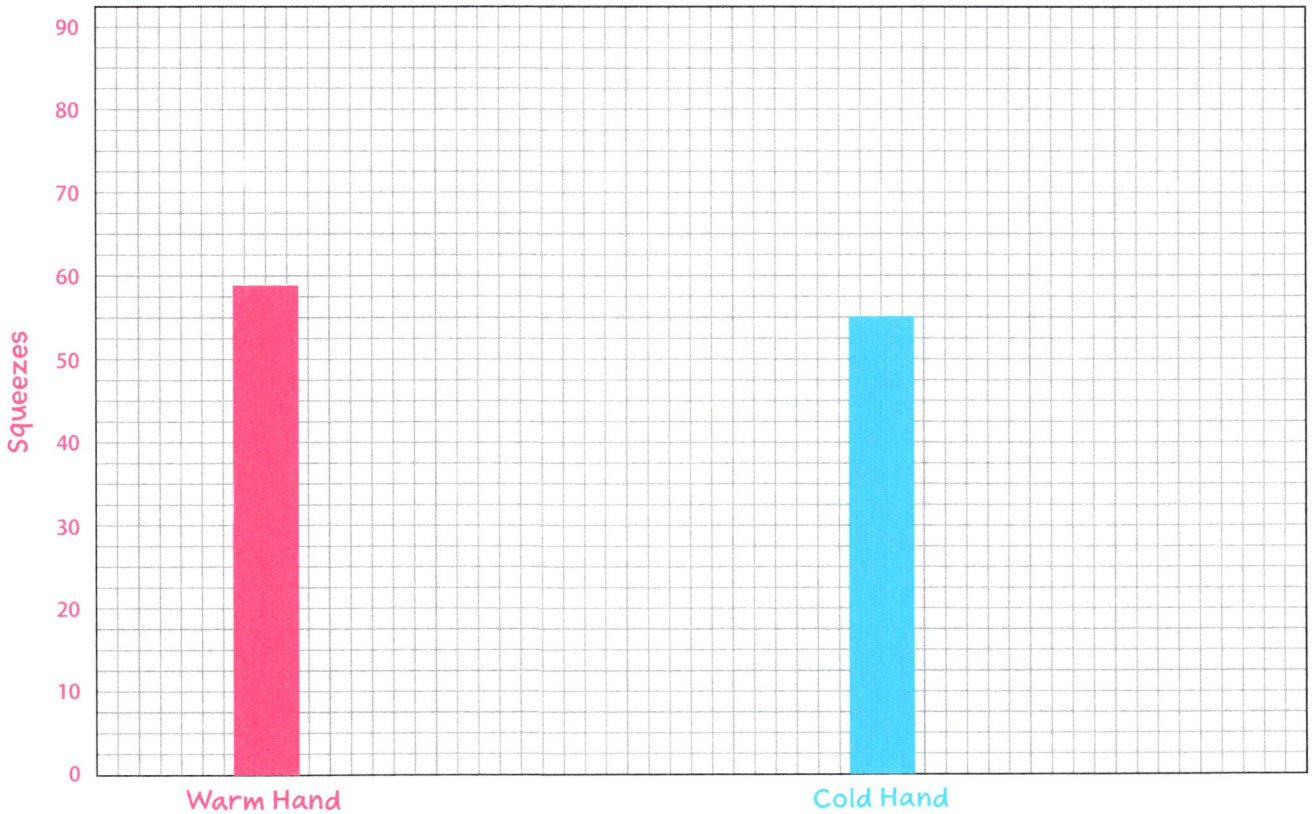

Squeezes

Warm Hand   Cold Hand

22A LAB

Relax and Take a Deep Breath

Exploring the Human Respiratory System

We've all been told to relax and take a deep breath at one time or another, usually right before we have to do something unpleasant or stressful—like share a lab report with our classmates! But exactly how deep a breath can a person take? And does your friend's deep breath take in the same amount of air as your deep breath? Is there even a way to tell? It turns out there is, and looking at breath volume is one of the things that you'll do during this activity.

We don't give too much thought to breathing. It's an involuntary activity, one that doesn't require conscious thought to control. That's a good thing—otherwise it might be difficult to sleep at night! Your brain's respiratory centers watch over every breath you take—all 600,000,000 or more over the course of an average lifetime. While those respiratory centers go about their business, let's take a look at the respiratory system.

How much air can an average person inhale in one breath?

❓

Equipment

lung volume bag
mouthpieces (3)
rubber band

stethoscope
isopropyl (rubbing) alcohol
tissues

paper towels
☠

QUESTIONS

• What is an average lung capacity?

• What are the structures of the respiratory system?

PROCEDURE

Measuring Tidal Volume

A Prepare the lung volume bag. Select one person from your group to be tested and another to act as assistant.

» Insert the mouthpiece partway into the open end of the lung volume bag and secure it with a rubber band.

» Have the assistant sit down.

» Slide the bag slowly across the assistant's knee while the assistant presses the bag with a paper towel. This will remove all air from the lung volume bag.

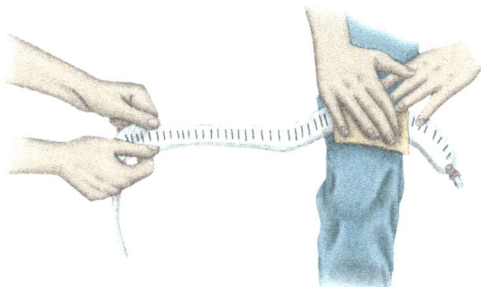

LAB 22A OBJECTIVES

• Measure lung capacity.

• Differentiate between tidal volume, reserve volume, and vital capacity.

• Describe the sounds made by different parts of the respiratory system.

• Identify the structures of the respiratory system.

✔ Introduction to Lab 22A

The first section of this lab activity can be done as a demonstration that takes about fifteen to twenty minutes. Various lung volume tests will supply the data to use in answering the questions.

This lab activity includes an exercise in using a stethoscope for listening to breathing sounds and to the voice. Following these activities is an exercise that deals with the structures of the respiratory system. This can be done outside of class.

⚗ Equipment Notes

A spirometer is a good piece of equipment to have, but it is very expensive. If you have access to one, you can modify the lung volume experiments to use it. A spirometer sensor is also available for use with some probeware systems.

Lung volume bags and kits are available from science equipment suppliers. Their instructions may differ slightly from those given here. Kits are typically intended for use with groups of three to four students students. If cost is an issue, consider purchasing one kit and selecting students to perform the activity as a demonstration for the rest of the class.

Lung volume bags are not usually designed for all the lung capacity measurements made in this exercise. Thus, some of the experimental results will not be quite as accurate as they would be if a spirometer and different techniques were used, but they will be sufficiently accurate for this activity.

Breathe Normally!

Demonstrate how to use the lung volume bag before you ask for volunteer "breathers" and "assistants."

Because students know that they are being tested, they often exaggerate their breathing. Emphasize that this is not a contest and that the data should be as representative of normal breathing as possible. Don't be surprised, however, if your students have "normal" breath volumes that are several times larger than average.

Students with respiratory issues may need accommodations for this lab activity.

More Data Yields Better Conclusions

Students will obtain more meaningful averages if you compile data for the entire class on the board and have them calculate their averages from the larger data set.

B The person being tested should breathe in a normal breath, pinch the nose, put the lung volume bag mouthpiece in the mouth, and breathe out a normal breath. This normal breath is known as the *tidal volume* of the lungs.

C The assistant should take the bag and hold it closed while sliding it across the knee and pressing it with a paper towel in order to force all the air to the closed end. Note how many liters of air are in the bag and record the data in the Tidal Volume section of Table 1.

D Empty the lung volume bag, using the procedure described at the end of Step A.

E Repeat Steps B through D for two more trials.

F Using new mouthpieces, repeat Steps B through E for the other two students. Students should hold on to their mouthpieces.

Measuring Expiratory Reserve Volume

G Prepare the lung volume bag using the first student's mouthpiece.

H The first student should next breathe in a normal breath, breathe out a normal breath, pinch the nose, put the lung volume bag mouthpiece in the mouth, and breathe out as much as possible. This extra bit of breath is known as the *expiratory reserve volume*.

I Using the procedure described in Step B, measure the air and record the volume in the Expiratory Reserve Volume section of Table 1.

J Repeat Steps G–I for two more trials.

K Repeat Steps G–J for the other two students. Make sure that students are using their own mouthpieces for their trials.

Measuring Vital Capacity

L Prepare the lung volume bag using the first student's mouthpiece.

M The person being tested should now breathe in as much as possible, pinch the nose, put the lung volume bag mouthpiece in the mouth, and breathe out as much as possible. This maximum-sized breath measures the lungs' *vital capacity*.

N Using the procedure described in Step B, measure the air and record the volume in the Vital Capacity section of Table 1.

O Repeat Steps L–N for two more trials.

P Repeat Steps L–O for the other two students.

Computing Average Lung Capacities

Q Determine the average for each of the volumes.

1. What is the average tidal volume for the students tested?

   approximately 500 mL

2. What is the average expiratory reserve volume?

approximately 1000 mL

3. What is the average vital capacity?

approximately 4500 mL

The additional air that can be forcibly inhaled after the inspiration of a normal tidal volume is known as the *inspiratory reserve volume.*

4. Using the answers to Questions 1–3, determine the average value for the inspiratory reserve volume for the students tested.

approximately 3000 mL

5. Assuming a residual volume—the amount of air left in the lungs after a person has forced as much out as possible—of 1000 mL, what would be the average *total lung capacity* of the people tested?

approximately 5500 mL

6. According to the information on page 496 of your textbook, are these lung volume results considered average? If not, what factors might account for the difference?

Answers will vary. A person's size, athletic ability, and degree of physical fitness, as well as diseases and disorders such as asthma, can all affect lung volume. Also, the testing procedure is not completely accurate.

Listening to Your Lungs

Air rushing into and out of healthy respiratory structures makes various sounds. Many respiratory problems cause abnormal sounds that a physician can hear using a stethoscope. Hopefully, you will not hear abnormal respiratory sounds when listening to your lab partner's breathing! But it can be interesting to listen to normal sounds of the respiratory system.

R Using a tissue and alcohol, clean the earpieces of the stethoscope.

S Place the stethoscope in your ears, allowing the tubes to hang freely and being careful not to hit the diaphragm on hard objects. (The noise can be very loud.)

T Place the diaphragm of the stethoscope on your lab partner's back and press lightly as he or she breathes normally.

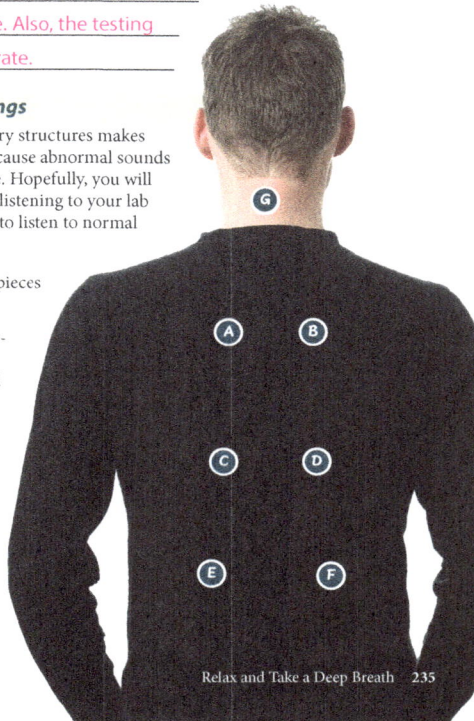

Calculations

Students may need some help arriving at the answers for Questions 4 and 5. The inspiratory reserve is equal to the vital capacity minus the tidal volume and expiratory reserve. The total lung capacity is the sum of all the various capacities.

7. Listen to a normal breath or two in Areas A–F as labeled on page 235. Describe the sounds you hear.

Answers will vary.

8. Are the sounds you hear different in different areas? Explain.

Answers will vary.

9. Listen to a deep breath in the same areas. Do you hear a difference? If so, describe the difference and tell what may account for it.

There should be little difference in a normal, healthy person.

10. Listen to Areas A and E while your partner coughs. Describe what you hear.

Answers will vary.

11. Listen to Areas A and E while your partner talks. Describe what you hear.

Answers will vary. Students should hear their partner's voice but with a different timbre.

U Now listen to your lab partner's breathing by placing the stethoscope in Area G. Listen to your lab partner's voice in the same area.

12. Describe what you hear.

Answers will vary. Students should hear little while their partners are breathing and should hear their voices but with a different timbre when they are speaking.

Structures of the Respiratory System

Label the diagram below as completely as possible. Under each label indicate the function of that structure (enclose this information in parentheses).

13. In the Greek New Testament the word *pneuma* is used for both the Holy Spirit and for breath. How are the work of the wind (breath) and the Holy Spirit compared in John 3:5–8?

Like wind, the Holy Spirit is invisible, but His work is undeniable. Wind moves trees and other objects even though it is invisible. The Spirit regenerates—that is, makes a person born of the Spirit or born again—and changes people even though He cannot be seen. ("Spirit" at the end of verse 6 and "wind" at the beginning of verse 8 are the same word: *pneuma*.)

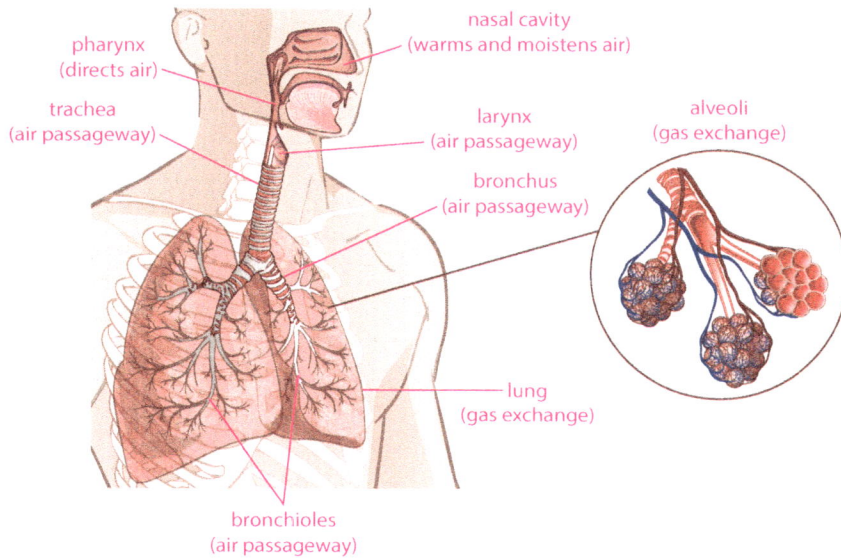

pharynx
(directs air)

trachea
(air passageway)

nasal cavity
(warms and moistens air)

larynx
(air passageway)

bronchus
(air passageway)

alveoli
(gas exchange)

lung
(gas exchange)

bronchioles
(air passageway)

TABLE 1 Lung Volume

PERSON A	PERSON B	PERSON C	AVERAGE
Tidal Volume			
1. _____ 2. _____ 3. _____	1. _____ 2. _____ 3. _____	1. _____ 2. _____ 3. _____	_____ mL or _____ L
Expiratory Reserve Volume			
1. _____ 2. _____ 3. _____	1. _____ 2. _____ 3. _____	1. _____ 2. _____ 3. _____	_____ mL or _____ L
Vital Capacity			
1. _____ 2. _____ 3. _____	1. _____ 2. _____ 3. _____	1. _____ 2. _____ 3. _____	_____ mL or _____ L

22B LAB

Feeling the Pressure

Investigating Blood Pressure and Hypertension

Shelana was a 36-year-old mother of four. She had spent a busy day planning for her daughter's tenth birthday, in spite of having had a terrible headache for the previous two weeks. Shelana suffered from migraines, so she hadn't thought too much about the headache, but that night over dinner she mentioned her intention to see her doctor the next day.

As Shelana was drifting off to sleep later that night, she felt what she described as a "weird" sensation. She tried to tell her husband about it, but her speech was slurred and incoherent. Her husband rushed her to an emergency room. After a night in the hospital and an MRA (a type of scan that provides pictures of blood vessels), a diagnosis was made. Shelana had suffered two ischemic strokes. Two weeks in intensive care and many months of physical therapy followed. Even two-and-a-half years later she still did not have full use of the right side of her body.

A stroke occurs when blood flow to the brain is reduced, either due to a clot (an ischemic stroke) or bleeding (a hemorrhagic stroke). In 2021 strokes were the fifth-leading cause of death in the United States. We sometimes think of a stroke as something that happens to older people, but as Shelana's case shows, a stroke can happen to anyone.

A *risk factor* is anything in a person's life that increases the risk of having a particular medical condition. The greatest risk factor for stroke is high blood pressure, or hypertension. Blood pressure measurement shows how much pressure is exerted by blood on the walls of a person's blood vessels. Let's take a closer look at this vital topic.

How can I lower my risk of having a stroke?

QUESTIONS
- What is blood pressure?
- What is hypertension?
- What are some health risks associated with high blood pressure?

Equipment
digital blood pressure cuff

PROCEDURE

Measuring Blood Pressure

To obtain a blood pressure measurement, follow the instructions provided with the digital blood pressure cuff given to you by your teacher. Blood pressure measurements are generally taken in a relaxed, seated position. At the conclusion of the test, the digital blood pressure cuff will display two numbers. The larger of the two numbers is the *systolic pressure*, the pressure within your arteries when your heart is contracting. This pressure is measured in millimeters of mercury (mmHg).

Feeling the Pressure   **239**

LAB 22B OBJECTIVES

- Take a blood pressure reading.
- Explain what the numbers in a blood pressure reading indicate.
- Explain the relationship between hypertension and stroke.
- List some risk factors for hypertension.
- Explain why Christians should be concerned about maintaining healthy blood pressure.

✓ Introduction to Lab 22B

Hypertension and stroke are serious medical issues in the United States. It is quite likely that one or more of your students have hypertension, have risk factors associated with hypertension, or know someone living with hypertension. They may also know someone who has had a stroke. As for any personal health issue, a good measure of tact is called for when discussing the subject with students. At the same time, though, raising awareness of issues like hypertension and stroke at this age might convince a student to make healthier lifestyle choices.

✓ Prelab Discussion

Consider polling the class for how many students know someone who has had a stroke. Invite students to share any personal experiences they may have had in coping with the effects of stroke in a family member.

⚗ Equipment Notes

Digital blood pressure cuffs, available at your local pharmacy, are inexpensive and much easier to use than manually operated sphygmomanometers. Two types are generally available, one that is used on the upper arm and another that is used on the wrist. The arm-style cuffs are usually more accurate and reliable. While digital blood pressure cuffs are generally similar in operation, differences between brands and models may exist. Make sure that you are familiar with your cuff before doing the activity, and consider demonstrating its use before students try using it themselves. This activity can also be done with a traditional sphygmomanometer.

Question 3 Answer

Although it is possible to get three identical measurements, it is far more typical to get three slightly varying measurements.

A Obtain a blood pressure measurement.

1. What is your systolic pressure?

   Answers will vary.

The smaller number is your *diastolic pressure*, the pressure within your arteries when your heart is between beats.

2. What is your diastolic pressure?

   Answers will vary.

B Obtain two additional blood pressure measurements.

3. Are the values for the second and third measurements the same as for the first measurement?

   See margin for answer.

4. Suggest a way in which to report a more reliable blood pressure measurement.

   Answers will vary. Students may suggest taking several readings

   and reporting the average.

Blood pressure measurements are categorized according to their indications for hypertension. The following chart reflects categories defined by the American Heart Association.

TABLE 1 Blood Pressure

BLOOD PRESSURE CATEGORY	SYSTOLIC (mmHg)		DIASTOLIC (mmHg)
Normal	less than 120	and	less than 80
Elevated	120–129	and	less than 80
High Blood Pressure Stage 1 Hypertension	130–139	or	80–89
High Blood Pressure Stage 2 Hypertension	140–180	or	90–120
Hypertension Crisis (emergency care needed)	higher than 180	and/or	higher than 120

5. Using the information in Table 1, indicate which blood pressure category you are in.

   Answers will vary.

6. What are some reasons why that category might not accurately describe your current blood pressure health?

Answers will vary. Students may suggest aspects of the test that were done incorrectly, doubts about the skill of the user, doubts about the accuracy of the equipment, physical or emotional factors that may have influenced the measurement, or the need to measure blood pressure over a period of time rather than to take just a single measurement.

Assessing Risk for High Blood Pressure

As we saw in the opener, having high blood pressure increases a person's chances for having a stroke. Do an internet search for high blood pressure risk factors, and then answer the following questions.

7. What are some risk factors for high blood pressure?

Answers will vary. *Examples:* age, ethnicity, family history, obesity, physical inactivity, tobacco use, excess salt intake, potassium deficiency, vitamin D deficiency, consumption of alcohol, stress, and certain chronic conditions such as kidney disease, diabetes, and sleep apnea

8. Can the cause of a person's high blood pressure always be determined? Explain.

No. Many people have high blood pressure without having any known risk factors.

9. Besides stroke, what are some other health risks associated with high blood pressure?

Answers will vary. *Examples:* increased risk of aneurysms, heart disease, kidney disease, impaired cognitive ability, and dementia

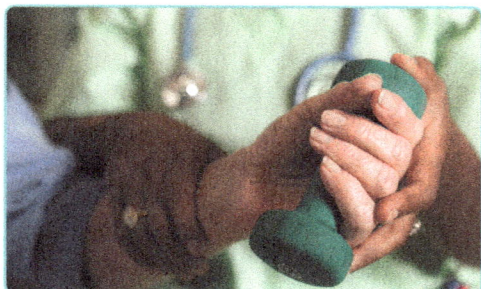

A severe stroke victim faces the possibility of a very long recovery period, perhaps including the need to relearn fine motor skills.

Elevated Readings

If any students are worried about their blood pressure readings, encourage them to discuss this with a parent or guardian.

✝ Caring for Our Bodies

A body isn't something we wind up with by chance, and it's not the end product of a long evolutionary process. It is a marvelous gift from God, something that allows us to perceive the rest of creation and to operate within it. Our bodies also allow us to serve God and demonstrate His love to others. Both gratitude to God and consideration of our usefulness for service suggest that we should give some attention to the care of our bodies.

10. What are some lifestyle choices that you can make right now to reduce your risk for high blood pressure?

    Answers will vary. *Examples:* making dietary changes, beginning an exercise program, reducing or learning to manage stress

11. Why should Christians be concerned about maintaining healthy blood pressure and reducing risks?

    Caring for our bodies demonstrates gratitude toward God for what He has given us. It can also be considered a good witness to others as well as a way of keeping us physically fit for God's service.

23A LAB

Calorimetry in a Can
Measuring the Energy in Food

It's break time at your school and time for a snack. But among the seemingly infinite variety of snack foods available today, which one will provide you with the biggest energy boost? Let's find out!

Which snack food delivers the most energy?

Equipment

laboratory balance (accurate to 0.01 g)
evaporating dish, 100 mL
ring stand
graduated cylinder, 50 mL
thermometer or temperature probe
iron ring
cork stopper

large straight pin
glass stirring rod
empty soda can with attached tab
grill lighter
tongs or mitts
various snack foods with nutrition facts labels
laboratory apron
goggles

QUESTIONS

- How is a soda can calorimeter used to determine the energy content of food?
- Which snack foods contain the most energy?
- How accurate is a soda can calorimeter?

PROCEDURE

Calorimetry is the technique that we'll use to determine the amount of energy in a sample of your favorite snack. A calorimeter burns a small sample of food to heat a known quantity of water. The resulting temperature change of the water is then used to calculate the energy released by the burning sample.

A Create a food holder by pushing the straight pin through the cork stopper as shown on the right. Push the pin through the center of the cork, if possible; otherwise, push it through from one side.

B Secure a small piece of the first snack food item to be tested on the point of the pin, then place the food holder on the evaporating dish.

C Use the balance to determine the mass of the evaporating dish, food holder, and food sample. Record this as the initial mass in Table 1.

D Place the evaporating dish on the base of the ring stand.

E Use the graduated cylinder to add 50.0 mL of water to the soda can.

F Measure the temperature of the water to the nearest 0.1 °C and record the result in Table 1.

- Calculate the energy content of snack foods using data obtained from a soda can calorimeter.
- Compare the energy densities of different snack foods.
- Evaluate the accuracy of a soda can calorimeter.

Prelab Discussion

This lab activity may be smelly, so you may want to warn your students. Be mindful of using snack foods that may contain common allergens (e.g., peanuts, tree nuts) since the burning process may vaporize them, potentially leading to serious allergic reactions. Have a variety of snack foods on hand to be prepared for this contingency.

Equipment Notes

The only requirement for any snack food candidate is that it can be secured to the mounting pin. You can boost student interest in this activity by giving them advance notice to allow them to select their own snacks to be tested. Snacks with high moisture content will be difficult to ignite.

The quantities of food burned in this procedure are typically very small. Your laboratory balances should be accurate to at least 0.01 g. The measurements produced by less precise instruments will not lead to meaningful results.

G Bend the tab of the soda can upward so that the stirring rod will easily pass through it. Slide the rod through the tab and suspend the can from the ring of the ring stand, leaving about 2 cm between the bottom of the can and the top of the food sample.

H Use the grill lighter to ignite the food sample. As soon as the sample finishes burning, measure the temperature of the water in the soda can, again to the nearest 0.1 °C. Record this final temperature value in Table 1.

I With a set of tongs or mitts, move the evaporating dish and food sample to the balance and determine the final dish and food mass. Record this result in Table 1.

J After making sure that the evaporating dish has cooled, clean the dish and food holder, then repeat Steps B–I for the remaining two snack food items.

ANALYSIS

Using the data that you collected, you can calculate the amount of heat energy that was released when the food was burned.

K For each snack food test, subtract the starting temperature of water from its final temperature and record the temperature change (Δt) in Table 2.

L For each snack food item, calculate the mass of food consumed during the test by subtracting the final mass of the dish, holder, and sample from its starting mass. Record the mass in Table 2.

The formula for heat energy is

$$Q = c_{sp}m\Delta t,$$

where Q is the heat energy in calories absorbed by the water, c_{sp} is the specific heat of water (1.0 cal/g °C), m is the mass of the water, and Δt is the change in temperature of the water.

M Calculate the heat energy absorbed by the water from each snack food sample. Nutrition fact panels typically use the Calorie (note the capital C) as a standard unit of heat energy. A Calorie is 1000 calories, so divide each value for Q by 1000 and record the results in Table 2.

Example:

$$Q = c_{sp}m\Delta t$$

$$= \left(\frac{1.0 \text{ cal}}{g \cdot {}^\circ C}\right)(50.0 \text{ g})(5.3 \,{}^\circ C)$$

$$= 265 \text{ cal}$$

$$= 0.265 \text{ Cal}$$

N Find each snack food's energy per unit of mass by dividing the value for Q from Step M by that snack food's change in mass. Record these values in Table 2.

1. Did any of the snack foods have a substantially higher energy content per gram than the others? If so, which one?

   Answers will vary.

© 2024 BJU Press. Reproduction prohibited.

O In Table 2 record the reported energy per unit of mass for each snack food in Calories per gram. You will probably need to calculate these values on the basis of the serving size and calories per serving found in each snack's nutrition facts label.

P Calculate the percent error of your calorimetry results by using the formula below.

$$\% \ error = \left(\frac{|value_{estimated} - value_{actual}|}{value_{actual}} \right) 100\%$$

The estimated value is the result obtained from your calorimetry test, and the actual value is the value obtained from the nutrition facts label on the snack food packaging. Record your results in Table 2.

GOING FURTHER

2. According to your data, is a soda can calorimeter reliable for determining the energy content of snack foods?

   Answers will vary. Students are likely to arrive at large percent errors for the foods they tested. The soda can calorimeter simply has too many avenues for heat energy to escape the system. At least some escapes into the can's surroundings, and some is used to heat the food holder and soda can.

3. According to your data, is a soda can calorimeter useful for comparing the energy contents of different snack foods?

   Answers will vary. Although the soda can calorimeter is not a precise enough instrument for determining the exact calorie contents of foods, that is, for quantitative measurements, it should, in theory, be able to make qualitative distinctions between samples if the procedure is done consistently.

4. How do you suppose snack food producers determine the reported energy contents of their products?

   Answers will vary. The reported energy contents of snack foods are determined by calorimetry but with equipment that is obviously far more precise than a soda can calorimeter.

5. How can knowing the energy content of a snack food help you make healthier living choices?

   Answers will vary. Since what we eat affects our health, using our knowledge of the energy content in foods can guide our snack choices so that we do not consume excess calories.

Making Wise Choices

Because our bodies are a gift from God, we should make wise choices about things that affect our health, such as our diet. A basic understanding of the nutritional value of various foodstuffs, including Calorie content, can help us make some of those wise choices.

Sample Data

The data shown in Tables 1 and 2 is typical. Students' results may be different.

TABLE 1 Calorimetry Data

Snack Food	Initial Mass (dish, food holder, and sample) (g)	Initial Water Temperature (°C)	Final Water Temperature (°C)	Final Mass (g)
cheese puff	17.78	20.0	25.3	17.68

TABLE 2 Analysis

Snack Food	Temperature Change (Δt) (°C)	Mass of Food Consumed (g)	Energy Released (Q) (Cal)	Energy per Unit of Mass, Experimental (Cal/g)	Energy per Unit of Mass, Reported (Cal/g)	% error
cheese puff	5.3	0.10	0.27	2.7	5.7	53%

23B LAB

What a Waste!

Modeling Dialysis

Do you remember your study of osmosis in Lab 5B? Osmosis occurs when the water from a solution passes through a semipermeable membrane but the solute cannot. But what happens when some of the solutes can also pass through the membrane? This process is called *dialysis*, and it is the basis for the cleansing action of the kidneys.

You will be working with a mixed solutes solution containing the protein albumin, the carbohydrates glucose and starch, and the salt sodium chloride.

How do the kidneys use dialysis to remove wastes from the blood?

?

Equipment

multimeter	disposable pipettes	iodine solution
beakers, 250 mL (3)	graduated cylinder, 100 mL	paper towels
test tubes (9)	hot water bath	laboratory apron
test tube rack	mixed solutes solution	nitrile gloves
test tube holder	distilled water	goggles
dialysis tubing strip	biuret reagent	
clamps (2)	Benedict's solution	

PROCEDURE

Dialysis Setup

A Label three 250 mL beakers A, B, and C.

B Label the nine test tubes A1, A2, A3, B1, B2, B3, C1, C2, and C3 and place them in the test tube rack.

C Obtain a 12 cm strip of dialysis tubing that has been soaked in water. Close one end of the strip with a clamp to form a bag with an open end.

D Using a pipette, place 10 mL of the mixed solutes solution in the dialysis bag.

E Clamp the other end of the tubing, leaving about 2 cm of air in the bag. Rinse the bag with distilled water and blot dry with a paper towel.

F Place the bag in Beaker A with 200 mL of water. Allow the bag to soak in the water overnight.

1. Form a hypothesis about which solutes will pass through the membrane.

Answers will vary.

QUESTIONS
- What is dialysis?
- Which solutes will pass through a semipermeable membrane?
- What factors determine which molecules can pass through a semipermeable membrane?

© 2024 BJU Press. Reproduction prohibited.

✓ Introduction to Lab 23B

This lab activity requires the dialysis tubing bag to soak in water overnight. You should plan to spend two half-hour periods on consecutive days with this activity and use the remainder of the period of each day for instruction.

✓ Prelab Discussion

Students may be familiar with the term *dialysis* as a means of treating patients with kidney disease. Explain to them that dialysis, the *treatment*, is based on dialysis, the *process*.

🧪 Equipment Notes

You should prepare 150 mL of the mixed solutes solution for each student. Dissolve 7.5 g of albumin, 0.75 g of starch, 7.5 g of glucose, and 7.5 g of NaCl in 150 mL of distilled water.

In many areas tap water will work fine, but some areas have water that has enough solutes to give a false positive for the salt test. If you live in such an area, you should use distilled water for this lab activity. If your school is in an area with soft water, you may be able to use tap water, although using any tap water supply may result in false positives for the conductivity test. If you are connected to a municipal water supply, your utility should have water quality information available.

Prepare the hot water bath before class. Set it to around 90 °C. The exact temperature is not critical.

Positive and Negative Tests

While the dialysis bag is soaking, you can examine what positive and negative tests look like for each of the solutes in the mixed solutes solution.

G Pour 100 mL of water into Beaker B.

H Pour 10 mL of water into each of Tubes B1, B2, and B3.

I Use the graduated cylinder to pour 100 mL of the mixed solutes solution into Beaker C.

J Use a pipette to transfer 10 mL of the mixed solutes solution into each of Tubes C1, C2, and C3.

Biuret Reagent for Albumin

Biuret reagent is a liquid that changes from blue to pink or purple in the presence of proteins such as albumin.

K Use a pipette to add 2–5 drops of biuret reagent to Tubes B1 and C1.

L Check whether the solutions test positive or negative for albumin and record your results in Table 1. Enter a "+" for a positive test and a "−" for a negative test.

Benedict's Solution for Glucose

Benedict's solution is a blue liquid. When mixed with monosaccharides such as glucose and heated, Benedict's solution changes to an orangish red.

M Use a pipette to add 2 mL of Benedict's solution to Tubes B2 and C2.

N Place both test tubes in a hot water bath for 10 minutes.

O Check whether the solutions test positive or negative for glucose and record your results in Table 1.

Iodine Solution for Starch

Iodine solution is a brown liquid that changes to blue or purple in the presence of starch.

P Use a pipette to add 2–5 drops of iodine solution to Tubes B3 and C3.

Q Check whether the solutions test positive or negative for starch and record your results in Table 1.

Electrolyte Test

Although pure water is a poor conductor of electricity, dissolved electrolytes, such as sodium and chloride ions, which result when NaCl is dissolved in water, make it a good conductor. You can take advantage of this fact to determine whether the sodium and chloride ions passed through the dialysis tubing into Beaker A. You will use a multimeter set to measure resistance to test whether the water will conduct an electric current. If the water has electrolytes in it, the multimeter will indicate a certain number of ohms (Ω) of resistance. If the water contains no electrolytes, the multimeter will indicate either 0 Ω or infinite Ω (depending on the multimeter).

R Use the multimeter to check whether the contents of Beakers B and C test positive or negative for the presence of electrolytes, and record your results in Table 1. Be sure to rinse the probes between tests.

Testing Dialysis

After your dialysis bag has soaked overnight, it's time to test the water from Beaker A.

S Remove the dialysis bag and set it aside.

T Use a clean pipette to place 10 mL of water from Beaker A into each of Tubes A1, A2, and A3.

U Following the steps from the previous procedures, check Tube A1 for albumin using biuret reagent, Tube A2 for glucose using Benedict's solution, and Tube A3 for starch using iodine solution. Record the results in Table 1.

V Check Beaker A for sodium and chloride ions using the multimeter. Record the results in Table 1.

2. Which solutes passed through the dialysis membrane?

   See margin for answer.

3. Why do you think that some solutes passed through the membrane while others did not?

   Answers will vary. Students should mention that solutes with smaller particles—either small molecules or ions—pass through the membrane, while the solutes with larger molecules like protein or starch cannot.

4. How do the kidneys use dialysis to cleanse the blood of wastes? You may need to reference your textbook or other sources.

   Answers will vary. Students should mention that blood plasma and smaller molecules pass into the Bowman's capsule while whole proteins and blood cells do not. In the proximal convoluted tubule, beneficial molecules move back into the surrounding capillaries while the wastes remain in the tubules to be excreted.

5. How do the kidneys show evidence of a Creator?

   Answers will vary. The extreme intricacies of the nephrons can operate only if all the parts are present. Such irreducible complexity requires a Designer to create; it could not have evolved through a series of incremental changes.

Question 2 Answer
Answers will vary. Students should indicate that glucose and NaCl passed through the membrane.

Evidence of Design

Evolutionists tell us that the human kidney is the product of millions of years of a blind process. But the kidney is another excellent example of an irreducibly complex system. If any part of the kidney system is dysfunctional or missing, the entire organ fails to work. The production of a functioning kidney by chance alone seems absurdly impossible. But just as the creation of something as complex as a mechanical filtration system requires a human designer, so our kidneys give evidence of a Creator as well, one whose handiwork never ceases to amaze.

TABLE 1 Indicator Results

	Albumin (biuret reagent)	Glucose (Benedict's solution)	Starch (iodine solution)	NaCl (multimeter)
Water	−	−	−	−
Mixed Solutes Solution	+	+	+	+
Beaker A Water	−	+	−	+

24A LAB

Sensational!

Exploring the Sensory Organs

We perceive the world around us through our senses. Seeing a flash of green at a traffic light indicates that it is safe to drive through the intersection. The smell of pot roast in the oven tells us that dinner will soon be ready. The sting of a cold winter wind on the cheek suggests that we may need to put on a heavier coat. Although we label some senses major and others minor, the truth is that the absence of any of them puts a person at a disadvantage. In this lab activity you will test some of both the major and minor senses.

? How do my senses help me gather information about my environment?

Equipment

beakers, 250 mL (2)
metal or glass rods (2)
washable ink pen
metric ruler
blindfold

metric tape measure
tuning forks of assorted frequencies (including 128 Hz and 256 Hz)
hot water

ice water
laboratory apron
goggles

PROCEDURE

Sensing Hot and Cold

Our skin receives numerous sensations. The sense of touch involves a particular set of nerves. Pain, temperature, and pressure (and possibly other sensations) are perceived by other nerve endings in the skin. In this portion of the activity you will determine whether hot and cold are different sensations of the same nerve or of two different nerves.

A Obtain a beaker of hot water and place a blunt metal or glass rod in it. (Make sure that the water is not hot enough to burn the skin.) Also obtain a beaker of ice water and place a similar rod in it.

B With a washable ink pen place six small dots on the back of your partner's hand separated by at least 1 cm. Diagram the hand on a piece of paper, draw in the dots, and assign each dot a letter.

C Blindfold your partner.

D Select either a hot or cold rod. After quickly drying the rod, randomly touch the rod to one of the dots on your partner's hand for three seconds. Your partner must then state whether the rod felt hot or cold. Mark a correct response in the appropriate column of Table 2. You do not need to record an incorrect response.

QUESTIONS

- How does the skin sense hot and cold?
- What aspects of vision can I test for?
- How do my ears determine the direction from which a sound is coming?

- **Perform simple sensory tests.**

✓ Options

You may choose to limit the number of sections you do in this lab activity. The least significant test in the activity is the test for astigmatism. As another alternative, have students do this exercise at home with their families.

Equipment Notes

The metal or glass rods used for the hot- and cold-sensing demonstration should be about half the width of a normal pencil. The procedure moves along more rapidly if each group has several rods for the hot and cold beakers. Have one person prepare the rods for another person to touch the hand of a third person. Sixteen-penny nails work well for the rods since they have a uniform size and composition and are readily available in sufficient quantities from a hardware or home improvement store. Dull the nails to reduce the risk of injury.

E Return the rod to its proper beaker between tests to keep it at the correct temperature.

F Repeat Steps D–E until every dot has been tested twice for hot and twice for cold.

1. Which did your partner more often correctly sense, hot or cold?

   Answers will vary.

2. Were each of the locations equally sensitive to hot and cold, or were some more prone to be correct about one temperature or the other?

   Answers will vary.

3. If different areas were sensitive to different temperatures, what might be the reason?

   There are separate receptors for hot and cold, and those

   receptors are not distributed evenly in the skin.

4. If all areas were equally sensitive to different temperatures, what might be the reason?

   The receptors for hot and cold are the same, and they are evenly

   distributed in the skin.

G With a washable ink pen mark a 2.5 cm square on the back of your partner's hand and then draw a 6 × 6 grid in the square like the ones shown below.

H Use the cold rod to test all the boxes within the grid. Mark in the "Cold" box below all the corresponding areas that were sensitive to the cold rod.

I Repeat Step H for the hot rod. Mark in the "Hot" box below all the corresponding areas that were sensitive to the hot rod.

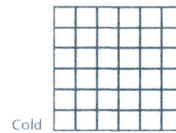

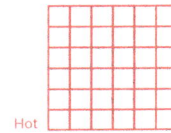

Cold   Hot

5. According to your experiment, are heat and cold receptors in identical locations on the hand? What can you conclude about temperature receptors?

   No. Heat and cold receptors must be different since they are

   located in different areas of the skin.

Vision

Vision is one of the senses on which humans rely most. It is also more prone to minor disorders than are other senses. A slight shortcoming of a person's vision or hearing may go unnoticed for years. For example, astigmatism in one eye may be compensated for by the other eye so that a person may not be aware of having blurred vision. Some people who have red-green colorblindness (one of the most common forms of colorblindness) do not realize they have it until they are adults. In the following activity you will perform some simple tests on your (or your lab partner's) vision. First, you will check yourself for astigmatism.

6. Using a dictionary or internet source, define astigmatism.

   Astigmatism is a condition characterized by an uneven area of

   the cornea or lens that leads to an area of blurred vision.

   _____

J Cover your left eye and hold the image below about 15 cm in front of your right eye. Look directly into the center of the empty circle of the diagram. Be sure that your pupil is looking directly at the central white area. If any of the lines appear blurred, you probably have astigmatism in the corresponding area of your right eye. Repeat the test on your left eye using the same procedure.

7. Do you have astigmatism in your right eye? If so, in which areas (indicated by the numbers at the ends of the lines)?

   Answers will vary.

8. Do you have astigmatism in your left eye? If so, in which areas (indicated by the numbers at the ends of the lines)?

   Answers will vary.

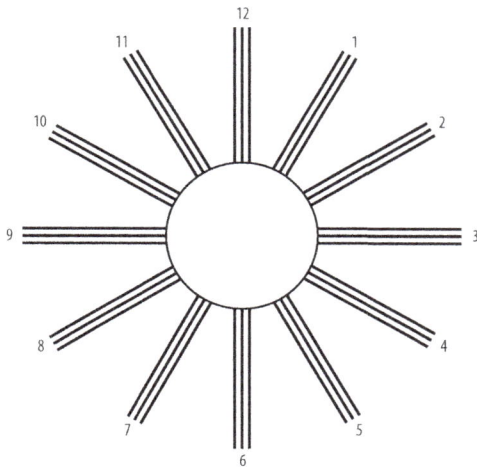

There are no photoreceptors in the eye where the optic nerve fibers leave the eye to form the optic nerve. This area is called the *blind spot*. In people with normal eyes the blind spot of each eye affects a different area of vision; therefore, the total field of vision is unbroken.

K Position the image below so that the dot is to the right of the plus sign. Hold the diagram about 45 cm from your eyes with the plus sign directly in front of your right eye. Cover your left eye.

L Move the diagram slowly toward you as you stare at the plus sign. At a certain point the dot will disappear. The dot is then in your blind spot. Have your partner measure the distance from your eye to the paper.

9. What is the distance from your right eye to the paper at your blind spot?

   usually 20–30 cm

M Repeat Steps K and L for your left eye.

10. What is the distance from your left eye to the paper at your blind spot?

    usually 20–30 cm

11. What effect do you think the blind spot would have on a person who has only one eye?

    Answers will vary. Most one-eyed (monocular) people are not
    aware of the blind spot because their brains fill in the image.

Now let's test your eyes' ability to focus.

N Have your partner look at an object across the room while you hold a pencil about 45 cm in front of his or her nose. Then ask your partner to look at the pencil.

12. How did the pupils change?

    They became smaller.

13. Did your partner's eyes move in any other way? If so, how?

    Yes. The eyes directed the pupils toward the nose.

Another important part of vision is *near-point accommodation*. The near point is the closest distance at which sharp focus is attained.

O Determine the near point of your eyes by covering your left eye and focusing with your right eye on the *D* at the beginning of this sentence. Move this page toward your right eye until the letter no longer appears sharp and clear. Move the page away until the letter is clear again.

P Have your partner measure the distance from the page to your eye to the nearest 0.1 cm.

✓ **Take Off the Glasses!**

Consider having students with corrective lenses test their near-point accommodation distance with and without their glasses or contact lenses.

14. What is your right eye's near-point accommodation distance?

Answers will vary.

Q Repeat Steps O and P for your left eye.

15. What is your left eye's near-point accommodation distance?

Answers will vary.

One's near-point accommodation distance increases with age because the lens of the eye loses its elasticity. In the table below notice the average near-point accommodation distance by age.

TABLE 1 Average Near-Point Accommodation by Age

AGE (years)	DISTANCE (cm)
15	8.9
25	11.4
35	17.2
45	52.1
55	83.8

16. Is your near-point accommodation distance normal for your age? Explain.

If a person's near-point accommodation distance isn't normal, it could be because of not wearing corrective lenses and needing them or because his or her corrective lens prescription needs to be updated. There may also be other causes.

17. When a person can no longer focus on the print of a book at a "comfortable" distance (about 50 cm), reading glasses are necessary. Considering your present near-point accommodation distance, at what age should you expect to need reading glasses?

Answers will vary. This is typically true around age 45.

Hearing

You depend greatly on your sense of hearing. The ear plays a vital role in sound perception. Your ears tell you more than just the kind and volume of sound, as this lab activity will demonstrate.

The tuning fork is an instrument used to produce a specific pitch. This instrument can be used to test for possible hearing problems. You will use it in the next few experiments. A tuning fork consists of a base and a forked pair of prongs. The base supports the prongs, allowing them to vibrate and produce sound when they are struck.

When using the tuning fork during a test, be careful to hold it at its base only; touching the fork along the prongs interferes with its vibrating and thus stops the sound. Also, strike the fork against the palm of your hand or your thigh only. Do not strike it on any hard surface since resulting dents may alter its frequency.

Using a Tuning Fork

Striking a tuning fork on hard surfaces such as a laboratory table can damage the tuning fork. Provide students with a hard rubber striker or have them strike the fork on the knee or the heel of the hand. It is beneficial to demonstrate how to properly strike a tuning fork.

R Sit in a quiet room with your eyes closed. Your partner will strike the fork on the hand and hold it 20–25 cm away from the sides, top, and then back of your head.

S As your partner moves the fork to different positions around your head, point to the direction from which the sound is coming. Your partner may need to strike the tuning fork each time it is moved.

18. Were you able to locate sound near the sides of your head? the back of your head? the top of your head?

Answers will vary.

T Place your index finger over the opening of one of your ears and repeat the experiment.

19. Was there any change in the results? Why or why not?

Answers will vary. Both ears are needed to establish sound direction.

20. In what ways do our senses equip us to be effective servants for God?

Answers will vary.

Serving with Our Senses

Students may need to give some serious thought to Question 20! Because of our sin nature, we tend to think of our senses only in terms of how they serve and please us. Becoming Christ's disciple includes dying to self and serving the needs of others.

TABLE 2 Correct Identification of Temperature for Various Areas of the Hand

	HOT CORRECTLY IDENTIFIED	COLD CORRECTLY IDENTIFIED
Dot A		
Dot B		
Dot C		
Dot D		
Dot E		
Dot F		

24B LAB

Rat Recap

Dissecting a Rat

Rats flourish around human habitation, often to the detriment of people living in the vicinity. Rats eat just about anything and spoil much that they leave behind. They also carry disease and are believed to be partly responsible for the spread of bubonic plague that killed an estimated twenty to fifty million Europeans in the fourteenth century.

But in the last hundred years, domesticated brown rats have been kept as pets and have earned a reputation for being easily trained and loyal to their owners. Other domesticated brown rats have been put to work laying computer cable or serving as therapy animals. Gambian pouched rats have been trained to detect land mines and tuberculosis. But probably the most common use for domesticated rats is as laboratory animals in medical and behavioral research. In this lab activity you will dissect a rat to see the structures typically found in mammals.

Scientists often study mammals in order to learn more about humans. Many medicines are tested in rats or mice to ensure their safety before they are made available for use by humans. Rats also have a body plan similar to that of humans. Think of this activity as the capstone of your study of human anatomy. Obviously, a rat is different from a human, but its internal structure is similar enough that it will allow you to visualize how human internal organs are arranged.

What typically mammalian structures can be observed in a rat?

Equipment

preserved rat	mounted rat skeleton
dissection pan	laboratory apron
dissection kit	nitrile gloves
pins	goggles
reference materials	

PROCEDURE

Getting Ready

Read all of Lab 24B before you come to class and review Section 19.2 in your textbook. Also feel free to read and consult other sources as you work on the lab activity. These outside sources can help you find information that will be helpful to know to complete this activity. Your teacher can suggest texts for you to consult, and you may also find some information on the internet. A search on rat dissection and rat anatomy may yield good results.

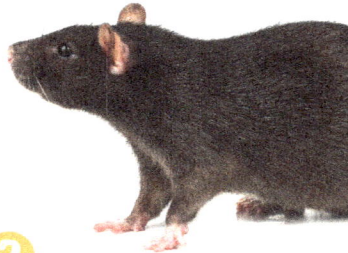

QUESTIONS

- What is the best way to dissect a vertebrate?
- How can a rat be used as a model of human anatomy and physiology?

LAB 24B OBJECTIVES

- Identify basic mammalian anatomical features.
- Explain the function of certain rat anatomical structures.

Introduction to Lab 24B

Lab 24B is designed to review the human body systems, since those of a mammal are similar to those of a human. This lab activity may take several class periods to complete. If at all possible, give students extra time to work, such as during study halls or after school.

Suggested References for Lab 24B

Hickman, Cleveland P. et al., *Integrated Principles of Zoology*, 14th ed. New York: McGraw-Hill, 2007.

Homberger, Dominique G., and Warren F. Walker, *Anatomy and Dissection of the Rat*, 3rd ed. New York: W. H. Freeman, 1997.

Olds, Ronald John, and Joan R. Olds, *A Colour Atlas of the Rat: Dissection Guide*. Boca Raton, FL: CRC Press, 1988.

Schenk, Michael P., and David G. Smith, *Dissection Guide & Atlas to the Rat*. Englewood, CO: Morton Publishing Company, 2001.

Wingerd, Bruce D., *Rat Anatomy and Dissection Guide*. Burlington, NC: Carolina Biological Supply, 2008.

Wingerd, Bruce D., *Rat Dissection Manual* (Johns Hopkins Laboratory Dissections Series), 1st ed. Baltimore, MD: The Johns Hopkins University Press, 1988.

Preparing for Lab 24B

Announce the date and time that this lab activity must be completed. Students will need to have all questions in the lab report completed by this time.

Virtual Rat Dissection

To prepare your students for the rat dissection, you might have them try a virtual dissection first. Do an internet or app search on virtual rat dissection. Be sure to try any yourself first before assigning it to students.

The Ethics of Dissection

A variety of opinions exist on the ethics of using animals for research in general and dissection in particular. Though there is value in being able to see real anatomical structures in a hands-on activity, some teachers, students, or both may not be comfortable with this activity. In such instances give careful consideration to substituting a virtual dissection.

Storing Rats Overnight

This lab activity may easily take two class periods, requiring students to store a rat overnight. In between periods, have them place the rat's internal structures in its body cavity and then wrap the entire rat in a damp paper towel. They should then place this in a plastic bag, removing as much air as possible, close the bag tightly, and tie or zip it shut. Have students record their names and class period on their bags to easily identify their rats the following day. These should then be placed in a refrigerator or cooler for the night.

Question 6 Answer
Answers will vary. It is covered with scales and a few bristlelike hairs.

External Structures

A Identify on your specimen each of the structures listed below. Check the box beside each structure after you locate it.

- ☐ auricles
- ☐ vibrissae (whiskers)
- ☐ external nares (nostrils)
- ☐ upper eyelid
- ☐ lower eyelid
- ☐ incisors
- ☐ forelimbs
- ☐ hind limbs
- ☐ head
- ☐ trunk
- ☐ anus
- ☐ tail
- ☐ (male:) penis and scrotal sacs (containing testes)
- ☐ (female:) vaginal opening and teats
- ☐ fur

1. Do the auricles contain bone?

   No. The auricles contain cartilage.

2. What is the function of the vibrissae?

   Answers will vary. The vibrissae are tactile organs that allow the animal to sense its immediate surroundings.

3. Why do rats have such prominent incisors?

   Answers will vary. Their incisors constantly grow, both allowing and requiring rats to continually gnaw.

4. Of the upper and lower incisors, which ones are longer?

   the lower incisors

5. How is a rat's upper lip different from that of a human?

   The upper lip is cleft in the middle.

6. How is the surface of the tail different from the rest of the body?

   See margin for answer.

7. How many toes does a rat have on each of its forelimbs? on each of its hind limbs?

   4; 5

8. What is the sex of your rat?

   Answers will vary.

Removing the Skin

Now you will remove the rat's skin.

B Place your rat in the dissection pan with its ventral side up.

C Pinch the skin in the thoracic region (over the sternum) enough to puncture it with the tip of your scissors. Carefully cut by continuing to raise the skin from the muscles and cut along the indicated lines with your scissors, penetrating only into the subcutaneous layer (see right). Do not cut into the muscles because you will need to observe them intact. Note that you will cut around each limb just proximal to each foot. You will also cut around the neck, the base of the tail, and the groin area, leaving the fur intact.

D After you have made all the indicated cuts, use your forceps to pull the skin away from the muscles beneath. It should peel back as flaps if you carefully use your scissors and perhaps the probe to separate the skin from the muscles.

E Turn your rat over so that the dorsal side is up. Now complete the skin removal by pulling the skin away from the back in one piece. If all cuts were made properly, this should be possible.

F Examine the skin that you have removed.

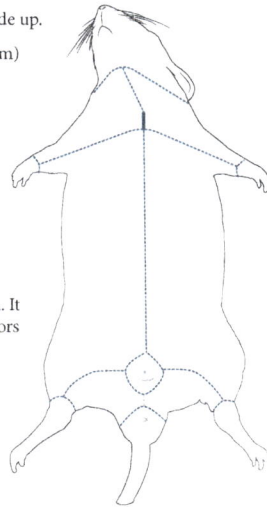

9. How many layers can you see?

   See margin for answer.

10. How many types of hair can you discern?

   See margin for answer.

Muscles and Bones

Now that you've skinned your rat, look for the muscle groups that you learned for humans in Chapter 21. Muscle functions in the rat are often different from those in the human because of the difference in body plans. The origins and insertions, though, are generally the same and are summarized in Table 1 (p. 262). You should be able to see all the muscle groups indicated without making any further cuts. You will probably be able to distinguish the superficial muscles shown on the next two pages by carefully separating them with your probe. Simply insert the probe under the loose edge of the muscle and slide along its margin. You can usually tell where one muscle separates from another by the direction of the fibers.

Question 9 Answer
Answers will vary. Students may be able to distinguish between the epidermis and the dermis.

Question 10 Answer
Answers will vary. Students should be able to distinguish between the longer guard hairs and the softer underhair.

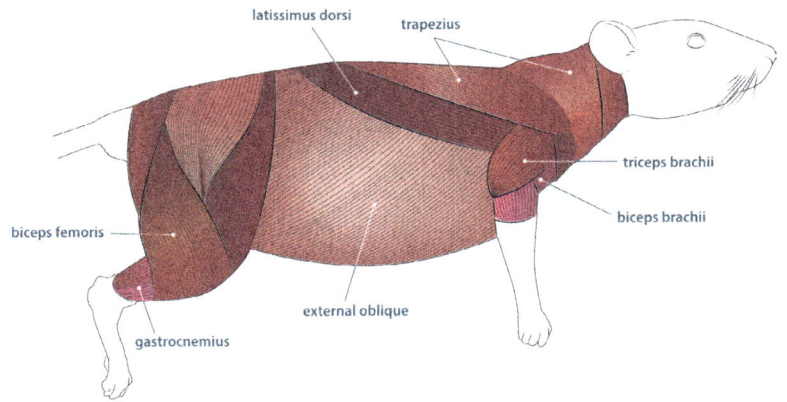

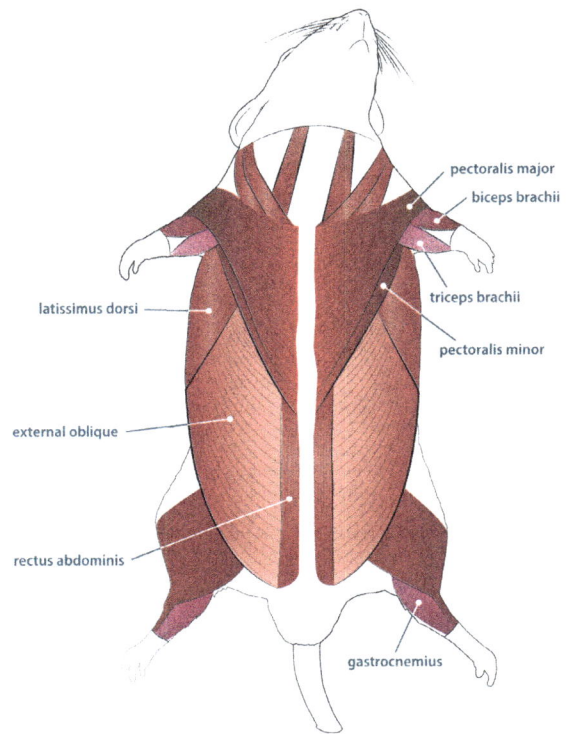

latissimus dorsi
trapezius
triceps brachii
biceps brachii
biceps femoris
external oblique
gastrocnemius

pectoralis major
biceps brachii
triceps brachii
pectoralis minor
latissimus dorsi
external oblique
rectus abdominis
gastrocnemius

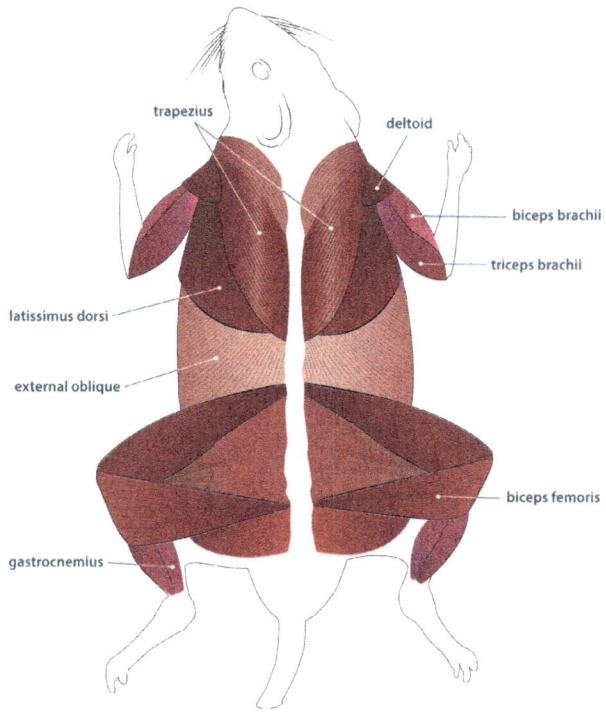

trapezius

deltoid

biceps brachii

triceps brachii

latissimus dorsi

external oblique

biceps femoris

gastrocnemius

G As you find each muscle or muscle group, check the box beside its name in Table 1. After mastering the identification of the muscles in the chart and the figures, choose one hind leg and carefully remove the biceps femoris, gastrocnemius, and any other necessary underlying muscles to reveal the tibia, fibula, and femur, as well as the patella.

TABLE 1 Muscles of the Rat

NAME	ORIGIN	INSERTION	FUNCTION
☐ trapezius	vertebrae	scapula	pulls scapula inward, up, and backward
☐ latissimus dorsi	vertebrae	humerus	moves humerus toward the back
☐ external oblique	muscles in region of last nine to ten ribs	pelvis	compresses abdomen
☐ gastrocnemius	lower femur	Achilles tendon	extends the foot
☐ biceps femoris	lower vertebrae	femur and tibia	flexes the lower rear leg
☐ biceps brachii	scapula	radius	flexes lower front leg
☐ triceps brachii	scapula and humerus	ulna	extends lower front leg
☐ deltoid (spinodeltoid)	spine near the scapula	humerus	pulls scapula back and outward
☐ pectoralis major	sternum	humerus	draws front leg forward, toward chest
☐ pectoralis minor	sternum	humerus and scapula	draws front leg forward, toward chest
☐ rectus abdominis	pelvis	sternum	compresses abdomen

Many of the bones in the human body can be found in a rat, though they may look somewhat different because of the different body plans that God has designed.

H Examine your skinned specimen to identify the bones shown below. As you find each one, check the box by its name.

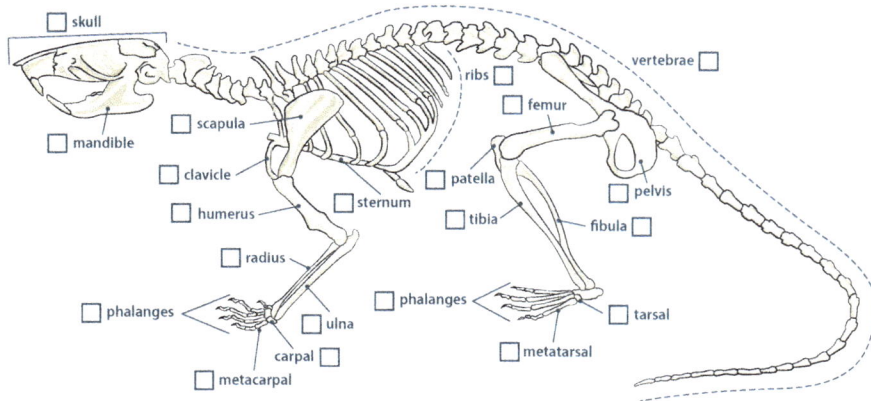

- [ ] skull
- [ ] mandible
- [ ] scapula
- [ ] clavicle
- [ ] humerus
- [ ] radius
- [ ] phalanges
- [ ] ulna
- [ ] carpal
- [ ] metacarpal
- [ ] ribs
- [ ] sternum
- [ ] patella
- [ ] tibia
- [ ] phalanges
- [ ] femur
- [ ] vertebrae
- [ ] pelvis
- [ ] fibula
- [ ] tarsal
- [ ] metatarsal

Opening the Body Cavity

There are many ways to open the body cavity of a rat. The method described below is recommended. It usually results in fewer damaged organs.

I Place your rat in the dissection pan with its ventral side up.

J Use your scissors to cut through just the muscle layers along the lines indicated on the right. It is best to make the longitudinal cut first and then the shorter transverse cuts. Be careful not to cut the internal organs as you make the incisions.

K After all the cuts are made, fold the flaps back and pin them to the pan so that the organs are more easily seen. Do not remove limbs or any other structures unless your teacher instructs you to do so.

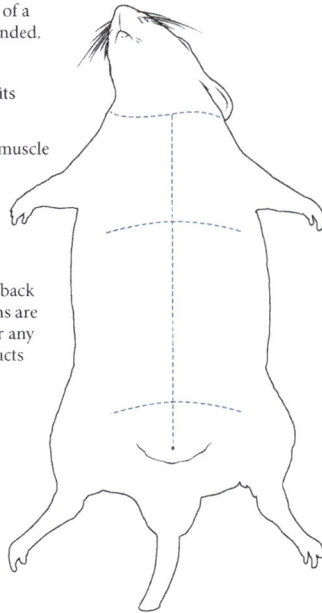

The Internal Organs

L Identify the labeled structures below. As you find each one, check the box beside its name. Mesenteries (thin, clear connective tissues with blood vessels running through them) hold the organs in place. Using a probe to loosen these mesenteries will make the internal organs more visible.

Note: If your rat is older, the thymus may not be distinguishable.

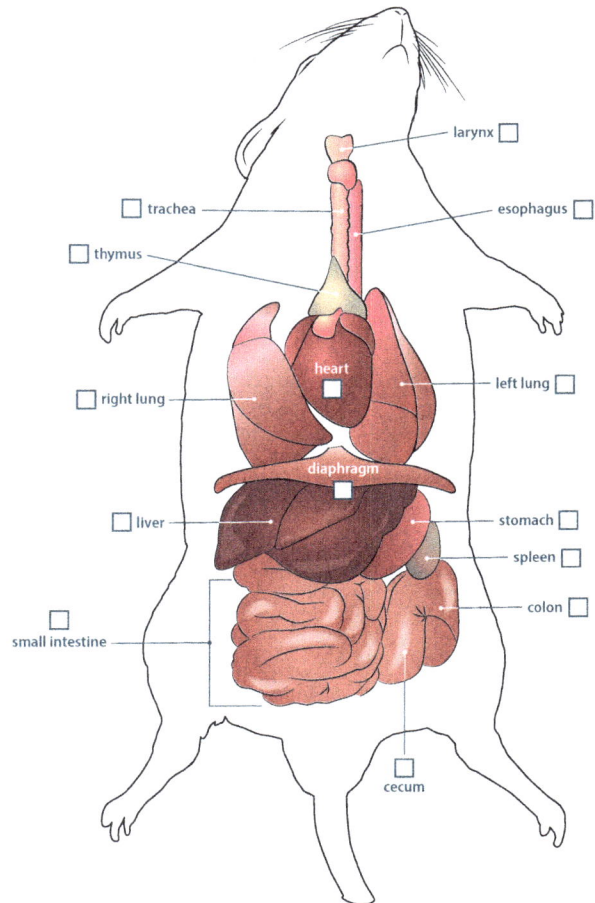

larynx ☐

☐ trachea   esophagus ☐

☐ thymus

heart ■

☐ right lung   left lung ☐

diaphragm ■

☐ liver   stomach ☐

spleen ☐

colon ☐

☐ small intestine

cecum ☐

M While the rat's heart is in its natural position, identify the parts of the heart listed on the right. Check the box by each structure as you find it.

11. Which structures of the heart contain deoxygenated blood?

the right atrium and the right ventricle

N Remove the heart.

THE RESPIRATORY SYSTEM

O Remove the respiratory system. Follow the trachea as near to the mouth cavity as possible. Include the larynx when you make the cut to remove this system.

12. How do the rat's diaphragm and rib cage function in breathing?

The diaphragm pushes the organs in the abdominal cavity downward, and the ribs expand outward. Both of these actions increase the space in the thoracic cavity, and the lungs expand to fill it, drawing in air from outside the body.

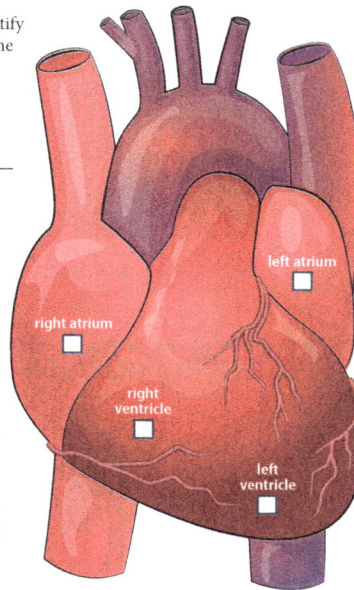

left atrium

right atrium

right ventricle

left ventricle

THE DIGESTIVE SYSTEM

P Remove the structures of the digestive system by carefully cutting the esophagus as close to the skull as possible and cutting across the rectum, posterior to the end of the colon.

Q Carefully separate any mesenteries connecting digestive organs to parts of the urogenital system and the vertebrae. The entire digestive system—with the liver, spleen, and pancreas intact—should now come out of the body cavity.

R Slowly tease the mesenteries with your probe so that the alimentary canal can be straightened. All accessory organs should remain attached to the alimentary canal.

13. (a) How long is the alimentary canal? (b) How long is it in relation to the length of the rat?

Answers will vary. (a) Larger rats will have larger alimentary canals. (b) It is much longer than the rat.

14. In regard to the rat and other vertebrates, what is the relationship between an animal's diet and the length of the intestine?

Animals that eat plants tend to have longer intestines than those that eat meat.

15. How many lobes does the liver have?

4

THE URINARY, ENDOCRINE, AND REPRODUCTIVE SYSTEMS

These structures in a rat are basically the same as in a human.

S Carefully locate the male or female urinary structures shown below and check the box beside the name of each structure as you find it. Do not remove the urinary structures from the rat body cavity.

The ureters are sometimes hard to find because of their small size, but if you wiggle the kidneys it may be easier to see them. Note that the adrenal glands, atop the kidneys, are considered part of the endocrine system.

T Carefully locate and expose the male or female reproductive structures shown below. If you are dissecting a male rat, you will need to cut the scrotal sac to access the testes. Leave them in the rat body cavity. As you find each structure, check the box beside its name.

Male Rat Urogenital System

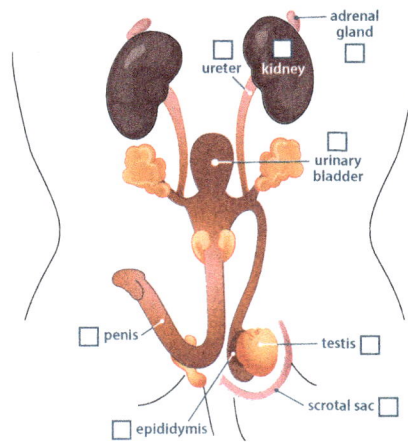

- adrenal gland ☐
- ureter
- kidney
- urinary bladder ☐
- penis ☐
- testis ☐
- scrotal sac ☐
- epididymis ☐

Female Rat Urogenital System

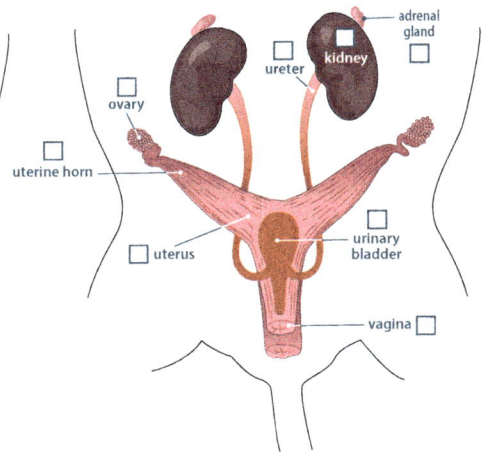

- adrenal gland ☐
- ureter
- kidney
- ovary ☐
- uterine horn ☐
- uterus ☐
- urinary bladder ☐
- vagina ☐

25A LAB

Unusual Development

Modeling the Amazing Growth of Robert Wadlow

You've learned that as you get older, you get taller. But you've also learned that at some point you stop getting taller even though you're getting older. It's a good thing too, otherwise you might be 16 feet tall when you retire!

Have you ever wondered how tall the tallest person in the world was? Did he grow like a normal person? If not, how was his growth different?

The famous Guinness Book of World Records has verified Robert Pershing Wadlow to be the tallest known human in history. Robert was born in 1918 at a normal baby weight of about 4 kg. His record-breaking growth soon began. When Robert was twelve years old, doctors concluded that he had an enlarged, hyperactive pituitary gland that produced abnormally high levels of growth hormone. Robert passed the height of an average adult male by the age of seven. His four younger siblings all matured to a normal size, and there is no record of unusual growth for any other family members. Average males stop growing at about age eighteen, but Robert continued growing throughout his life.

In this lab activity you will graph and compare data from Robert Wadlow's amazing growth with that of a normal male to see how different Robert's growth was.

? **How does the growth of the tallest man to ever live compare with normal male development?**

Equipment
colored pencils, 4 different colors
calculator

PROCEDURE

Graphing the Data

Let's see what you can learn from Robert's incredible growth. You will be using the data in Table 1 (see next page) to create two "dual-line" graphs.

A Create a graph of male height versus age for both Wadlow and a normal male in Graphing Area A. Be sure to label the resulting lines.

B Create a graph of male mass versus age for both Wadlow and a normal male in Graphing Area B. Be sure to label the resulting lines.

QUESTIONS

- What is the relationship between Robert Wadlow's growth and normal growth?
- How can you interpret trends in data?
- Can data be extrapolated from existing information?

- Plot growth rates when given tabular data on height and mass.
- Make a prediction that is based on extrapolating data.
- Assess the health of an individual by calculating his or her BMI.

✓ Software Suggestion

If students are familiar with a spreadsheet program or other graphing software, suggest that they use it to graph the data in Table 1 and attach printouts in the graphing areas.

TABLE 1 Growth Data for Robert Wadlow and Normal Males

Age (y)	ROBERT WADLOW'S GROWTH		NORMAL MALE GROWTH	
	Height (cm)	Mass (kg)	Height (cm)	Mass (kg)
Birth	51.0	3.8	49.8	3.3
1	107.0	20.0	75.7	9.6
2	137.0	34.0	86.8	12.5
3	150.0	40.0	95.2	14.0
4	160.0	48.0	102.3	16.3
5	169.0	64.0	109.2	18.4
6	170.0	66.0	115.5	20.6
7	178.0	72.0	121.9	22.9
8	183.0	77.0	128.0	25.6
9	188.0	82.0	133.3	28.6
10	196.0	95.0	138.4	32.0
11	211.0	109.0	143.5	35.6
12	218.0	130.0	149.1	39.9
13	224.0	137.0	156.2	45.3
14	226.0	150.0	163.8	50.8
15	239.0	161.0	170.1	56.0
16	248.0	170.0	173.4	60.8
17	251.0	173.0	175.2	64.4
18	254.0	177.0	175.7	66.9
19	259.0	220.0	176.5	68.9
20	262.0	221.0	177.0	70.3
21	264.0	223.0		
22.4*	272.0	199.0		

*Age of Wadlow at his death

CONCLUSIONS

Answer the following questions on the basis of your graphs and the data in Table 1.

1. In what year did Wadlow's height increase the most? When does the average male grow the most?

   first year; first year

2. What was the average amount that Robert grew in height per year? What is the average amount that the average male grows in height in one year?

   10.0 cm per year; 6.4 cm per year

3. How does Robert's increase in height compare to other males?

   Robert's growth rate was about 56% greater than that of a

   normal male.

4. In what year did Wadlow's mass increase the most? When does the average male increase in mass the most?

   during his nineteenth year; during his first year

5. What was the average amount that Robert increased in mass per year? What is the average amount that the typical male increases in mass per year?

   8.8 kg per year; about 3.4 kg per year

6. How does Robert's growth in mass compare to other males?

   Robert's gain in mass happened at over twice the rate of a

   normal male.

7. Scientists often extrapolate data to make predictions. To do this, they extend a graph along the same slope, above or below the measured data. If Robert had lived until the age of thirty and had continued growing at the same rate that you previously calculated, what height might he have attained? What mass might he have grown to?

   about 354 cm; about 302 kg

The formula for calculating BMI—body mass index, a calculation used to assess the relationship between a person's height and mass—is

$$\text{weight in kg} \div (\text{height in cm})^2 \times 10\,000.$$

8. According to the US Centers for Disease Control, a healthy BMI is 18.5–24.9 kg/cm². Calculate Robert's BMI at the ages of five, ten, fifteen, and twenty. Would Robert have been considered healthy?

   Answers will vary. Robert's BMI was outside the "healthy" range

   for most of his life, but only at age 20 would he have been

   considered obese according to CDC guidelines.

© 2024 BJU Press. Reproduction prohibited.

Extrapolation

Visual learners will benefit from having this concept explained with a brief demonstration at the whiteboard. An internet search on extrapolation will yield many video resources on the topic.

Calculating BMI

Students might be interested in calculating their own BMIs. They might *not* be thrilled about reporting what they discover! Suggest that they consider this subject heading into Lab 25B.

Unusual Development   269

Unusual Development   **269**

9. After consulting the internet or other resources, list some other health problems that might be faced by a true "giant" other than the ones mentioned in this lab activity.

   Adults with acromegaly (gigantism) are at increased risk for type 2 diabetes, hypertension, cardiovascular disease, and arthritis.

10. What kinds of additional modifications might a person of this size need for daily life?

    Answers will vary.

GOING FURTHER

Most pediatricians do a rough estimate of how tall a boy will grow using the formula shown below.

$$(\textit{mother's height} + 5 \text{ in.} + \textit{father's height})/2$$

Robert's father was 5 ft 11 in. tall. There is not an easily obtainable recorded height for his mother, but by comparing photos, we can surmise that she was about 5 ft 5 in. tall.

11. Using these measurements of Robert's parents, how tall would a pediatrician predict one of their sons to be?

    5 ft 10.5 in.

12. What is the significance of the pituitary gland?

    The pituitary gland produces the hormones that regulate human growth.

13. What might cause the pituitary to become enlarged?

    Pituitary enlargement is normally caused by a tumor in the pituitary gland.

14. If Robert Wadlow had been born in 2020, how might doctors have treated him?

    Acromegaly is usually treated with medications, surgery, or a combination of the two.

15. Considering the health issues associated with acromegaly, how does the function of the pituitary gland demonstrate the wisdom of God's design for the human body?

    By limiting the maximum height that a person can normally attain, God's original design for the pituitary gland spared people from the health problems associated with acromegaly.

✝ Fighting the Effects of the Fall

The mechanisms that God has built into the human body for governing and regulating its growth are indeed marvelous, and they provide a powerful testimony of God's meticulous care and craftsmanship. But because of the Curse, sometimes these same mechanisms don't function as they should, often with tragic consequences. For now, researching ways to combat the effects of disorders such as acromegaly provides Christians with the opportunity to exercise wise dominion and display God's concern for the afflicted.

Growth (Height) Robert Wadlow vs. Typical Male

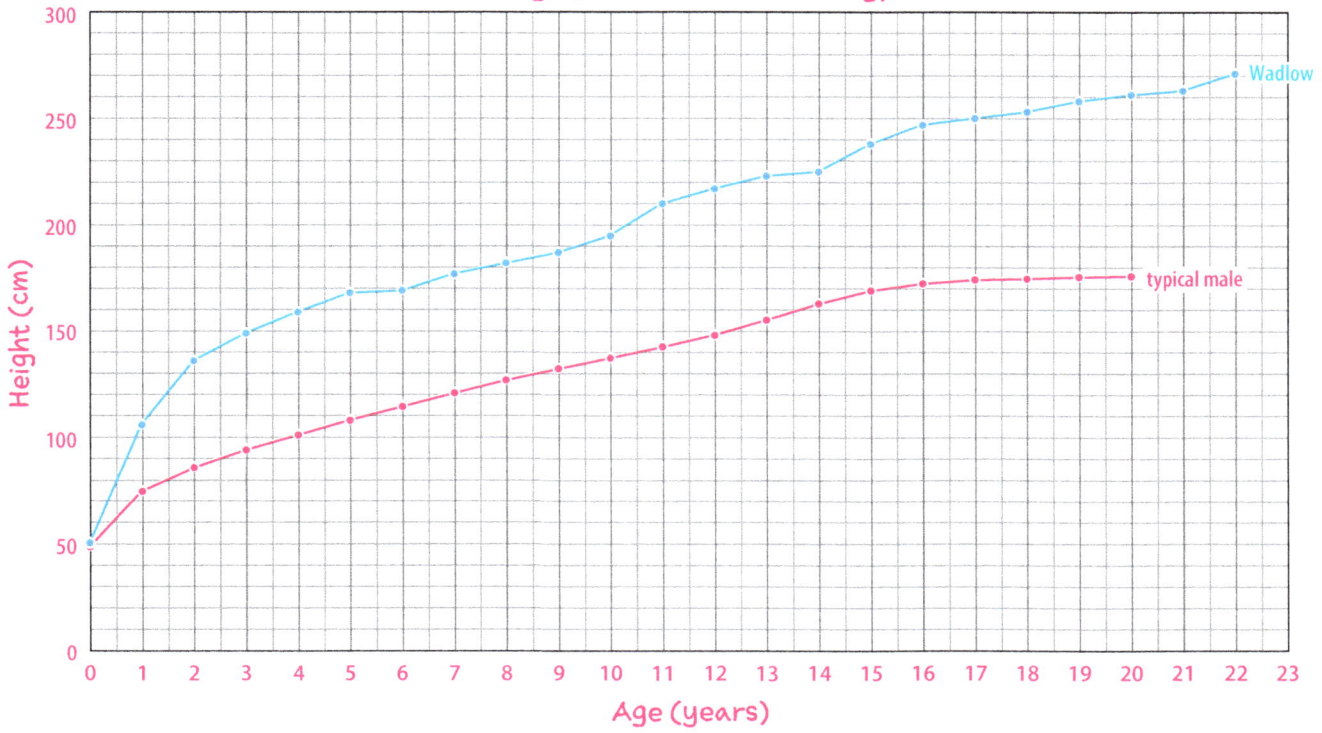

Height (cm) vs. Age (years). Curves labeled "Wadlow" and "typical male."

Growth (Mass) Robert Wadlow vs. Typical Male

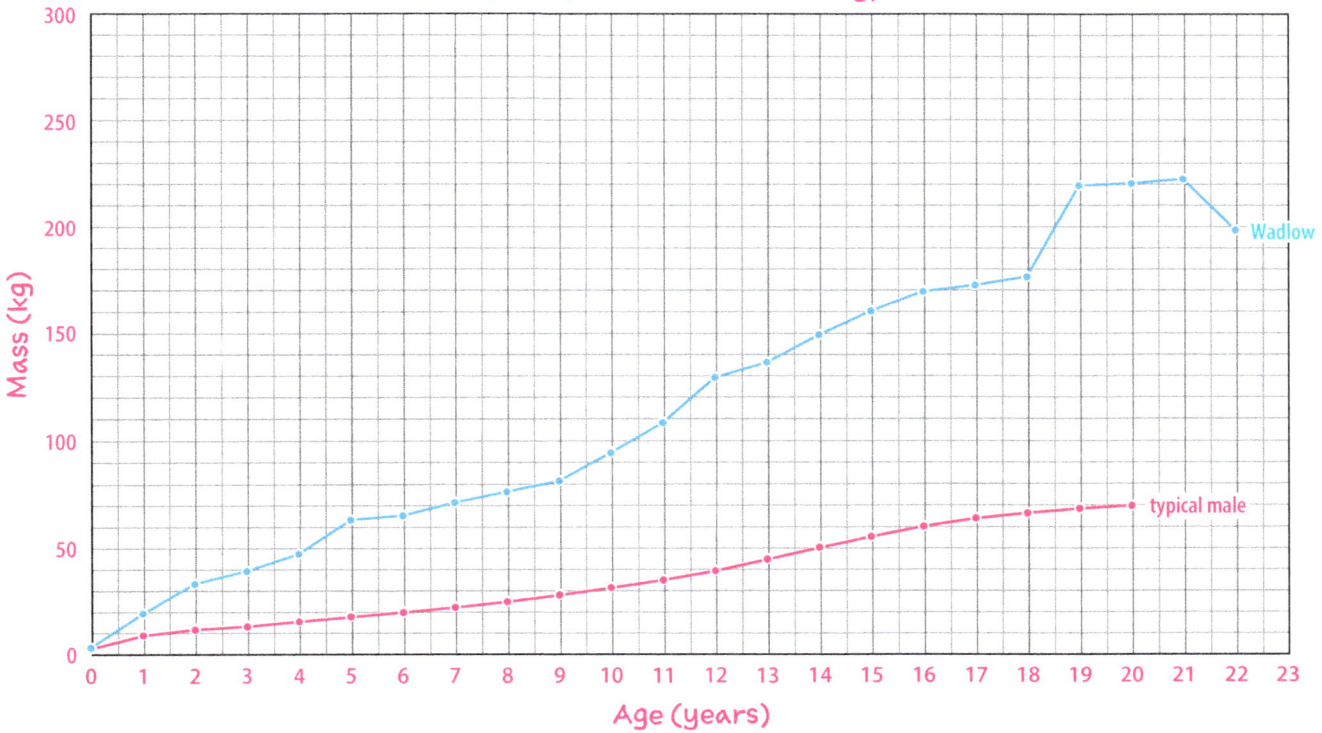

Mass (kg) vs. Age (years). Curves labeled "Wadlow" and "typical male."

25B LAB

Fast Food Fact-Finding

Exploring the Perception of Fast Food versus Reality

Fast food is big business! In 2015 estimated US fast food sales generated nearly five times the amount of revenue earned by movies and professional sports combined. In fact, during that same year Americans spent an amount of money on fast food greater than the gross domestic products of three-quarters of the world's countries. That's a lot of money—and a lot of food! Yes, American fast food is popular, both here and around the world, but you would probably have to be a hermit not to have heard some of the troubling accusations being made about fast food. Various studies, news outlets, and talk-show doctors have linked fast food to obesity, heart disease, diabetes, and a host of other maladies. But is fast food really that bad for you? In this activity you will examine fast food, and you'll make use of a surprising source of information—the fast food franchises themselves.

Is eating fast food a healthy lifestyle option?

Equipment
none

PROCEDURE

Road Trip!

Imagine this scenario: you just got your driver's license, and you've been given use of the family car for the whole day! So you head out early, deciding that you'll stop somewhere along the way for meals. With no one along on the trip to tell you something's bad for you or costs too much, you get to eat whatever you want at whatever restaurants you want. Ah, but here's the catch:

» You'll need to eat three meals (breakfast, lunch, and dinner) plus dessert.

» For each meal you'll need to record the name of the restaurant where you eat. You must include an itemized list of everything you eat and drink at that restaurant, using the menu names as given by the restaurant. If you order a combo, you'll need to list separately everything included in the combo.

» The only restriction on where you can eat is that it must be a restaurant that publishes nutrition information that you can access either at the restaurant, in a brochure, or online.

Don't overthink what you'll order. Remember, it's your road trip, so order whatever you want.

QUESTIONS
- What are my fast food preferences?
- Do my fast food choices increase my health risks?
- Do I need to change my fast food eating habits?

- Analyze a meal for its nutrient content.
- Compare the nutrient contents of fast foods with the dietary guidelines provided by the USDA.
- Self-assess fast food habits.

Introduction to Lab 25B

Fast food is part of the American landscape, and teenagers are the demographic that eats fast food the most frequently and regularly. The goal of this lab activity is to get teens to think about just how nutrient-dense fast food fare really is.

Prelab Discussion

Most students love fast food and eating out. You should have no difficulty polling them about their attitudes toward and experiences with fast food dining.

Planning a Menu

Although most teens will have no trouble planning where and what they would like to eat, there are of course students who don't eat out for various reasons or who have dietary needs that preclude eating fast food. In such instances it is not necessary for a student to map out what he himself would eat; he can just as readily imagine what a typical teenager might eat.

It is important that students consider what they would actually eat on their imaginary trip since we're trying to get them to think about their actual habits. Make sure that they're not trying to "cheat the system" by underestimating what they would eat. Nor should they be trying to create a menu with an unrealistic amount of food in an attempt to create the most shocking caloric intake possible.

Although not truly fast food restaurants, many popular restaurant chains also have menus dominated by entrée items with high calorie, fat, carbohydrate, and sodium contents. Consider allowing students to include "visits" to these restaurants on their road trip.

Big C or Little C?

Recall from Lab 23A that the energy values for foods are generally given in kilocalories, for which the correct unit abbreviation is either Cal (uppercase *C*) or kcal. However, in everyday language we commonly use the term *calorie* even when kilocalories are meant. Similarly, caloric intake guidelines are also expressed in calories when, scientifically speaking, the indicated numerical quantities are actually kilocalories.

Too Much Can Be As Unhealthy As Too Little

In Western culture the abundance of food in general, and fast food in particular, poses health issues that most of humanity throughout history could have scarcely comprehended. Since the Fall, most of humanity has struggled with having enough to eat, and the Bible contains many examples of famines and the devastating consequences they had on families and societies. It takes a certain steadfastness and godly wisdom to resist worldly eating habits.

Sample Data

The sample data in this lab activity, the product of an imaginary road trip, is derived from real restaurant menus. The meals itemized are not significantly different from what an average teen might eat, so some students will undoubtedly experience "calorie shock." Encourage these students to create a "game plan" to help make healthier eating choices.

A Record what you ate at each meal in the Menu Item column of Tables 1–4.

1. Think about the nutrients that you've learned about this year. Describe whether you feel that the meals you've chosen for the day will satisfy your body's nutritional needs.

   Answers will vary.
   _____
   _____
   _____

Breaking Down the Nutrition Information

Most restaurants now post nutritional information for their menus online. To find the nutritional information for each restaurant that you visited on your road trip, do an internet search for the name of the restaurant plus the keyword "nutrition."

B Consult each restaurant's online nutrition information, then fill in the number of calories and the amount of fat, carbohydrates, protein, and sodium for each item that you listed in Tables 1–4.

C Add up the total amount of each nutrient and record the results in the last row of each table.

D Now add up the totals from each meal to find the total amount of each nutrient consumed during the day and record the results in Table 5.

Measuring Up

Now let's find out whether your day of dining out on the road provided you with adequate nutrition. First, you'll need some idea of what the basic nutritional needs of a teenager are. Not all sources agree on the nutritional needs of teenagers, but for this lab activity we'll use the guidelines published by the US Department of Agriculture (USDA) for 2015.

E Consult Table 6, the USDA Estimated Calorie Needs per Day by Age, Sex, and Physical Activity Level, to determine your daily calorie need.

2. What is your daily calorie need?

   Answers will vary between 1800 and 3200 calories.

3. Express your caloric intake for the day from Table 5 as a percentage of your estimated calorie need from Table 6.

   Answers will vary.

4. How does your caloric intake for the day compare to the calorie need suggested by the USDA?

   Answers will vary.
   _____

Table 7 shows the USDA recommended daily intake amounts for teens for several nutrients.

5. How does your intake for the day for each of these nutrients compare to the recommended amounts in Table 7?

Answers will vary.

Assessing Risk

6. Considering your road trip data and the USDA recommended calorie needs for teens, do you think eating fast food increases your risk for weight gain? Explain.

Answers will vary. Students should observe that even a modest consumption of fast foods can easily result in consuming far more than the recommended number of calories, with weight gain a consequent result.

Type 2 diabetes is a disease that results in the body being unable to properly use and store sugars and fats. There is no cure for type 2 diabetes. Once a person has diabetes, the disease must be managed for the rest of his or her life. One of the risk factors for developing type 2 diabetes is being overweight.

7. Considering your road trip data, do you think eating fast food increases a person's risk of developing type 2 diabetes? Explain.

See margin for answer.

As you learned in Lab 22B, hypertension (high blood pressure) is a risk factor for stroke. Your chances of developing hypertension are increased by being overweight and by eating a high-sodium diet.

8. Considering your road trip data, do you think eating fast food increases a person's risk of developing hypertension?

Answers will vary. In addition to the potential for causing weight gain, fast food typically has a high sodium content as well. Therefore, a diet rich in fast foods has the potential to contribute to developing hypertension.

Question 7 Answer
Answers will vary. Fast food tends to be high in carbohydrates and fats, resulting in calorie-dense menu items. Even a modest intake of fast foods can easily result in consuming more calories than are recommended by the USDA, with weight gain a consequent result.

The risk factors for heart disease include being overweight and having hypertension, diabetes, or both.

9. Considering your road trip data, do you think eating fast food increases a person's risk of developing heart disease?

   Answers will vary. Since a diet rich in fast foods can contribute to developing the risk factors for heart disease, then by extension that same diet can be a factor in increasing the risk of heart disease.

Think about It

10. Suggest some alternatives to eating fast food during your road trip.

    Answers will vary. Students may suggest packing healthier meals for the trip or selecting only healthier items from restaurant menus.

11. Are there ways to eat healthier even at a fast food restaurant? Explain.

    Answers will vary. Many restaurants offer menu items aimed at health-conscious consumers. These items are usually lower in fat, carbohydrates, and sodium. Customers may also request that menu items be prepared without certain ingredients, or they may ask for substitutions.

12. Fast food advertising often attempts to portray fast food as fun, tasty, convenient, and inexpensive—just think about the commercials you have seen on TV showing fit young people or families smiling and enjoying a hamburger or fried chicken strips. Suggest a more balanced perception of fast food.

    Answers will vary. Frequent consumption of fast food is most definitely more expensive than dining in. In addition, fast food fare normally contains unhealthy amounts of calories, fats, carbohydrates, and sodium. Fast food should be consumed in moderation and with careful consideration given to selecting healthier menu options.

One slice: 435 Cal, 25 g of fat, and 883 mg of sodium

13. Why should Christians give careful consideration to their fast food dining habits?

Answers will vary. A diet rich in fast foods can contribute to many significant health problems. Thus, eating whatever we like without regard to health risks jeopardizes our own health, which in turn affects those who depend on us, such as a spouse or children. Eating out is also not very cost effective, so frequent dining out is not the wisest use of the financial resources that God gives us. Lastly, our bodies are a gift to us from God and are intended to be used in His service; it is therefore our responsibility to take care of what God has entrusted to us.

GOING FURTHER

14. Obesity, diabetes, hypertension, and heart disease are only a few of the problems that have been linked to fast food. Do an internet search on the topic of fast food health risks, then briefly discuss some of the additional risks associated with fast food.

Answers will vary. Fast food, or the nutrients found in high quantities in fast food, has been linked to kidney disease, stomach cancer, asthma, sleep apnea, depression, memory loss, acne, eczema, tooth decay, osteoporosis, and headache.

TABLE 1 Breakfast

NAME OF RESTAURANT: _____					
Menu Item	Calories (kcal)	Fat (g)	Carbohydrates (g)	Protein (g)	Sodium (mg)
two sausage biscuits with egg	1020	68	70	34	2160
hash browns	150	9	15	1	310
latte	170	9	14	9	115
Totals	1340	86	99	44	2585

TABLE 2 Lunch

NAME OF RESTAURANT: _____					
Menu Item	Calories (kcal)	Fat (g)	Carbohydrates (g)	Protein (g)	Sodium (mg)
double cheeseburger	670	41	39	37	1440
fries	395	18	54	7	245
chocolate shake	590	29	72	10	320
Totals	1655	88	165	54	2005

TABLE 3 Dinner

Menu Item	Calories (kcal)	Fat (g)	Carbohydrates (g)	Protein (g)	Sodium (mg)
NAME OF RESTAURANT: _____					
pepperoni pizza (4 slices)	1740	97	144	72	3530
20 oz cola	250	0	69	0	55
Totals	1990	97	213	72	3585

TABLE 4 Dessert

Menu Item	Calories (kcal)	Fat (g)	Carbohydrates (g)	Protein (g)	Sodium (mg)
NAME OF RESTAURANT: _____					
ice cream	530	18	66	8	85
Totals	530	18	66	8	85

TABLE 5 Daily Total

Calories (kcal)	Fat (g)	Carbohydrates (g)	Protein (g)	Sodium (mg)
5515	289	543	178	8260

TABLE 6
USDA Estimated Calorie Needs per Day by Age, Sex, and Physical Activity Level

	MALES				FEMALES		
Age	Sedentary	Moderately Active	Active	Age	Sedentary	Moderately Active	Active
14	2000	2400	2800	14	1800	2000	2400
15	2200	2600	3000	15	1800	2000	2400
16	2400	2800	3200	16	1800	2000	2400

Sedentary means a lifestyle that includes only the physical activity of independent living. *Moderately Active* means a lifestyle that includes physical activity equivalent to walking about 1.5 to 3 miles per day at 3 to 4 miles per hour, in addition to the activities of independent living. *Active* means a lifestyle that includes physical activity equivalent to walking more than 3 miles per day at 3 to 4 miles per hour, in addition to the activities of independent living.

TABLE 7 USDA Nutritional Goals

MACRONUTRIENT OR MINERAL	MALE 14–18	FEMALE 14–18
Fat (g)*	70–100	55–75
Carbohydrates (g)	130	130
Protein (g)	52	46
Sodium (mg)	2300	2300

* Fat amounts are calculated according to recommended caloric intake of fats for moderately active teens.

APPENDIX A

Laboratory and First Aid Rules

Laboratory Rules

Safety Rules
Be sure to review Safety in the Laboratory on page x of To the Teacher in this Teacher Lab Manual.

1. Never perform an unauthorized experiment or change any assigned experiment without your teacher's permission.

2. Avoid playful, distracting, or boisterous behavior.

3. Never work alone. Students must not conduct lab activities without supervision.

4. Work at your own lab station.

5. Always work in a well-ventilated area. Use the fume hood when working with toxic vapors. Never put your head in the fume hood.

6. Always wear safety goggles when working with chemicals, glassware, projectiles, and other materials or objects that are potentially hazardous to the eyes.

7. Wear protective clothing and gloves when working with corrosive or staining chemicals.

8. While working in the laboratory, tie back long hair and avoid wearing loose clothing such as scarves or ties.

9. Never taste any chemical, eat or drink out of laboratory glassware, or eat or drink in the laboratory.

10. Always use appropriate instruments for cutting and handle them carefully. Always cut away from yourself.

11. When handling live organisms, follow instructions and do not cause them undue harm or discomfort.

12. Thoroughly wash your hands with soap after handling any live organisms, cultures containing organisms, or chemicals.

13. To smell a substance, gently fan its vapor toward you.

14. Never leave a flame or heater unattended. Keep combustible materials away from heat sources.

15. When diluting acid solutions, always add the acid to water slowly. ***Never add water to an acid!***

16. When heating a test tube, point the open end away from you. ***Never heat a closed or stoppered container!***

17. Dispose of waste as instructed by your teacher.

18. Do not return unused chemicals to a container. Dispose of them properly.

19. Notify the teacher of any injuries, spills, or breakages.

20. Know the locations of the fire extinguisher, safety shower, eyewash station, fire blanket, first-aid kit, and Safety Data Sheets (SDSs).

APPENDIX A

First Aid Rules

BURNS

Flush the area with cold water for several minutes. Do not apply ice.

CHEMICAL SPILLS

Notify your teacher of all chemical spills.

ON A LABORATORY DESK

1. If the material is not particularly volatile, toxic, or flammable, your teacher may have you clean the spill. For liquids, use an absorbent material that will soak up the chemical. For solids, use the designated dustpan and brush. Dispose of chemicals and cleaning materials properly. Then clean the area with soap and water.

2. If the material is volatile, toxic, or flammable, and not a large spill, ask your teacher for help. If it is a large spill, you may need to evacuate the laboratory.

3. If a highly reactive material, such as hydrochloric acid, is spilled, your teacher will clean it up.

ON A PERSON

1. If the spill covers a large area, begin rinsing in the chemical shower, then remove all contaminated clothing and remain under the safety shower. Flood the affected body area for fifteen minutes. Obtain medical help immediately.

2. If the spill covers a small area, immediately flush the affected area with cold water for several minutes. Then wash the area with soap and water. Get medical attention.

3. If the chemical splashes in the eyes, immediately wash them in the nearest eyewash fountain for at least fifteen minutes. Get medical attention.

4. If the spill is an acid, rinse the affected area with sodium bicarbonate or sodium carbonate solution; if it is a base, use citric or ascorbic acid solution.

FIRE

1. For any fire other than a contained fire, do *not* attempt to put it out on your own. ***Understand that some fires can't be put out with water.***

2. Smother a small fire in a container by covering it.

3. If a person's clothes are on fire, remember to stop, drop, and roll—roll the person on the floor and use a fire blanket to extinguish the flames. The safety shower may also be used. ***Do not use a fire extinguisher!***

SWALLOWING CHEMICALS

Determine the specific substance ingested and follow the instructions on the SDS. Contact the Poison Control Center in your area immediately.

CUTS

If the wound is superficial, clean it with a disinfectant, apply triple antibiotic ointment, and cover with a bandage. If it is a deep cut, apply pressure and seek immediate medical attention.

BITES AND STINGS

For minor bites (that do not break the skin) and stings, wash the affected area with soap and water and cover with antibiotic ointment and a clean bandage. Communicate with your teacher about any allergies that you may have and whether you have an up-to-date tetanus shot.

If the bite is from an animal, wash the affected area with soap and water and cover with antibiotic ointment and a clean bandage. See a doctor immediately.

RASHES FROM PLANTS

Rashes caused by exposure to poison ivy, oak, or sumac will not appear immediately after contact; it usually takes about twenty-four hours for symptoms to show. If you suspect that you have contacted these plants, wash the area with soap and water within ten minutes of contact. Be sure to clean any clothing that has contacted plants since it may also contain the toxins.

If a rash is visible, treat with calamine lotion, hydrocortisone cream, or oral antihistamines. If conditions worsen, see a doctor.

APPENDIX B

Laboratory Equipment

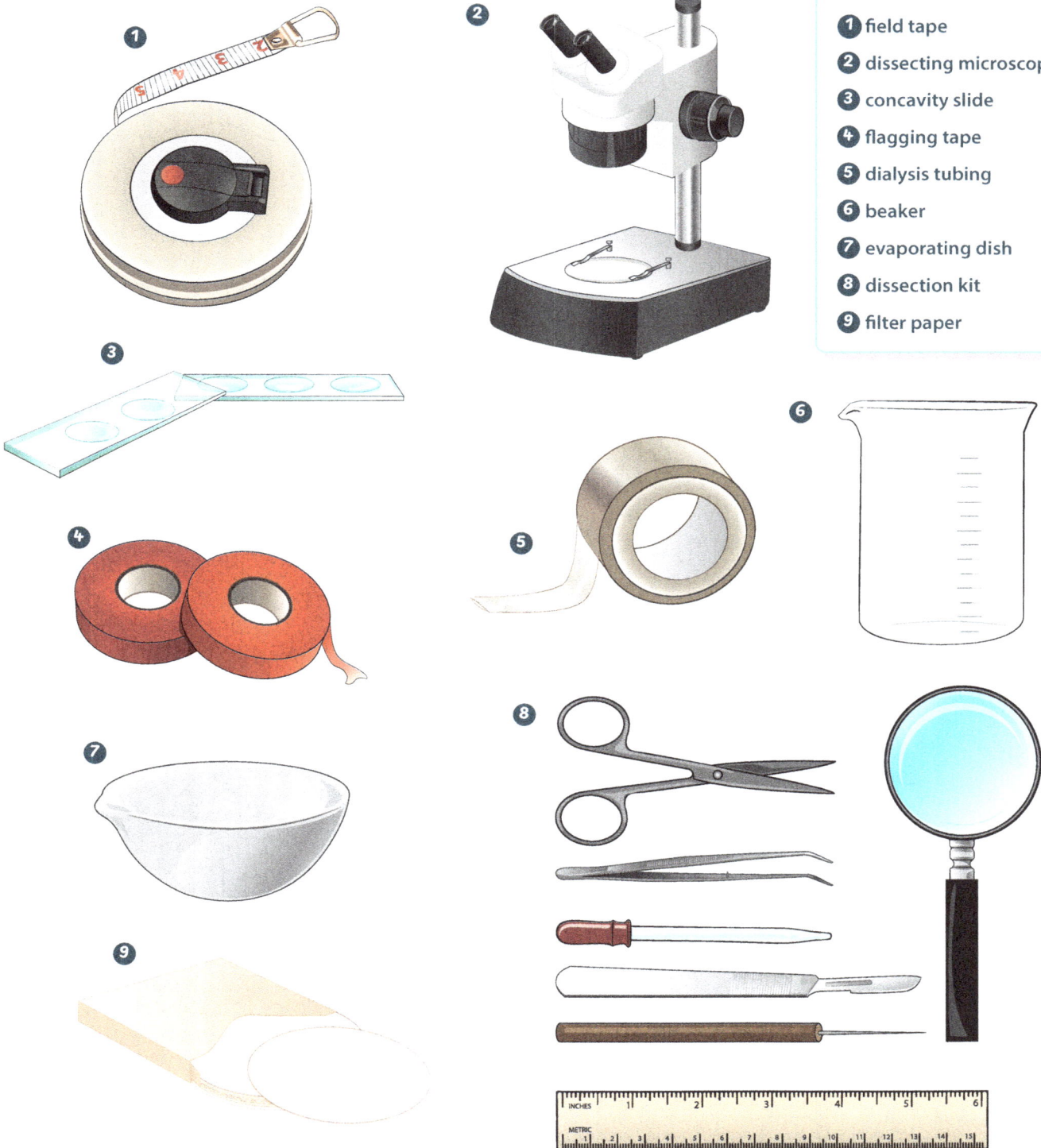

1 field tape
2 dissecting microscope
3 concavity slide
4 flagging tape
5 dialysis tubing
6 beaker
7 evaporating dish
8 dissection kit
9 filter paper

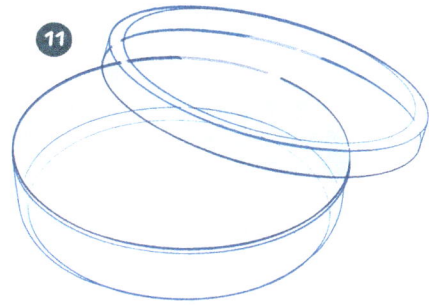

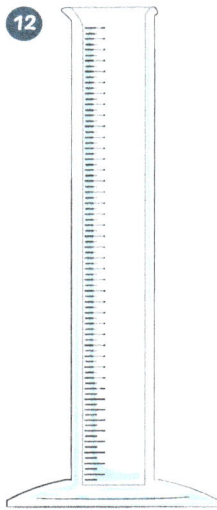

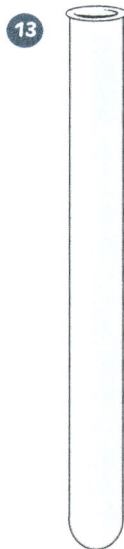

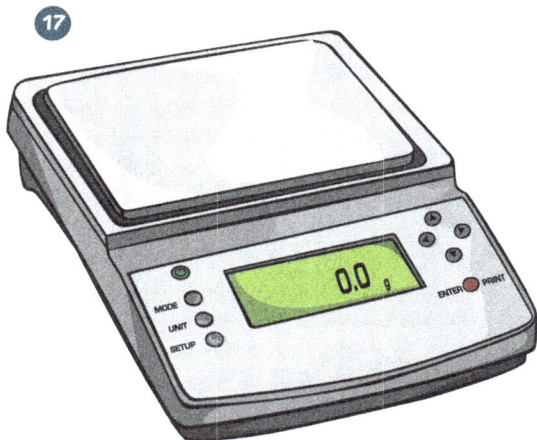

0.0 g

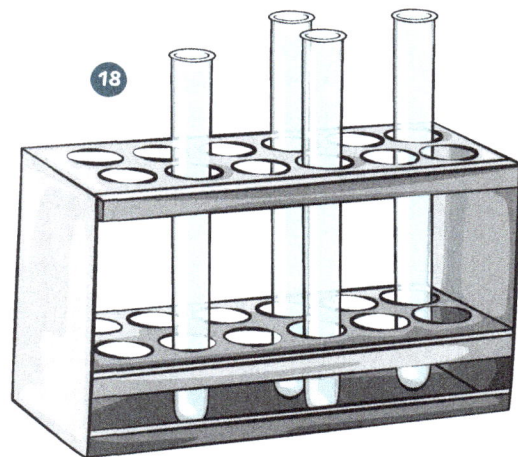

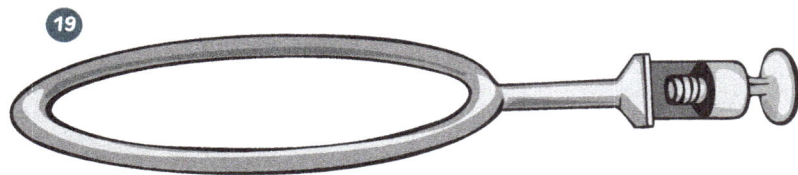

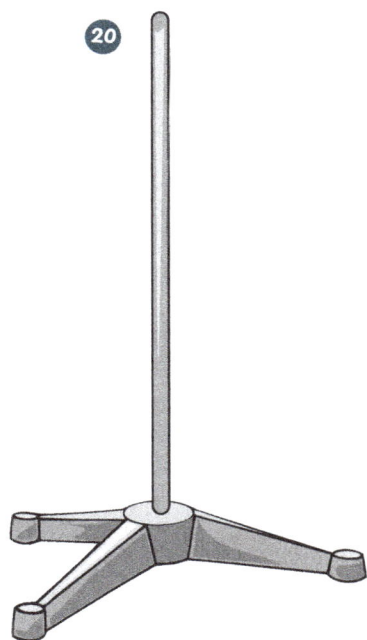

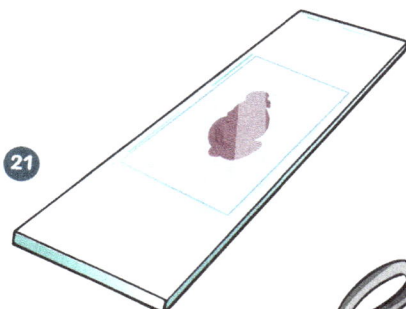

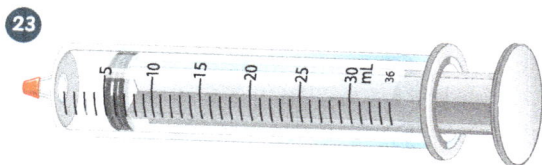

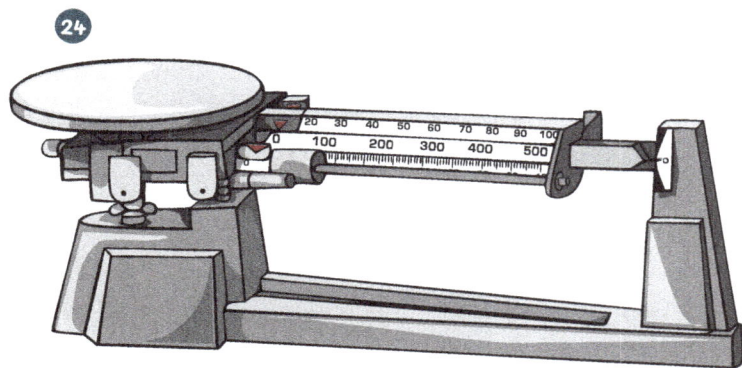

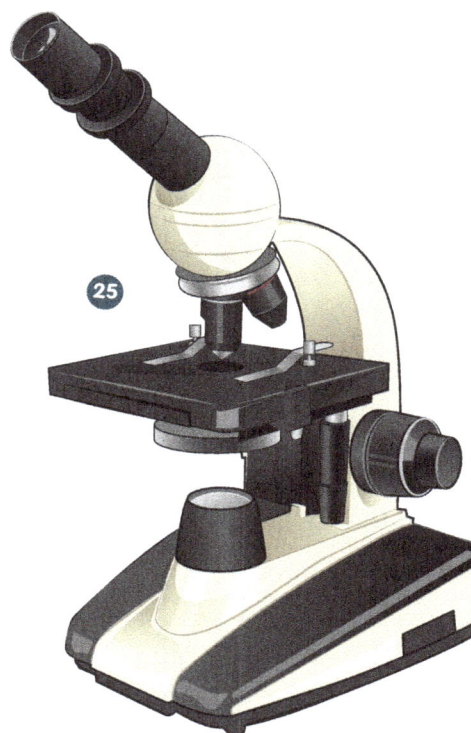

10 stoppers
11 petri dish
12 graduated cylinder
13 test tube
14 pipette
15 spatula
16 test tube brush
17 laboratory balance
18 test tube rack
19 iron ring
20 ring stand
21 microscope slide and cover slip
22 test tube holder
23 syringe
24 triple-beam balance
25 microscope

Anatomical Terms

cephalic—related to the head

anterior—toward the front

transverse plane—divides into front and back

distal—away from the trunk

proximal—closer to the trunk

dorsal, superior—related to the upper side

posterior—toward the rear

frontal or coronal plane—
divides into top and bottom

superficial—
toward the surface

deep—toward
the inside

ventral, inferior—related
to the underside

caudal—related to the tail

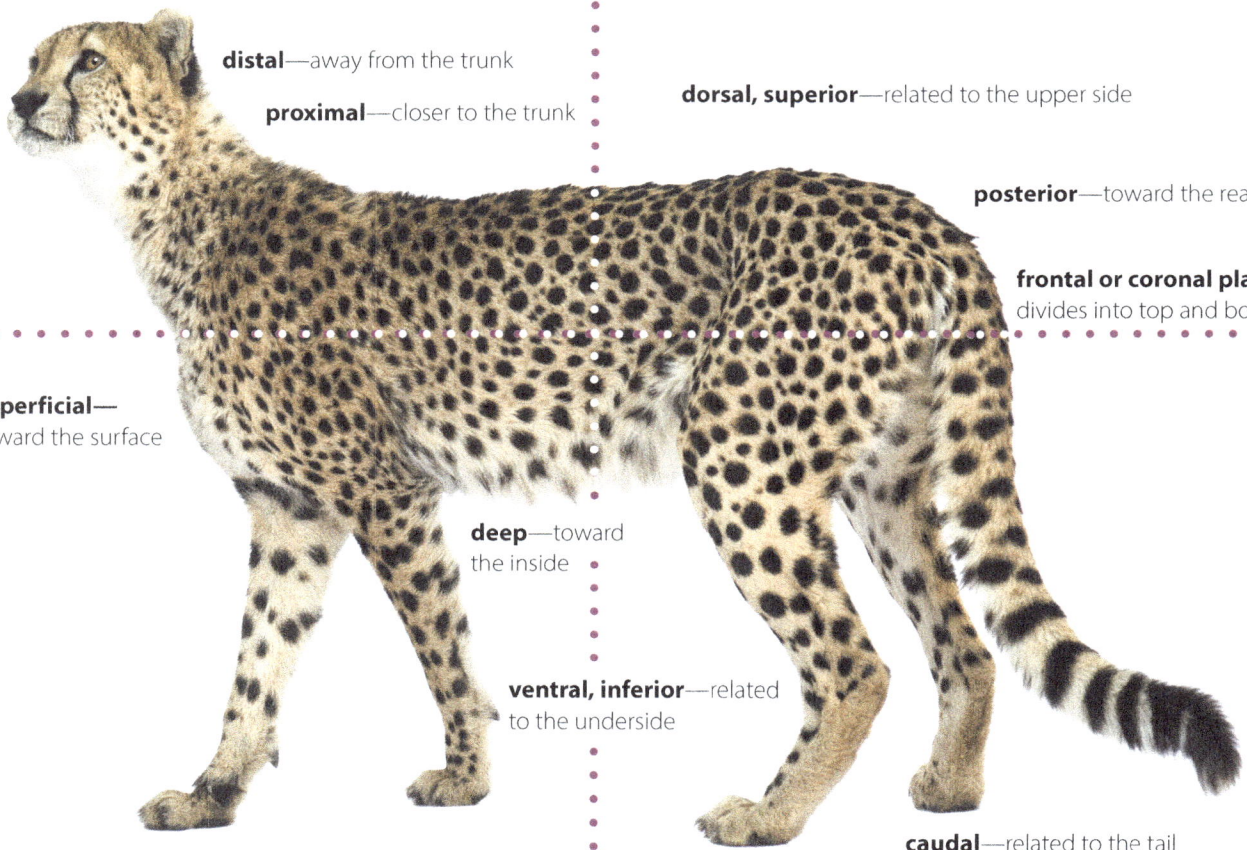

midline (also median or sagittal plane)—divides into left and right

dorsal, superior—related to the top

lateral—related to the side

right side of organism

left side of organism

ventral, inferior—related to the underside

medial—close to the midline

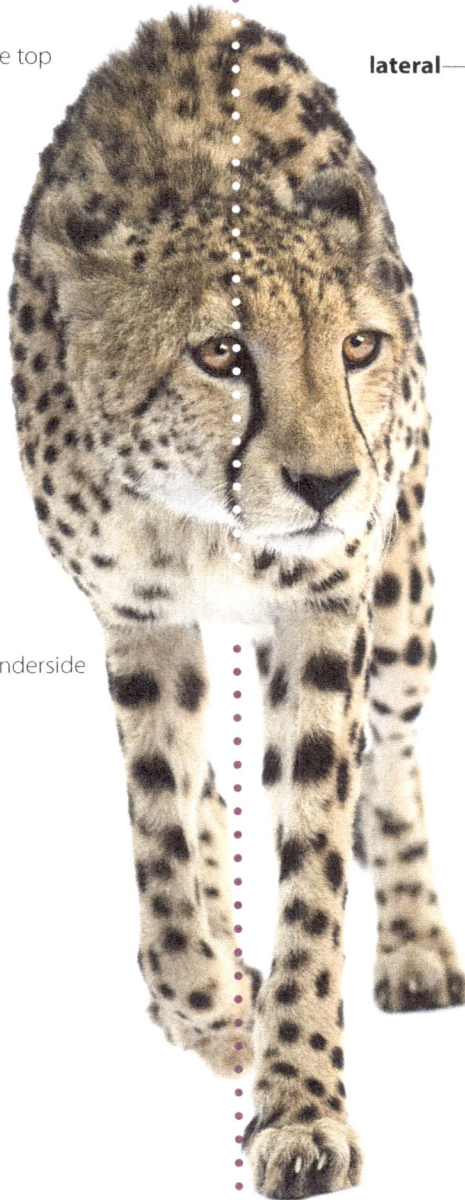

APPENDIX D

Laboratory Techniques

Making Biological Drawings

During some lab activities you will be asked to draw what you observe. Drawing scientific specimens is one of the best ways to learn some of the complex biological structures and processes that you will be observing. Drawing is also a great field skill to have. As you draw, concentrate on the shape and textures of the different structures you observe. By the time you have finished, you should better understand the structures that you have drawn.

Drawing areas have been provided for you. But your teacher may also allow you to take a photo, print it out, and include it in the drawing area.

A Center the name of the specimen above the drawing. If the drawing is of a portion of an organism, indicate the portion in the title or under the title. For example, "Fruit Fly, leg" indicates that the drawing is of the leg of a fruit fly. The title "Fruit Fly" alone indicates a drawing of the entire fly.

B If the specimen or structure has been prepared as a wet mount or a longitudinal mount before you observe it, indicate this under the name of the specimen.

C Make the drawing large and center it in the space available.

D You can use just a pencil for most drawings so that you can erase anything you want to change. If it makes your drawing clearer, use colored pencils, though a colorful drawing should not be a goal in itself.

E Add labels after the drawing is complete; label what you can identify and only what you observe.

F If you used a microscope, indicate the power in the lower right-hand corner. If you used some other type of magnification (such as a hand lens), write "magnified" or "enlarged" in that corner. Don't draw the microscope field or the container that the specimen is in.

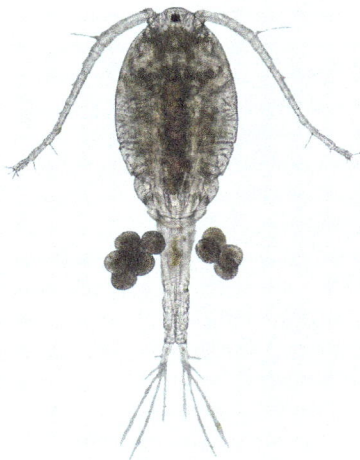

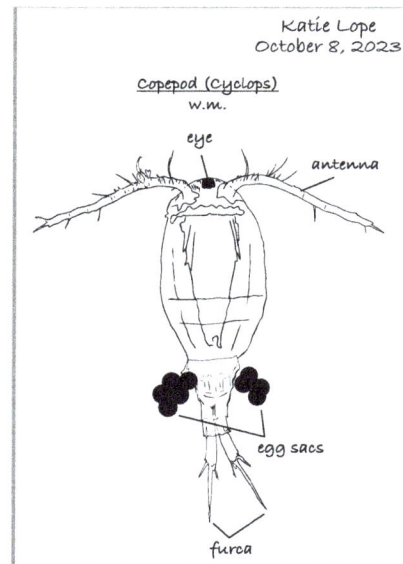

Preparing a Wet Mount

When you use a microscope to look at specimens, sometimes you will use slides that have been prepared ahead of time. But for some lab activities you will need to make your own temporary slide. Placing a specimen in a drop of water to observe it is called *making a wet mount*. To make a wet mount, you need a microscope slide and a cover slip as well as the specimen that you want to observe.

A Handle glass slides and cover slips by their edges so that you do not leave fingerprints on them.

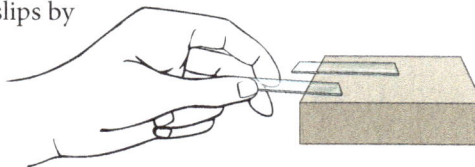

◀ Handling glass slides

B Be careful not to bend plastic cover slips. If you are using a glass cover slip, handle it gently; splinters from shattered glass cover slips easily enter the fingers and may require surgical removal. Inspect plastic cover slips for excessive scratches. If too many scratches appear, discard the cover slip.

C Use only water when washing the slides and cover slips. Soap film may kill or damage living specimens. Shake off excess water and dry the slide with a tissue.

D Place a drop of water on the clean slide.

E Using forceps, place the specimen in the drop of water. In some cases your teacher may provide a special concavity slide that has a shallow well to hold larger specimens.

F Place the cover slip on top of the specimen so that one edge is touching the slide and the cover slip is held at a 45° angle above the drop of water. Slowly lower the cover slip down on top of the water and specimen. If bubbles appear in the area that you are going to view, tap the cover slip with the tip of a probe to remove the bubbles.

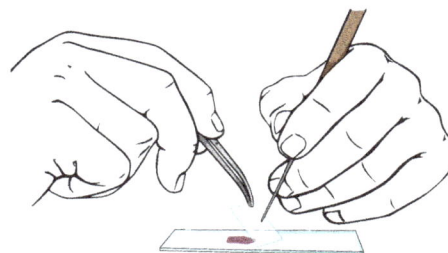

▲

Creating a wet mount

G When finished observing, dispose of the specimen and wash the slide and cover slip, placing them on a paper towel to air dry.

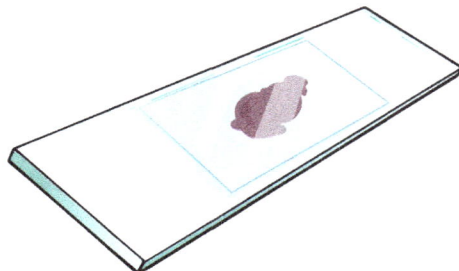

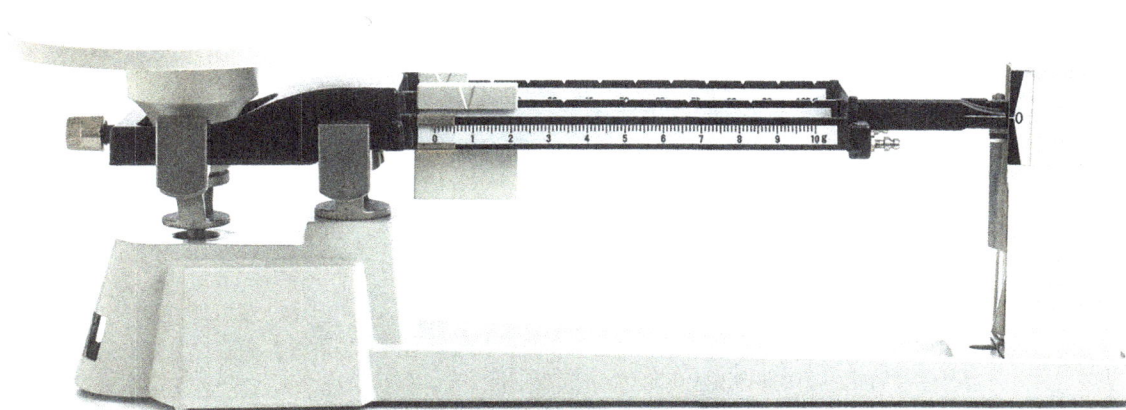

Using Mechanical Balances

The mass of a substance can be determined in the laboratory with the use of a mechanical balance. Several kinds of mechanical balances are common, but all of them operate on the same principles. To use a mechanical balance properly, follow the steps given below.

A Place the balance on a smooth, level surface.

B Keep the balance pan(s) clean and dry. Never put chemicals directly on the metal surface of the pan(s). Always place materials on a sheet of weighing paper, on a weighing boat, or in a container.

C Check the rest point of the empty balance. To do this, remove all weight from the pans and slide all movable masses to their zero positions. If the balance beam swings back and forth, note the central point of the swing. You do not have to wait until the beam stops swinging completely. If the central point on the balance arm does not align with the marked zero point on the post, have your teacher adjust the balance. *Do not adjust the balance yourself!*

D Place the substance on the pan and adjust the sliding masses. Move the largest mass first, and then make final adjustments with the smaller masses. The sum of all the readings is the mass of the object.

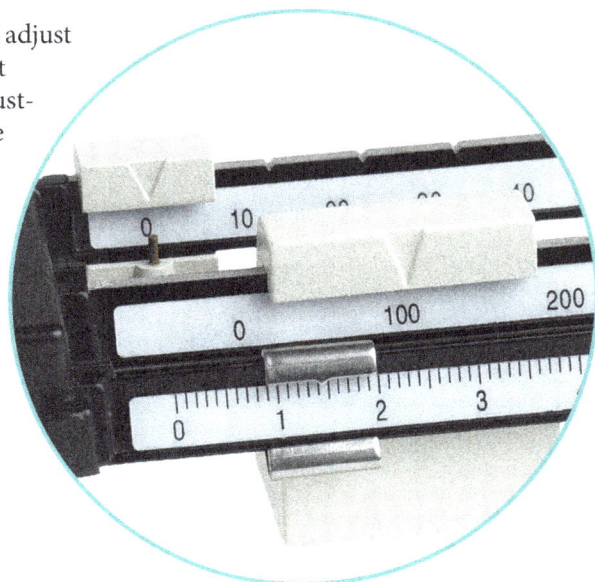

Using an Electronic Balance

Electronic balances are generally faster and easier to use than their mechanical counterparts. To use an electronic balance properly, follow the instructions given below.

A Place the balance on a smooth, level surface.

B Keep the balance pan(s) clean and dry. Never put chemicals directly on the metal surface of the pan(s). Always place materials on a sheet of weighing paper, on a weighing boat, or in a container.

C Turn on the balance. Place the container or weighing paper that will hold the substance whose mass you are trying to find and make sure that there is a reading of 0 by pushing the *Tare* button.

D Place your substance on the paper or in the container and read the mass on the display. If measuring out a predetermined amount of a substance, add the substance until you have reached the appropriate mass.

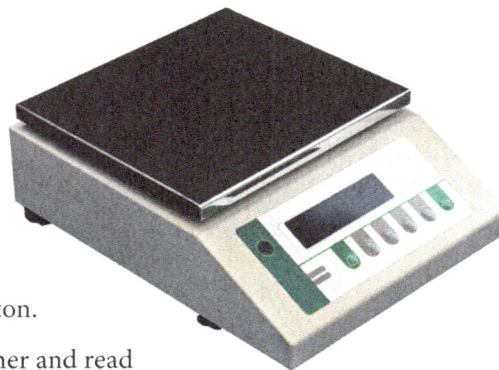

Using a Thermometer

When using a thermometer in lab activities, make sure that you are using one that has the proper temperature range for the experiment that you will be doing. Support the thermometer in a one-hole rubber stopper when necessary to contain the substance whose temperature the thermometer is measuring. To avoid breaking the thermometer and cutting your hand while inserting it in the stopper, lubricate the thermometer and the stopper hole with soap or glycerol. Then protect your hands with puncture-resistant gloves or several layers of paper towels. Hold the thermometer near the stopper and gently twist it into the hole. If you have to use a great amount of force, ask your teacher to enlarge the hole.

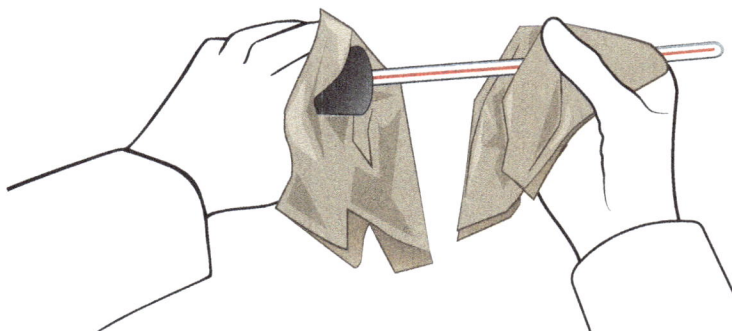

Position the thermometer bulb just above the bottom of the container. If the bulb touches the container, your readings will be inaccurate. Always hold or secure a thermometer in a container; never leave an unsecured thermometer sticking out the top of a container. If a thermometer breaks, alert your teacher and do not touch the inner contents. Some thermometers still in use contain mercury. The spilled mercury may look fascinating, but it is toxic and can be absorbed through the skin.

Handling Liquids

Proper technique for handling liquids is essential if you are to remain safe, keep reagents pure, and obtain accurate measurements. For increased safety do not splash or splatter liquids when pouring. Pour them slowly down the insides of test tubes and beakers. If anything is spilled, wipe it up quickly. (See First Aid Rules on pages 281–82.)

To keep the liquid chemicals pure, keep stirring rods out of the stock supply. Do not let the stoppers and lids become contaminated while you are pouring. Instead, hold the stopper between your fingers. If you must put a lid down, keep the inside surface from touching the surface of the table.

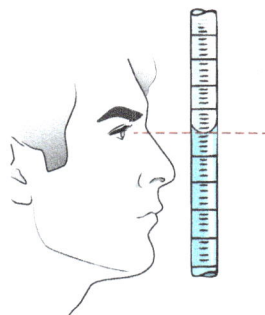

Accurate measurements of liquids can be made in burettes, graduated cylinders, and volumetric flasks. You should measure volumes in these pieces of glassware unless you need only a rough approximation. When reading the level of a liquid, look at the meniscus (curved surface) at the center of the upper surface of the liquid along a horizontal line of sight. For most liquids you will measure at the bottom of the meniscus.

*QUICK NOTE
Measuring Out Powdered Solids

1. Scoop out a little of the sample with a spatula.

2. Gently tap the spatula until the desired amount falls off onto a sheet of weighing paper.

3. Cup the paper to pour the powdered solid into a test tube or other container.

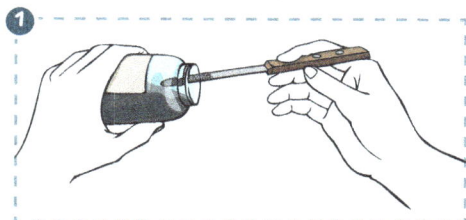

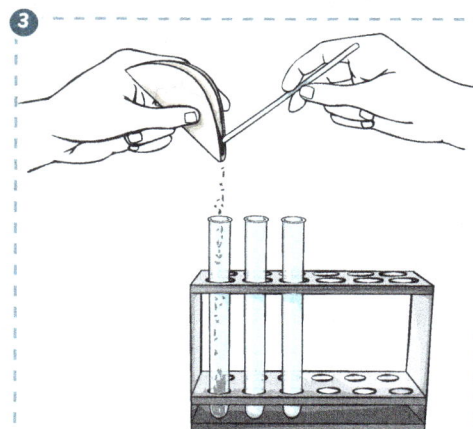

Slowing Protozoan Movement

In some lab activities you will be working with protozoans—microscopic organisms that move. This makes them hard to observe! You can slow down protozoans and make them easier to observe in a wet mount by using a cover slip, cotton fibers, or a thicker medium.

As the culture medium of a wet mount evaporates, the cover slip will press on the organism and slow its movement. Use a paper towel on the edge of the cover slip to speed the process. Your lab partner can draw off small portions of water while you chase the organism. Do not permit the medium to evaporate completely. Replenish it by placing a drop of medium beside the cover slip and letting some of it seep under the cover slip.

Before you place the cover slip on the medium containing protozoans in a wet mount, place a small quantity of cotton fibers on the medium. These serve as obstacles, blocking the path of protozoans and thus localizing their activities.

Use special media (glycerin, methyl cellulose, or commercially prepared products) to slow protozoans. Because these media are thicker than water, the protozoans move more slowly through them.

Dissection Techniques

In this course there are two lab activities in which you will be dissecting specimens. Before you do a dissection, make sure that you read the directions. Your teacher may also have you practice with a virtual dissection. Keep in mind that once you make a cut, you can't undo it!

Make sure that you have identified the correct structures by comparing them to drawings before you cut. Handle the specimens delicately. Preserved structures often tear and break easily.

If you will be finishing your dissection on a following lab day, you will need to label and preserve your specimen. Put your name and lab hour on a plastic bag with a permanent marker. Wrap the organism in a wet paper towel and put it in the bag. Gently squeeze out most of the air and tie or zip the bag closed. Organisms wrapped in this manner may be kept for a few days with good results.

APPENDIX E

Graphing Techniques
Constructing Graphs

When data is recorded in tables, it is difficult to see the relationship that exists between sets of numbers. To make trends and patterns easier to see, you will often put your data on a graph.

In experiments that search for a cause-effect relationship between two variables, you will cause one variable (the independent variable) to change and observe the effect on the second (the dependent variable). For example, if you were to investigate how the level of glucose in the blood changes with time after a meal, time would be the independent variable and blood glucose level would be the dependent variable. Traditionally, the independent variable is plotted on the x-axis of the graph and the dependent variable is plotted on the y-axis.

As you construct a graph, choose appropriate scales. Do not make the graph so small that the data cannot be clearly seen or so large that the graph will not fit on a single sheet of paper. Pick scales that include the entire range of each variable. Keep in mind that the scales on each axis do not have to be the same. For instance, the scale on the x-axis might be 1 h for every line, while the scale on the y-axis could be 50 mg/100 mL for every line. Also, your scale should be easy to subdivide. Subdivisions of 1, 2, 5, and 10 units are the most convenient. Once you have settled on a scale, label the increments on the x- and y-axes. It is not necessary to label every increment. Labeling every second, fifth, or tenth increment will often make your axes look less cluttered.

When you have decided which variable will be plotted on which axis, include a label of each quantity and the units used to measure it on the appropriate axis.

Blood Glucose Level vs. Time

HOURS AFTER EATING (h)	BLOOD GLUCOSE (mg/100 mL)
1	140
2	175
3	120
4	110
5	80
6	75

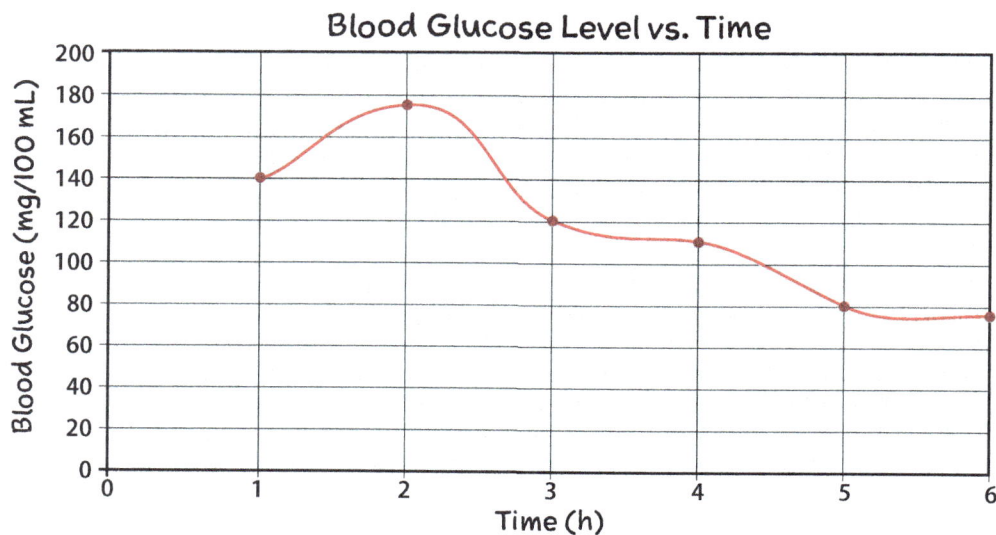

Blood Glucose Level vs. Time

The title of the graph should state the variables (e.g., "Blood Glucose Level vs. Time") and be printed at the top of the graph. If more than one line will be sketched on the same graph, include a legend that identifies each line. Plot each of your data points by making small dots. Follow the specific lab guidelines for the proper way to handle these data points. You will usually draw a smooth curve through the data points. The graph on the facing page illustrates these techniques.

In some cases you will want to draw a straight line even though your data points do not fall precisely in a line. If this occurs, draw a line that shows the general relationship. Be sure to make the line go through the average values of the plotted points. In the first graph at right, the line is incorrect because it lies above the cluster of points near the bottom of the graph and below the cluster of points at the top. The second graph shows the correct method of fitting a straight line to a series of points.

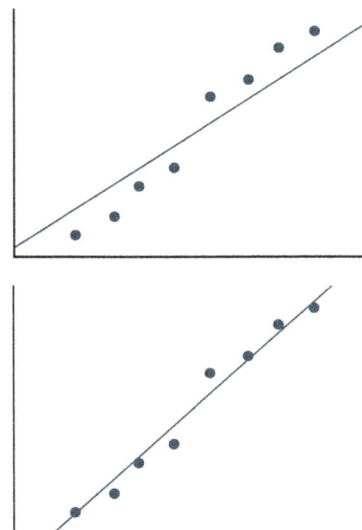

Interpreting Graphs

The shape of a graph tells much about the relationship between the variables. When data appears to be arranged in a straight line, the *x*- and *y*-variables are related in a way that can be expressed as a linear equation. If this line goes through the origin, the linear equation will express the *x*-variable as a multiple of the *y*-variable (a direct variation). A positive slope means that the *y*-variable increases with the *x*-variable; a negative slope indicates that the *y*-variable decreases as the *x*-variable increases.

Data points that curve up (or down) from left to right indicate that data may be best modeled by some nonlinear equation. The equation relating the two variables may contain an exponent. Other possible relationships are inverse, exponential, or logarithmic; some can be modeled by a polynomial.

Graphs can be used to predict additional data points that have not been experimentally determined. Predicting points between data points on the basis of the graphed line is called *interpolation*. From the graph of blood glucose levels on the previous page it is reasonable to assume that the blood glucose level would be near 150 mg/100 mL at 2.5 h. Predicting values past the data points by extending the graphed line in either direction is called *extrapolation*. The graph of blood glucose level indicates that it could fall to 60 mg/100 mL at 7 h. This extrapolation is reasonable. The further from the data points we attempt to extrapolate, the less reasonable our predictions are likely to be. For example, if we tried to use the data provided to extrapolate out to 24 h, we would not make a reasonable prediction.

Predictions from a Graph
Extrapolation always introduces the element of uncertainty in scientific studies. Use this fact to discuss the ways in which extrapolation may be misused in our modern culture—in particular, projected climate change and age-of-the-earth theories that are based on dating methods with limited reliability.

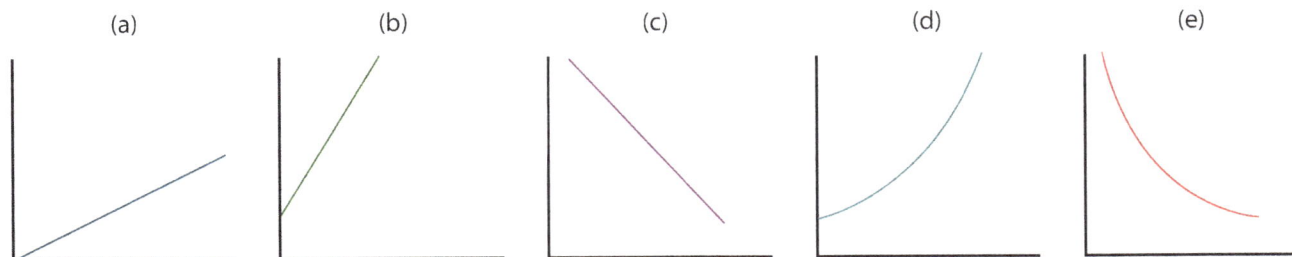

The graphs in this lab manual can have different shapes, showing linear relationships (a, b, c) or exponential relationships (d, e).

Writing Formal Lab Reports

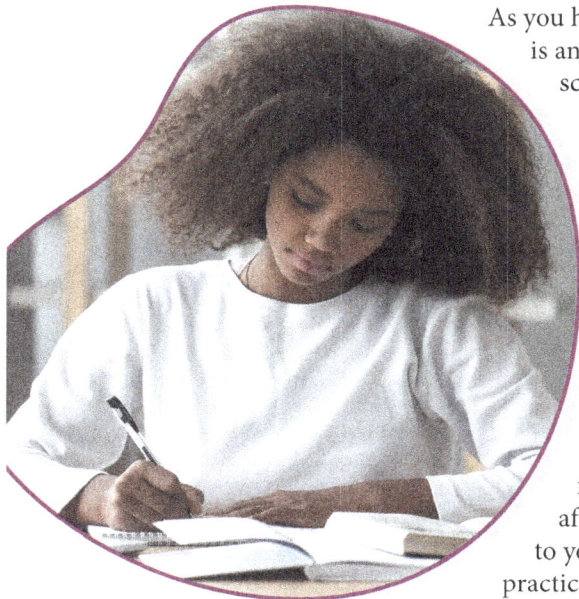

As you have learned in your study of science, communication is an important part of the scientific process. Part of what scientists do is share what they have learned with other scientists. One example of this is the writing of formal lab reports. In a lab report a scientist presents a problem that was investigated, describes how the investigation was conducted, and summarizes what was learned during the investigation. Many college science courses require students to present all their laboratory work in formal lab reports.

To better prepare you for college, your teacher may choose to have you submit your lab assignments in formal lab reports rather than use the fill-in-the-blank format found in this lab manual. You may be more familiar with the latter form, but there really is little to be afraid of about writing formal reports. They may be new to you, and the word "formal" may sound alarming, but with practice such reporting can become second nature.

There really isn't a standardized way of writing and compiling a formal lab report. Your teacher may require you to do all your reporting in a bound journal, or you may be asked to submit individual typed reports. Regardless of which method your teacher prefers, there are some properties of lab reports that teachers generally agree upon. Let's look at the style and mechanics of a formal lab report and the type of content that is normally included.

Style and Mechanics

The first thing to keep in mind about a formal lab report is that it is a type of formal writing. That means that a formal lab report is like the final draft of an essay or research paper. It should have a neat, orderly appearance. To that end, formal lab reports normally have the following elements of style and mechanics.

» Unbound lab reports should be typed in black ink. Reports in a journal should be neatly written.

» Lab reports should be written in complete sentences and be checked for proper spelling, punctuation, and grammar.

» Formal lab reports should be free of erasures, whiteout, or scratched-out words.

» Since objectivity is a key aspect of the scientific process, lab reports usually avoid first-person narrative and instead use an impersonal third-person form.

Example:

INFORMAL: "I learned that there aren't any juvenile Chinook salmon upstream of Misty Falls."

FORMAL: "The survey indicated an absence of juvenile Chinook salmon upstream from Misty Falls."

Format

Scientists (and science teachers) may disagree on the exact number of components in a formal lab report, but there is a general consensus on what is presented and the order in which it is presented. Keep in mind as you write that a formal lab report answers three basic questions.

» What problem was investigated?

» What procedure was used during the investigation?

» What was learned from the investigation?

To answer those questions, a formal lab report generally includes the elements shown below.

» title

» synopsis

» introduction

» list of any materials used and a description of the procedure

» data yielded by the procedure

» analysis of the data and a discussion of the findings

» sources

Let's take a closer look at each of these.

TITLE

A formal lab report title should succinctly describe what the lab activity was about. It should be matter-of-fact rather than creative. Consider the following possible titles for a report on a Chinook salmon survey.

Example:

INFORMAL: Here, Fishy, Fishy!

FORMAL: The Distribution of Chinook Salmon within Boulder Creek

It's OK for a formal lab report title to sound dry—save the creative writing for your language arts class! Imagine yourself as another scientist who is reading your report as part of research and wants to quickly know what your report is about without having to read deep into the main body of the text.

SYNOPSIS

A synopsis is a very brief summary of the entire lab report, that is, two or three sentences. If the title of your report catches the reader's attention, the synopsis should then provide just enough detail to decide whether to read further.

Example (based on a presence/absence survey for juvenile Chinook salmon):

Three-pass electrofishing was used to sample five reaches within Boulder Creek to evaluate the distribution of Chinook salmon (*Oncorhynchus tshawytscha*). Juvenile Chinook salmon were found at only one sampling site, downstream from Misty Falls, suggesting that the falls act as a complete barrier to the upstream migration of Chinook salmon.

APPENDIX F

INTRODUCTION

Your introduction should expand on your synopsis by providing more details about the experiment. An introduction should include the elements listed below.

» a description of the purpose of the experiment

» the hypothesis that will be tested

» a summary of the methods that will be used to test the hypothesis (Be brief. A more detailed description should be reserved for the Materials and Methods section that follows.)

» a brief description of the results of the procedure

» a summary of the conclusions that you have drawn on the basis of the results

MATERIALS AND METHODS

The goal of this section of your lab report is to describe your investigation in enough detail to allow another person to assess its validity or even replicate it. That is why there are two parts to this section: materials and methods. You should first list your materials, including names and quantities where appropriate (e.g., for chemicals). Your list should be just that—a list. Think about how you normally see such lists in this lab manual.

Next, describe your methods. You do not need to give instructions for how to do common tasks, such as using an electronic balance. Nor do you need to explain how to do common procedures if your audience is expected to be familiar with them. For example, in the population-sampling example used above it would not be necessary to explain how electrofishing is done; your audience probably already knows how to do this. What your audience would need to know in order to replicate your experiment is the type of electrofishing gear that you used along with the particular sampling methodology that was followed. As you write this section, ask yourself whether someone reading your descriptions would be able to replicate your experiment exactly. If the answer is no, then you need to be more specific. It is often helpful to include a diagram of any experimental setup used.

DATA

Your data section should be reserved for raw data only. This is where you report the results of your tests. Your data may be qualitative (i.e., based on observations) or quantitative (i.e., based on measurements). Often, a good way to report quantitative data is to use a data table. The format of data tables will vary depending on the data that you collected. An excellent technique, especially if you are using spreadsheets, is to record the units of measure in the column header and record only numerical data in the cells. This also allows you to use the computational features of the software. All quantitative data should be reported with the appropriate number of significant figures to indicate the precision of the measurements.

ANALYSIS AND DISCUSSION

» This is where the fun part begins! Now you get to interpret the data that you presented in the previous section. First, describe how the data was analyzed and show any formulas or equations that were used to do so. Sample calculations should be included. Calculated values should be reported with the proper number of significant figures. Descriptions of the results should include appropriate statistical parameters, such as the mean, median, and range of values. Where appropriate, use graphs to show trends in the data. All graphs should be properly scaled and include titles and axis labels with units.

» Once you have presented the analysis of your data, you may then address the all-important question: What does it mean? Think back to your stated hypothesis—did your data and analysis support your hypothesis? Were your results in line with what you expected? How did your results compare with those of others in your class? Be sure to support any claims you make with the relevant data and analyses. Your reader—your teacher—expects to see logical arguments that are consistent with your experimental results.

» Your discussion needs to be frank and honest. Don't try to make your data say something that it isn't actually saying. If your data does not support your hypothesis, then say so! Don't say that it does just because you think that might be the answer that your teacher is expecting. An important aspect of nurturing science skills is learning how to discuss in-conclusive or contrary data. Examine what may have gone wrong during the experiment, make suggestions for improving future investigations, or propose a modified hypothesis for future testing.

» It is also appropriate during your discussion to suggest possible practical applications of your work or to put forward additional related questions that might be investigated in the future. In doing so, you might be laying the groundwork for a future scientist. By this you can see how the process of science has links both to those who have investigated a question in the past and to those who may investigate it in the future.

SOURCES CITED

» Much like any formal research paper, a lab report should cite sources within the body of the report and include a bibliography at the end. There are different formats for doing this. Your teacher will advise you on which format to use.

APPENDIX G

Grading Rubric for Formal Lab Reports

Excellent—Student's improved design demonstrates thorough understanding of the task. Work may include minor flaws in design or execution that do not affect completion of the task.

Good—Student's improved design demonstrates good understanding of the task. Work may have some flaws.

Fair—Student's improved design demonstrates some understanding of the task. Work may have major flaws.

Needs Work—Student's improved design demonstrates little or no understanding of the task.

Point values are left blank to allow teachers to assign values within their grading system.

	Excellent (_____ pts)	Good (_____ pts)	Fair (_____ pts)	Needs Work (_____ pts)
CONCEPT				
Problem	clear and concise statement of problem or question to be addressed	statement of problem or question to be addressed provided but unclear on minor details	partial statement of problem or question to be addressed; details lacking or confusing	no statement of problem or question to be addressed
Hypothesis	clearly stated and testable hypothesis	hypothesis stated but unclear on minor details	unclear or untestable hypothesis	no hypothesis
PRODUCT				
Format	report complete, contains all expected components, and follows expected format	report mostly complete, contains most expected components, and generally follows expected format	report incomplete, contains only some expected components, and substantially deviates from expected format	report largely incomplete, contains few or none of the expected components; no attempt made to follow the expected format
Workmanship	report neat and free of mechanical errors	report mostly neat, contains minor mechanical errors that do not hinder understanding	report messy, contains some mechanical errors that hinder understanding	report messy, contains many mechanical errors that hinder understanding
Discussion	thorough discussion, provides evidence of whether the hypothesis was supported	discussion lacks detail or does not provide sufficient evidence of whether the hypothesis was supported	discussion confusing or incomplete, may lack evidence for whether the hypothesis was supported	little or no discussion
QUALITY				
Timeliness	on time			late

APPENDIX H

Equipment and Materials List (alphabetical)

Amount required is the number of that item needed to equip one lab group for the entire course of activities.

ITEM	SIZE	# REQUIRED	LAB	REMARKS
10-10-10 fertilizer			**12B**	
acetic acid (CH_3COOH)			**12B, 16A**	Lab 12B: 5 mL, 1 M solution Lab 16A: dilute (~5%) or may substitute vinegar
active dry yeast			**7B**	
banana purée, ripe			**15A**	
banana purée, unripe			**15A**	
beaker	50 mL	1	**6B**	
	150 mL	2	**6B**	
	250 mL	3	**23B, 24A**	
Benedict's solution			**15A, 23B**	
birding app		1	**19A**	downloaded onto smart device; may substitute field guide
biuret reagent			**23B**	
blender		1	**6B**	
blindfold		1	**24A**	
blood pressure cuff, digital		1	**22B**	may substitute manual sphygmomanometer
calculator		1	**4B, 25A**	
camera		1	**4A, 14A, 17A, 19A**	optional
carmine powder			**13A**	may substitute yeast
clamp		6	**23B**	
clear plastic cup			**7A**	Number of cups depends on how many trial groups created.
colored pencils			**7A, 25A**	1 set
computer with internet connection		1	**19B**	
concavity slide		1	**16A**	
container	small to medium	5	**9A, 15B**	
cooking oil	60 mL		**12B**	
cork stopper		1	**5A, 23A**	
cornmeal	1 tsp		**7B**	
correction fluid pen		1	**3A**	
cotton ball, sterile			**2, 12B**	Lab 12B: at least 4 per lab group
cotton fiber			**13A**	

APPENDIX H

ITEM	SIZE	# REQUIRED	LAB	REMARKS
cotton swab, sterile		4	**12A**	
cover slip		1	**5A, 6B, 13A, 16A**	Lab 5A quantity: 3 Lab 13A quantity: 5
culture dish		1	**16A, 16B, 17A**	
diagram, skeletal			**21A**	may substitute for skeleton model
dialysis tubing	12 in.	1	**23B**	
dip net	small	1	**3A**	may substitute small cup or other small container
dish detergent			**6B**	liquid
dishpan		1	**21B**	
dissecting microscope		1	**13B, 16B**	may substitute hand lens
dissection kit		1	**5A, 17A, 24B**	
dissection pan		1	**17A, 24B**	
duckweed			**3B**	Amount will vary according to students' procedures; expect ~50 plants per lab group.
Epsom salt			**16B**	
evaporating dish	100 mL	1	**23A**	
field guide		1	**14A**	
field tape		1	**4A, 4B**	may substitute tape measure or marked rope
flat toothpick		1	**5A**	
flower-head pin		8	**17A**	numbered 1–8 with permanent marker
forceps		1	**7A, 16B**	
garden shears		1 pair	**14A**	
gloves, garden		1 pair	**14A**	
gloves, nitrile			**6A, 6B, 12B, 15A, 17A, 23B, 24B**	1 pair per student per activity
glucose syrup	1 tsp		**7B**	may substitute Karo Light syrup
glue			**17A**	may substitute rubber cement or tape
glycerin			**13A**	may substitute methyl cellulose
goggles			**5A, 5B, 6A, 6B, 7B, 12A, 12B, 13B, 15A, 17A, 23A, 23B, 24A, 24B**	1 pair per student
graduated cylinder	10 mL	1	**12B, 15A**	
	50 mL	4	**7B, 12B, 23A**	
	100 mL	1	**6B, 23B**	

ITEM	SIZE	# REQUIRED	LAB	REMARKS
grill lighter		1	23A	
hand gripper		1	21B	
hand lens		1	14A, 14B, 16A, 16B	Lab 16A: may substitute dissecting microscope
hand sanitizer			12A	
hand soap			12A	
hexagonal metal nut	large	1	5A	
hole punch		1	7A	
hot water bath		1	15A, 23B	
immersion oil			1B	Omit if microscope does not have an oil immersion objective.
index card			16B	may substitute small piece of cardboard
iodine solution			15A, 23B	
iron ring		1	23A	
isopropyl alcohol			6B, 22A	Lab 6B: 90 mL
Key for Selected Fishes of North America			11A	included in Lab 11A
kitchen knife	large	1	14B	
laboratory apron			5A, 5B, 6A, 6B, 12A, 12B, 13B, 15A, 17A, 23A, 23B, 24A, 24B	1 per student
laboratory balance		1	12B, 23A	accurate to 0.01 g
laboratory spatula		1	15A	
laser pointer		1	2	may substitute flashlight
lens paper			1B	
lentils, dry			6B	uncooked
light source			7A	
live cricket	large		3A, 17B	Lab 3A: 30–50 needed Lab 17B: Number depends on the number of students and their experimental designs.
live fish			18A	
live lizard		1	18B	may substitute any small, carnivorous lizard
living culture, amoeba			13A	
living culture, brine shrimp			16A	may substitute *Daphnia*
living culture, euglena			13A	
living culture, hydra			16A	

APPENDIX H

ITEM	SIZE	# REQUIRED	LAB	REMARKS
living culture, paramecium			**13A**	
living culture, *Penicillium notatum*			**13B**	
living culture, planarian			**16B**	
living culture, *Rhizopus stolonifer*			**13B**	synonym *Rhizopus nigricans*
lung volume bag		1	**22A**	
marshmallow		1	**23A**	
measuring spoon	set	1	**7B**	
meter stick		1	**1A**	
methylene blue			**5A, 6B**	
microscope		1	**1B, 5A, 6B, 8A, 12A, 13A, 13B, 16A, 21A**	
microscope, dissecting		1	**16B**	optional
microscope slide, glass		3	**5A, 6B, 13A**	Lab 5A quantity: 3 Lab 13A quantity: 5
milk of magnesia	8 mL		**12B**	
mixed solutes solution	150 mL		**23B**	
mouthpiece		3	**22A**	
multimeter		1	**23B**	may substitute conductivity probe
notebook, field	small	1	**4A, 14A, 18A, 19A**	Labs 4A, 19A: optional
nutrient agar			**2**	
oil-eating bacteria suspension	80 mL		**12B**	
onion		1	**5A**	
oral syringe		1	**7A**	10 mL or larger
osmometer		1	**5B**	
paper sack		1	**15B**	
pen, washable ink		1	**24A**	
pencil		1	**18A**	
permanent marker		1	**7B, 12A, 14A**	
permanent marker, fine-tip		1	**17A**	
petri dish		1	**2, 12A, 15A, 16B**	sterile Lab 15A quantity: 4
phenol red			**6B**	pH indicator

ITEM	SIZE	# REQUIRED	LAB	REMARKS
pipette		several	**5A, 5B, 12B, 13A, 15A, 16A, 16B, 23B**	disposable plastic Lab 13A quantity: 5 Lab 16A quantity: 3
plastic bag	small		**14A, 17A**	Lab 14A: several Lab 17A: 1 bag
	medium	1	**24B**	
	large	3	**12A**	
plastic bag, resealable	snack-size	3	**7B**	
plastic bead, red		100	**9A**	
plastic bead, white		100	**9A**	
plastic jug	1 gal	1	**21B**	filled with water
preserved crayfish, injected	medium	1	**17A**	optional
preserved rat, injected		1	**24B**	
preserved slide, beef tapeworm bladders in meat		1	**16B**	optional
preserved slide, beef tapeworm proglottid		1	**16B**	optional; may substitute other species of tapeworm
preserved slide, beef tapeworm scolex		1	**16B**	optional; may substitute other species of tapeworm
preserved slide, budding hydra		1	**16A**	
preserved slide, *Coprinus*, c.s.		1	**13B**	
preserved slide, cup fungus apothecium		1	**13B**	
preserved slide, desmids or diatoms		1	**1B**	
preserved slide, dry ground bone, c.s.		1	**21A**	
preserved slide, fish embryos prepared for viewing mitosis		1	**8A**	
preserved slide, fish prepared for viewing meiosis		1	**8A**	
preserved slide, human liver fluke		1	**16B**	optional
preserved slide, hydra c.s.		1	**16A**	
preserved slide, *Rhizopus stolonifer*, w.m.		1	**13B**	
probe		1	**16A, 16B**	usually part of dissecting kit; may substitute toothpick
rangefinder		1	**4B**	optional
rat dissection reference material			**24B**	
rat skeleton, mounted		1	**24B**	

ITEM	SIZE	# REQUIRED	LAB	REMARKS
raw beef liver			**16B**	may substitute boiled egg yolk or flaked or pelleted fish food
razor blade, single-edged		1	**5A, 14B**	
reference book for protozoans			**13A**	or key
ring stand		1	**23A**	
rod, glass or metal		2	**24A**	may substitute dulled 16-p nail
rubber band		1	**22A**	
ruler		1	**5B**	
ruler, metric		1	**12A, 24A**	
salt solution	15 mL		**6B**	6% solution
scalpel		1	**14B**	usually included in dissecting kit
science periodicals			**10B**	current issues; may substitute digital copies of online articles
seawater			**15B**	
seed, plant			**15B**	
skeleton, human		1	**21A**	model
snack foods, various			**23A**	with nutrition fact labels
soda can, empty		1	**23A**	with tab attached
sodium bicarbonate solution			**7A**	2.5% solution
specimen, fresh flower		several	**14B**	
specimen, fresh fruit		several	**14B**	
specimen, preserved, crayfish		1	**17A**	showing crayfish life cycle; optional
specimen, preserved, mushroom		1	**13B**	
specimen, preserved, puffball		1	**13B**	
specimen, preserved, shelf fungus		1	**13B**	
specimen, seed		several	**14B**	
spinach or ivy leaves, fresh		1	**7A**	
spray bottle		1	**15B**	
squeeze bottle		1	**5B**	
stakes, or flagging tape			**4B**	
stethoscope		1	**22A**	
stirring rod, glass		1	**6B, 12B, 23A**	Lab 12B: 3 per student
stopwatch		1	**7A, 21B**	
straight pin	large	1	**23A**	
sucrose solution		2	**5B**	0.5 molal (Solution B) and 1.0 molal (Solution A)

ITEM	SIZE	# REQUIRED	LAB	REMARKS
tape measure, metric		1	24A	
thermometer or temperature probe		1	23A	
terrarium	medium to large	1	3A, 18B	Lab 3A: may substitute large bucket
test tube	medium	9	15A, 23B	Lab 15A quantity: 2
test tube holder		1	23B	
test tube rack		1	23B	
tissue			1B, 22A	
tongs		1	23A	may substitute mitts
toothpick		1	13A	per student
tuning fork, assorted			24A	include 128 Hz and 256 Hz
universal indicator paper			2, 12B	Lab 12B: may use a pH meter
water sample		2	2, 6A	obtained from local water source Lab 2: untreated
water test kit		1	6A	LaMotte 5860-01; may substitute similar kit
water, distilled			5B, 15B, 23B	
water, hot			24A	
water, ice			24A	
water, pond			13A	
water, spring			16A, 16B	may substitute treated tapwater
water, warm			7B	~40 °C
wax pencil		1	5B, 12B	
weighing paper			12B	
wood shavings			7B	may substitute grass clippings

APPENDIX H

Equipment and Materials List (by lab activity)

Amount required is the number of that item needed to equip one lab group for the entire course of activities.

LAB	ITEM	SIZE	# REQUIRED	REMARKS
1A	meter stick		1	
1B	immersion oil			Omit if microscope does not have an oil immersion objective.
	lens paper			
	microscope		1	
	preserved slide, desmids or diatoms		1	
	tissue			
2	cotton ball, sterile			
	laser pointer		1	may substitute flashlight
	nutrient agar			
	petri dish and lid		1	sterile
	universal indicator paper			
	water sample		2	untreated, obtained from local water source
3A	correction fluid pen		1	
	dip net	small	1	may substitute small cup or other small container
	live cricket	large		30–50 needed
	terrarium	medium to large	1	may substitute large bucket
3B	duckweed			Amount will vary according to students' procedures; expect ~50 plants per lab group.
4A	camera		1	optional
	field tape		1	may substitute tape measure or marked rope
	notebook, field	small	1	optional
4B	calculator		1	
	field tape		1	may substitute tape measure or marked rope
	rangefinder		1	optional
	stakes, or flagging tape			

LAB	ITEM	SIZE	# REQUIRED	REMARKS
	cork stopper		1	
	cover slip		3	
	dissection kit		1	
	flat toothpick		1	
	goggles			1 pair per student
	hexagonal metal nut	large	1	
5A	laboratory apron			1 per student
	methylene blue			
	microscope		1	
	microscope slide, glass		3	
	onion		1	
	pipette		1	
	razor blade, single-edged		1	
	goggles			1 pair per student
	laboratory apron			1 per student
	osmometer		1	
	pipette		several	disposable plastic
5B	ruler		1	
	squeeze bottle		1	
	sucrose solution		2	0.5 molal (Solution B) and 1.0 molal (Solution A)
	water, distilled			
	wax pencil		1	
	gloves, nitrile			1 pair per student per activity
	goggles			1 pair per student
6A	laboratory apron			1 per student
	water sample		2	obtained from local water source
	water test kit		1	LaMotte 5860-01; may substitute similar kit

LAB	ITEM	SIZE	# REQUIRED	REMARKS
6B	beaker	50 mL	1	
		150 mL	2	
	blender		1	
	cover slip		1	
	dish detergent			liquid
	gloves, nitrile			1 pair per student per activity
	goggles			1 pair per student
	graduated cylinder	100 mL	1	
	isopropyl alcohol	90 mL		
	laboratory apron			1 per student
	lentils, dry			uncooked
	methylene blue			
	microscope		1	
	microscope slide, glass		1	
	phenol red			pH indicator
	salt solution	15 mL		6% solution
	stirring rod, glass		1	
7A	clear plastic cup			Number of cups depends on how many trial groups created.
	colored pencils			1 set
	forceps		1	
	hole punch		1	
	light source			
	oral syringe		1	10 mL or larger
	sodium bicarbonate solution			2.5% solution
	spinach or ivy leaves, fresh		1	
	stopwatch		1	

LAB	ITEM	SIZE	# REQUIRED	REMARKS
7B	active dry yeast			
	cornmeal	1 tsp		
	glucose syrup	1 tsp		may substitute Karo Light syrup
	goggles			1 pair per student
	graduated cylinder		4	50 mL or larger
	measuring spoon	set	1	
	permanent marker		1	
	plastic bag, resealable	snack-size	3	
	water, warm			~40 °C
	wood shavings			may substitute grass clippings
8A	microscope		1	
	preserved slide, fish embryos prepared for viewing mitosis		1	
	preserved slide, fish prepared for viewing meiosis		1	
8B	none			
9A	container	small to medium	5	
	plastic bead, red		100	
	plastic bead, white		100	
9B	none			
10A	none			
10B	science periodicals			current issues; may substitute digital copies of online articles
11A	Key for Selected Fishes of North America			included in Lab 11A
11B	none			

APPENDIX H

LAB	ITEM	SIZE	# REQUIRED	REMARKS
12A	cotton swab, sterile		4	
	goggles			1 pair per student
	hand sanitizer			
	hand soap			
	laboratory apron			1 per student
	microscope		1	
	permanent marker		1	
	petri dish and lid		1	sterile
	plastic bag	large	3	
	ruler, metric		1	
12B	10-10-10 fertilizer			
	acetic acid (CH_3COOH)	5 mL		1 M solution
	cooking oil	60 mL		
	cotton ball, sterile			at least 4 per lab group
	gloves, nitrile			1 pair per student per activity
	goggles			1 pair per student
	graduated cylinder	10 mL	1	
		50 mL	4	
	laboratory apron			1 per student
	laboratory balance		1	accurate to 0.01 g
	milk of magnesia	8 mL		
	oil-eating bacteria suspension	80 mL		
	pipette		several	disposable plastic
	stirring rod, glass		3	quantity per student
	universal indicator paper			may use a pH meter
	wax pencil		1	
	weighing paper			

LAB	ITEM	SIZE	# REQUIRED	REMARKS
13A	carmine powder			may substitute yeast
	cotton fiber			
	cover slip		5	
	glycerin			may substitute methyl cellulose
	living culture, amoeba			
	living culture, euglena			
	living culture, paramecium			
	microscope		1	
	microscope slide, glass		5	
	pipette		5	disposable plastic
	reference book for protozoans			or key
	toothpick		1	per student
	water, pond			
13B	dissecting microscope		1	may substitute hand lens
	goggles			1 pair per student
	laboratory apron			1 per student
	living culture, *Penicillium notatum*			
	living culture, *Rhizopus stolonifer*			synonym *Rhizopus nigricans*
	microscope		1	
	preserved slide, *Coprinus*, c.s.		1	
	preserved slide, cup fungus apothecium		1	
	preserved slide, *Rhizopus stolonifer*, w.m.		1	
	specimen, preserved, mushroom		1	
	specimen, preserved, puffball		1	
	specimen, preserved, shelf fungus		1	

LAB	ITEM	SIZE	# REQUIRED	REMARKS
14A	camera		1	
	field guide		1	
	garden shears		1 pair	
	gloves, garden		1 pair	
	hand lens		1	
	notebook, field	small	1	
	permanent marker		1	
	plastic bag	small	several	
14B	hand lens		1	
	kitchen knife	large	1	
	razor blade, single-edged		1	
	scalpel		1	usually included in dissecting kit
	specimen, fresh flower		several	
	specimen, fresh fruit		several	
	specimen, seed		several	
15A	banana purée, ripe			
	banana purée, unripe			
	Benedict's solution			
	gloves, nitrile			1 pair per student per activity
	goggles			1 pair per student
	graduated cylinder	10 mL	1	
	hot water bath		1	
	iodine solution			
	laboratory apron			1 per student
	laboratory spatula		1	
	petri dish and lid		4	sterile
	pipette		several	disposable plastic
	test tube	medium	2	

LAB	ITEM	SIZE	# REQUIRED	REMARKS
15B	container	small to medium	5	
	paper sack		1	
	seawater			
	seeds, plant			
	spray bottle		1	
	water, distilled			
16A	acetic acid (CH_3COOH)			dilute (~5%) or may substitute vinegar
	concavity slide		1	
	cover slip		1	
	petri dish		1	may substitute a culture dish
	hand lens		1	may substitute dissecting microscope
	living culture, brine shrimp			may substitute *Daphnia*
	living culture, hydra			
	microscope		1	
	pipette		3	disposable plastic
	preserved slide, budding hydra		1	
	preserved slide, hydra c.s.		1	
	probe		1	usually part of dissecting kit; may substitute toothpick
	water, spring			may substitute treated tapwater
16B	Epsom salt			
	forceps		1	
	hand lens		1	may substitute dissecting microscope
	index card			may substitute small piece of cardboard
	living culture, planarian			
	microscope, dissecting		1	optional
	petri dish		1	sterile; may substitute a culture dish
	pipette		several	disposable plastic
	probe		1	usually part of dissecting kit; may substitute toothpick
	raw beef liver			may substitute boiled egg yolk or flaked or pelleted fish food
	water, spring			may substitute treated tapwater

APPENDIX H

LAB	ITEM	SIZE	# REQUIRED	REMARKS
17A	camera		1	
	culture dish		1	
	dissection kit			
	dissection pan		1	
	flower-head pin		8	numbered 1–8 with permanent marker
	gloves, nitrile			1 pair per student per activity
	glue			may substitute rubber cement or tape
	goggles			1 pair per student
	laboratory apron			1 per student
	permanent marker, fine-tip		1	
	plastic bag	small	1	
	preserved crayfish, injected	medium	1	
	specimen, preserved, crayfish		1	showing crayfish life cycle; optional
17B	live cricket	large		Number depends on the number of students and their experimental designs.
18A	live fish			
	notebook, field	small	1	
	pencil		1	
18B	live lizard		1	may substitute any small, carnivorous lizard
	terrarium	medium to large	1	
19A	birding app		1	downloaded onto smart device; may substitute field guide
	camera		1	optional
	notebook, field	small	1	optional
19B	computer with internet connection		1	
20A	to be determined by students' investigations			
20B	none			
21A	diagram, skeletal			may substitute for skeleton model
	microscope		1	
	preserved slide, dry ground bone, c.s.		1	
	skeleton, human		1	model

LAB	ITEM	SIZE	# REQUIRED	REMARKS
21B	dishpan		1	
	hand gripper		1	
	plastic jug	1 gal	1	filled with water
	stopwatch		1	
22A	isopropyl alcohol			
	lung volume bag		1	
	mouthpiece		3	
	rubber band		1	
	stethoscope		1	
	tissue			
22B	blood pressure cuff, digital		1	may substitute manual sphygmomanometer
23A	cork stopper		1	
	evaporating dish	100 mL	1	
	goggles			1 pair per student
	graduated cylinder	50 mL	4	
	grill lighter		1	
	iron ring		1	
	laboratory apron			1 per student
	laboratory balance		1	accurate to 0.01 g
	marshmallow		1	
	ring stand		1	
	snack foods, various			with nutrition facts labels
	soda can, empty		1	with tab attached
	stirring rod, glass		1	
	straight pin	large	1	
	thermometer or temperature probe		1	
	tongs		1	may substitute mitts

APPENDIX H

LAB	ITEM	SIZE	# REQUIRED	REMARKS
23B	beaker	250 mL	3	
	Benedict's solution			
	biuret reagent			
	clamp		6	
	dialysis tubing			one 12 in. strip
	gloves, nitrile			1 pair per student per activity
	goggles			1 pair per student
	graduated cylinder	100 mL	1	
	hot water bath		1	
	iodine solution			
	laboratory apron			1 per student
	mixed solutes solution	150 mL		
	multimeter		1	may substitute conductivity probe
	pipette		several	disposable plastic
	test tube	medium	9	
	test tube holder		1	
	test tube rack		1	
	water, distilled			
24A	beaker	250 mL	3	
	blindfold		1	
	goggles			1 pair per student
	laboratory apron			1 per student
	pen, washable ink		1	
	rod, glass or metal		2	may substitute dulled 16-p nail
	ruler, metric		1	
	tape measure, metric		1	
	tuning fork, assorted			include 128 Hz and 256 Hz
	water, hot			
	water, ice			

LAB	ITEM	SIZE	# REQUIRED	REMARKS
24B	dissection kit		1	
	dissection pan		1	
	gloves, nitrile			1 pair per student per activity
	goggles			1 pair per student
	laboratory apron			1 per student
	plastic bag	medium	1	
	preserved rat, injected		1	
	rat dissection reference material			
	rat skeleton, mounted		1	
25A	calculator		1	
	colored pencils			1 set
25B	none			

Nonconsumable Equipment List

ITEM	SIZE	# REQUIRED	LAB	REMARKS
beaker	50 mL	1	**6B**	
	150 mL	2	**6B**	
	250 mL	3	**23B, 24A**	
birding app		1	**19A**	downloaded onto smart device; may substitute field guide
blender		1	**6B**	
blindfold		1	**24A**	
blood pressure cuff, digital		1	**22B**	may substitute manual sphygmomanometer
calculator		1	**4B, 25A**	
camera		1	**4A, 14A, 17A, 19A**	optional
clamp		6	**23B**	
concavity slide		1	**16A**	
container	small to medium	5	**9A, 15B**	
cover slip			**5A, 6B, 13A, 16A**	Lab 5A quantity: 3 Lab 13A quantity: 5
culture dish		1	**16A, 16B, 17A**	
diagram, skeletal			**21A**	may substitute for skeleton model
dip net	small	1	**3A**	may substitute small cup or other small container
dishpan		1	**21B**	
dissecting microscope		1	**13B, 16B**	may substitute hand lens
dissection kit		1	**5A, 17A, 24B**	
dissection pan		1	**17A, 24B**	
evaporating dish	100 mL	1	**23A**	
field guide		1	**14A**	
field tape		1	**4A, 4B**	may substitute tape measure or marked rope
flashlight or laser pointer		1	**2**	
flower-head pin		8	**17A**	numbered 1–8 with permanent marker
forceps		1	**7A, 16B**	
garden shears		1 pair	**14A**	

ITEM	SIZE	# REQUIRED	LAB	REMARKS
gloves, garden		1 pair	**14A**	
goggles		1	**5A, 5B, 6A, 6B, 7B, 12A, 12B, 13B, 15A, 17A, 23A, 23B, 24A, 24B**	1 pair per student
graduated cylinder	10 mL	1	**12B, 15A**	
	50 mL	4	**7B, 12B, 23A**	
	100 mL	1	**6B, 23B**	
grill lighter		1	**23A**	
hand gripper		1	**21B**	
hand lens		1	**14A, 14B, 16A, 16B**	Lab 16A: may substitute dissecting microscope
hexagonal metal nut	large	1	**5A**	
hole punch		1	**7A**	
hot water bath		1	**15A, 23B**	
iron ring		1	**23A**	
Key for Selected Fishes of North America			**11A**	included in Lab 11A
kitchen knife	large	1	**14B**	
laboratory apron			**5A, 5B, 6A, 6B, 12A, 12B, 13B, 15A, 17A, 23A, 23B, 24A, 24B**	1 per student
laboratory balance		1	**12B, 23A**	accurate to 0.01 g
laboratory spatula		1	**15A**	
light source			**7A**	
lung volume bag		1	**22A**	
measuring spoon	set	1	**7B**	
meter stick		1	**1A**	
microscope		1	**1B, 5A, 6B, 8A, 12A, 13A, 16A, 21A**	

ITEM	SIZE	# REQUIRED	LAB	REMARKS
microscope, dissecting		1	**16B**	optional
microscope slide, glass			**5A, 6B, 13A**	Lab 5A quantity: 3 Lab 13A quantity: 5
mouthpiece		3	**22A**	
multimeter		1	**23B**	may substitute conductivity probe
oral syringe		1	**7A**	10 mL or larger
osmometer		1	**5B**	
petri dish		4	**2, 12A, 15A, 16B**	sterile Lab 15A quantity: 4
plastic bead, red		100	**9A**	
plastic bead, white		100	**9A**	
preserved slide, beef tapeworm bladders in meat		1	**16B**	optional
preserved slide, beef tapeworm proglottid		1	**16B**	optional; may substitute other species of tapeworm
preserved slide, beef tapeworm scolex		1	**16B**	optional; may substitute other species of tapeworm
preserved slide, budding hydra		1	**16A**	
preserved slide, *Coprinus*, c.s.		1	**13B**	
preserved slide, cup fungus apothecium		1	**13B**	
preserved slide, desmids or diatoms		1	**1B**	
preserved slide, dry ground bone, c.s.		1	**21A**	
preserved slide, fish embryos prepared for viewing mitosis		1	**8A**	
preserved slide, fish prepared for viewing meiosis		1	**8A**	
preserved slide, human liver fluke		1	**16B**	optional
preserved slide, hydra c.s.		1	**16A**	
preserved slide, *Rhizopus stolonifer*, w.m.		1	**13B**	
probe		1	**16A, 16B**	usually part of dissecting kit; may substitute toothpick
rangefinder		1	**4B**	optional
rat dissection reference material			**24B**	

ITEM	SIZE	# REQUIRED	LAB	REMARKS
rat skeleton, mounted		1	24B	
reference book for protozoans			13A	or key
ring stand		1	23A	
rod, glass or metal		2	24A	may substitute dulled 16-p nail
ruler		1	5B	
ruler, metric		1	12A, 24A	
scalpel		1	14B	usually included in dissection kit
science periodicals			10B	current issues; may substitute digital copies of online articles
skeleton, human		1	21A	model
specimen, preserved, crayfish		1	17A	showing crayfish life cycle; optional
specimen, preserved, mushroom		1	13B	
specimen, preserved, puffball		1	13B	
specimen, preserved, shelf fungus		1	13B	
spray bottle		1	15B	
squeeze bottle		1	5B	
stethoscope		1	22A	
stirring rod, glass		3	6B, 12B, 23A	Lab 12B: 3 per student
stopwatch		1	7A, 21B	
straight pin	large	1	23A	
tape measure, metric		1	24A	
terrarium	medium to large	1	3A, 18B	Lab 3A: may substitute large bucket
test tube	medium	9	15A, 23B	Lab 15A quantity: 2
test tube holder		1	23B	
test tube rack		1	23B	
thermometer or temperature probe		1	23A	
tongs		1	23A	may substitute mitts
tuning fork, assorted			24A	include 128 Hz and 256 Hz
water test kit		1	6A	LaMotte 5860-01; may substitute similar kit

APPENDIX H

Consumable Materials List

ITEM	SIZE	# REQUIRED	LAB	REMARKS
10-10-10 fertilizer			**12B**	
acetic acid (CH_3COOH)			**12B, 16A**	Lab 12B: 5 mL, 1 M solution Lab 16A: dilute (~5%) or may substitute vinegar
active dry yeast			**7B**	
banana purée, ripe			**15A**	
banana purée, unripe			**15A**	
Benedict's solution			**15A, 23B**	
biuret reagent			**23B**	
carmine powder			**13A**	may substitute yeast
clear plastic cup			**7A**	Number of cups depends on how many trial groups created.
colored pencils			**7A, 25A**	1 set
cooking oil	60 mL		**12B**	
cork stopper		1	**5A, 23A**	
cornmeal	1 tsp		**7B**	
correction fluid pen		1	**3A**	
cotton ball, sterile			**2, 12B**	Lab 12B: at least 4 per lab group
cotton fiber			**13A**	
cotton swab, sterile		4	**12A**	
dialysis tubing	12 in.	1	**23B**	
dish detergent			**6B**	liquid
Epsom salt			**16B**	
flat toothpick		1	**5A**	
gloves, nitrile			**6A, 6B, 12B, 15A, 17A, 23B, 24B**	1 pair per student per activity
glucose syrup	1 tsp		**7B**	may substitute Karo Light syrup
glue			**17A**	may substitute rubber cement or tape
glycerin			**13A**	may substitute methyl cellulose
hand sanitizer			**12A**	
hand soap			**12A**	

ITEM	SIZE	# REQUIRED	LAB	REMARKS
immersion oil			**1B**	Omit if microscope does not have an oil immersion objective.
index card			**16B**	may substitute small piece of cardboard
iodine solution			**15A, 23B**	
isopropyl alcohol			**6B, 22A**	Lab 6B: 90 mL
lens paper			**1B**	
lentils, dry			**6B**	uncooked
marshmallow		1	**23A**	
methylene blue			**5A, 6B**	
milk of magnesia	8 mL		**12B**	
mixed solutes solution	150 mL		**23B**	
notebook, field	small	1	**4A, 14A, 18A, 19A**	Labs 4A, 19A: optional
nutrient agar			**2**	
onion		1	**5A**	
paper sack		1	**15B**	
pen, washable ink		1	**24A**	
pencil		1	**18A**	
permanent marker		1	**7B, 12A, 14A**	
permanent marker, fine-tip		1	**17A**	
phenol red			**6B**	pH indicator
pipette		several	**5A, 5B, 12B, 13A, 15A, 16A, 16B, 23B**	disposable plastic Lab 13A quantity: 5 Lab 16A quantity: 3
plastic bag	small		**14A, 17A**	Lab 14A: several Lab 17A: 1 bag
	medium	1	**24B**	
	large	3	**12A**	
plastic bag, resealable	snack-size	3	**7B**	
plastic jug	1 gal	1	**21B**	filled with water
preserved crayfish, injected	medium	1	**17A**	optional
preserved rat, injected		1	**24B**	

ITEM	SIZE	# REQUIRED	LAB	REMARKS
raw beef liver			16B	may substitute boiled egg yolk or flaked or pelleted fish food
razor blade, single-edged		1	5A, 14B	
rubber band		1	22A	
salt solution	15 mL		6B	6% solution
seawater			15B	
seed, plant			15B	
snack foods, various			23A	with nutrition fact labels
soda can, empty		1	23A	with tab attached
sodium bicarbonate solution			7A	2.5% solution
specimen, fresh flower		several	14B	
specimen, fresh fruit		several	14B	
specimen, seed		several	14B	
spinach or ivy leaves, fresh		1	7A	
stakes or flagging tape			4B	
sucrose solution		2	5B	0.5 molal (Solution B) and 1.0 molal (Solution A)
tissue			1B, 22A	
toothpick		1	13A	per student
universal indicator paper			2, 12B	Lab 12B: may use a pH meter
water sample			2, 6A	obtained from local water source Lab 2: untreated
water, distilled			5B, 15B, 23B	
water, hot			24A	
water, ice			24A	
water, pond			13A	
water, sea			15B	
water, spring			16A, 16B	may substitute treated tapwater
water, warm			7B	~40 °C
wax pencil		1	5B, 12B	
weighing paper			12B	
wood shavings			7B	may substitute grass clippings

Live Materials

ITEM	SIZE	# REQUIRED	LAB	REMARKS
duckweed			3B	Amount will vary according to students' procedures; expect ~50 plants per lab group.
live cricket	large		3A, 17B	Lab 3A: 30–50 needed Lab 17B: Number depends on the number of students and their experimental designs.
live fish			18A	
live lizard		1	18B	may susbstitute any small, carnivorous lizard
living culture, amoeba			13A	
living culture, brine shrimp			16A	may substitute *Daphnia*
living culture, euglena			13A	
living culture, hydra			16A	
living culture, paramecium			13A	
living culture, *Penicillium notatum*			13B	
living culture, planarian			16B	
living culture, *Rhizopus stolonifer*			13B	synonym *Rhizopus nigricans*
oil-eating bacteria suspension	80 mL		12B	

This list includes only those items that must be ordered or planned for well in advance of an activity. It does not include items that are commonly available year-round at your local grocery. See specific lab activity for details.

LAB MANUAL PHOTO CREDITS

Key: (t) top; (c) center; (b) bottom; (l) left; (r) right

COVER

Tim Platt/Stone via Getty Images

FRONT MATTER

i Eric Isselee/Shutterstock.com; **iii** "Robert Hooke, Micrographia, detail: microscope"/Wellcome Images/Wikimedia Commons /modified/CC By 4.0; **iv** bonchan/iStock/Getty Images Plus via Getty Images; **v** GMVozd/E+ via Getty Images; **vit** GL Archive /Alamy Stock Photo; **vib** Olha Rohulya/Shutterstock.com

CHAPTER 1

1 Motortion Films/Shutterstock.com; **7** "Portrait of Anthonie van Leeuwenhoek (1632-1723)" by Jan Verkolje/Wikimedia Commons /Public Domain; **8** 3DMI/Shutterstock.com; **10l, r** Rattiya Thongdumhyu/Shutterstock.com

CHAPTER 2

15 michaeljung/Shutterstock.com

CHAPTER 3

19t Rostislav Stefanek/Shutterstock.com; **19b** Morphart Creation /Shutterstock.com; **23** Vicki Jauron, Babylon and Beyond Photography/Moment via Getty Images; **25** dugdax/Shutterstock .com

CHAPTER 4

27t Mark A Paulda/Moment via Getty Images; **27b** © iStock.com /JulianneGentry; **33** Phynart Studio/E+ via Getty Images; **34** ANN PATCHANAN/Shutterstock.com; **37** (all) Adobe Stock/Calvin

CHAPTER 5

39 "Robert Hooke, Micrographia, detail: microscope"/Wellcome Images/Wikimedia Commons/modified/CC BY 4.0

CHAPTER 6

49t Brian A Jackson/Shutterstock.com; **49b** (art reference) © iStock .com/Bill Oxford; **55** isak55/Shutterstock.com; **56** Luis Line /Shutterstock.com; **57** "Phenol red pH 6,0 - 8,0" by Max schwalbe /Wikimedia Commons/modified/CC By-SA 4.0; **58** koto_feja /iStock/Getty Images Plus via Getty Images

CHAPTER 7

59 Brent Hofacker/Shutterstock.com; **65** Photo 12/Alamy Stock Photo

CHAPTER 8

69 Kateryna Kon/Shutterstock.com; **75t** duncan1890/DigitalVision Vectors via Getty Images; **75b** Photoongraphy/Shutterstock.com; **77** Barbara Rich/Moment via Getty Images; **80** Jurgis Mankauskas /Shutterstock.com; **81t** LittleMiss/Shutterstock.com; **81ct, b** JIANG HONGYAN/Shutterstock.com; **81cb, 83** Grant Heilman Photography /Alamy Stock Photo; **82t** GoodFocused/Shutterstock.com; **82b** ARTSILENSE/Shutterstock.com

CHAPTER 9

85l Eric Isselee/Shutterstock.com; **85r** pandapaw/Shutterstock.com; **88t** Jim West/Alamy Stock Photo; **88b** "Seedbank" by R. C. Johnson /Wikimedia Commons/CC By 2.5; **91** D-Keine/E+ via Getty Images

CHAPTER 10

97 Chronicle/Alamy Stock Photo; **98** Edward Westmacott /Shutterstock.com; **99** The Natural History Museum/Alamy Stock Photo; **100** "Charles Darwin 1865 signature"/Wikimedia Commons /Public Domain; **103t** Jean-Philippe Offord/Alamy Stock Photo; **103b** © Dr. Seth Shostak/Science Source; **104** kenkistler /Shutterstock.com; **106** Jeff Topping/Stringer/Getty Images Sport via Getty Images

CHAPTER 11

109t vkbhat/iStock/Getty Images Plus via Getty Images; **109b** "Meadow deathcamas - Zigadenus venenosus" by Neal Herbert /National Park Service/Wikimedia Commons/Public Domain; **112tl** "Channel catfish1" by Ellen Edmonson and Hugh Chrisp/Wikimedia Commons/Public Domain; **112tr** "Bluespotted Sunfish" by Ellen Edmonson and Hugh Chrisp/Wikimedia Commons/Public Domain; **112bl** "Rainbow Trout"/Wikimedia Commons/Public Domain; **112br** "Bowfin" by Ellen Edmonson and Hugh Chrisp/Wikimedia Commons/Public Domain; **113tl** Raver Duane, U.S. Fish and Wildlife Service/Public Domain Images; **113tr** "Moxostoma anisurum" by Ellen Edmonson and Hugh Chrisp/Wikimedia Commons/Public Domain; **113bl** "Bonneville Cutthroat trout" /UDWR/Flickr/Wikimedia Commons/CC By 2.0; **113br** "Muskellunge USFWS" by Knepp, Timothy - U.S. Fish and Wildlife Service/Wikimedia Commons/Public Domain; **114tl** "Lepomis auritus" by Duane Raver/Wikimedia Commons/Public Domain; **114tr** "Lake whitefish1" by Ellen Edmonson and Hugh Chrisp /Wikimedia Commons/Public Domain; **114bl** Raver Duane, U.S. Fish and Wildlife Service/Public Domain Images; **114br** "Longnose gar" by Ellen Edmonson and Hugh Chrisp/Wikimedia Commons/ Public Domain; **115tl** "Micropterus dolomieu smallmouth bass fish" by Raver Duane, U.S. Fish and Wildlife Service/Wikimedia Commons/Public Domain; **115tr** Raver Duane, U.S. Fish and Wildlife Service/Public Domain Images; **115bl** "Pumpkinseed fish lepomis gibbosus" by Raver Duane, U.S. Fish and Wildlife Service /Wikimedia Commons/Public Domain; **115br** "Black bullhead" by Ellen Edmonson and Hugh Chrisp/Wikimedia Commons/Public Domain; **116tl** "Esox lucius1" by Timothy Knepp/Wikimedia Commons/Public Domain; **116tr** "Black Crappie" by Ellen Edmonson and Hugh Chrisp/Wikimedia Commons/Public Domain; **116bl** "The Shad (Clupea Sapidissima)" by Shermon Foote Denton /Wikimedia Commons/Public Domain; **116br** "Largemouth bass fish art work micropterus salmoides" by Raver Duane, U.S. Fish and Wildlife Service/Wikimedia Commons/Public Domain; **117tl** Raver Duane, U.S. Fish and Wildlife Service/Public Domain Images; **117tr** Raver Duane, U.S. Fish and Wildlife Service/Public Domain Images; **117bl** "Yellow perch fish perca flavescens" by Raver Duane, U.S. Fish and Wildlife Service/Wikimedia Commons/Public Domain; **117br** "Brindled madtom" by Ellen Edmonson and Hugh Chrisp/Wikimedia Commons/Public Domain; **118tl** "White perch morone chrysops" by Raver Duane/U.S. Fish and Wildlife Service /Wikimedia Commons/Public Domain; **118tr** "White crappie pomoxis annularis" by Raver Duane, U.S. Fish and Wildlife Service /Wikimedia Commons/Public Domain; **118bl** "Brook trout" by Raver, Duane/Wikimedia Commons/Public Domain; **118br** "Acipenser oxyrhynchus" by Raver Duane/U.S. Fish and Wildlife Service/Wikimedia Commons/Public Domain; **121** "Whirling disease pathology"/NOAA/Wikimedia Commons/Public Domain; **122** "Carl von Linné" by Alexander Roslin/Wikimedia Commons /Public Domain/Modified

CHAPTER 12
127 monkeybusinessimages/iStock/Getty Images Plus via Getty Images; **128** © iStock.com/ksass; **129** Schira/Shutterstock.com; **135** Bloomberg via Getty Images; **136** © STEVE GSCHMEISSNER/Science Source; **140** "Deepwater Horizon oil spill NOAA map"/NOAA/Wikimedia Commons/Public Domain

CHAPTER 13
151 Adobe Stock/伊藤ヨシユキ

CHAPTER 14
157 Aygul Bulte/iStock/Getty Images Plus via Getty Images; **158l** epantha/iStock/Getty Images Plus via Getty Images; **158r** North Wind Picture Archives/Alamy Stock Photo; **159t** saje/iStock/Getty Images Plus via Getty Images; **159b** Kypros/Moment via Getty Images; **163** CHALERMCHAI99/Shutterstock.com; **164l** MNStudio/iStock/Getty Images Plus via Getty Images; **164r** Gerald A. DeBoer/Shutterstock.com; **166** SeDmi/Shutterstock.com

CHAPTER 15
171 WoraponL/Shutterstock.com; **173** Photoongraphy/Shutterstock.com; **175** Martin Lisner/Shutterstock.com

CHAPTER 16
181 Lebendkulturen.de/Shutterstock.com; **184** Ground Picture/Shutterstock.com; **187** mirceax/iStock/Getty Images Plus via Getty Images

CHAPTER 17
191 bonchan/iStock/Getty Images Plus via Getty Images; **199** chuyuss/Shutterstock.com

CHAPTER 18
201t © iStock.com/NERYX; **201b** ac productions/Tetra images via Getty Images; **202** ryasick/E+ via Getty Images; **204** "Lamprologusstappersimalemcl"/Haplochromis/Wikimedia Commons/modified/CC BY-SA 3.0; **205** AP Photo/Chris O'Meara; **207** Antagain/E+ via Getty Images

CHAPTER 19
209 Erkki Alvenmod/Shutterstock.com; **215** fizkes/Shutterstock.com

CHAPTER 20
217 Jaromir Chalabala/Shutterstock.com; **219** Zay Nyi Nyi/Shutterstock.com; **220** New Africa/Shutterstock.com

CHAPTER 21
225l, r SciePro/Shutterstock.com; **227** Capuski/E+ via Getty Images; **228** Photology1971/iStock/Getty Images Plus via Getty Images

CHAPTER 22
235 pink_cotton_candy/E+ via Getty Images; **237** Ivanna Oliinyk/iStock/Getty Images Plus via Getty Images; **239** Ake Ngiamsanguan/iStock/Getty Images Plus via Getty Images; **241** fstop123/E+ via Getty Images

CHAPTER 23
243 Cook Shoots Food/Shutterstock.com

CHAPTER 24
251 agrobacter/E+ via Getty Images; **255** TPopova/iStock/Getty Images Plus via Getty Images; **257** Andregric/iStock/Getty Images Plus via Getty Images

CHAPTER 25
267l Bettmann via Getty Images; **267rt, rb** Vintage_Space/Alamy Stock Photo; **273** SolStock/E+ via Getty Images; **276** GMVozd/E+ via Getty Images

BACK MATTER
281 kali9/E+ via Getty Images; **283** (dissecting microscope art reference) Zern Liew/Shutterstock.com; **286, 287** Eric Isselee/Shutterstock.com; **288** Choksawatdikorn/Shutterstock.com; **290t** Science Photo Library via Getty Images; **290b** Martin Shields/Alamy Stock Photo; **291** Nordroden/Shutterstock.com; **296** fizkes/Shutterstock.com

PERIODIC TABLE OF THE ELEMENTS

1

1 Hydrogen
H
1.01
1

atomic number — **86** — radioactive

name — Radon

symbol — **Rn**

electron structure by energy level — **222.02** — atomic mass, rounded to hundredths place

2, 8, 18, 32, 18, 8

2

3 Lithium **Li** 6.94 *2, 1*
4 Beryllium **Be** 9.01 *2, 2*

11 Sodium **Na** 22.99 *2, 8, 1*
12 Magnesium **Mg** 24.31 *2, 8, 2*

3

21 Scandium **Sc** 44.96 *2, 8, 9, 2*

4

22 Titanium **Ti** 47.87 *2, 8, 10, 2*

5

23 Vanadium **V** 50.94 *2, 8, 11, 2*

6

24 Chromium **Cr** 52.00 *2, 8, 13, 1*

7

25 Manganese **Mn** 54.94 *2, 8, 13, 2*

8

26 Iron **Fe** 55.85 *2, 8, 14, 2*

9

27 Cobalt **Co** 58.93 *2, 8, 15, 2*

19 Potassium **K** 39.10 *2, 8, 8, 1*
20 Calcium **Ca** 40.08 *2, 8, 8, 2*

37 Rubidium **Rb** 85.47 *2, 8, 18, 8, 1*
38 Strontium **Sr** 87.62 *2, 8, 18, 8, 2*
39 Yttrium **Y** 88.91 *2, 8, 18, 9, 2*
40 Zirconium **Zr** 91.22 *2, 8, 18, 10, 2*
41 Niobium **Nb** 92.91 *2, 8, 18, 12, 1*
42 Molybdenum **Mo** 95.95 *2, 8, 18, 13, 1*
43 Technetium **Tc** 96.91 *2, 8, 18, 13, 2*
44 Ruthenium **Ru** 101.07 *2, 8, 18, 15, 1*
45 Rhodium **Rh** 102.91 *2, 8, 18, 16, 1*

55 Cesium **Cs** 132.91 *2, 8, 18, 18, 1*
56 Barium **Ba** 137.33 *2, 8, 18, 18, 2*
57 Lanthanum **La** 138.91 *2, 8, 18, 18, 9, 2*
72 Hafnium **Hf** 178.49 *2, 8, 18, 32, 10, 2*
73 Tantalum **Ta** 180.95 *2, 8, 18, 32, 11, 2*
74 Tungsten **W** 183.84 *2, 8, 18, 32, 12, 2*
75 Rhenium **Re** 186.21 *2, 8, 18, 32, 13, 2*
76 Osmium **Os** 190.23 *2, 8, 18, 32, 14, 2*
77 Iridium **Ir** 192.22 *2, 8, 18, 32, 15, 2*

87 Francium **Fr** 223.02 *2, 8, 18, 32, 18, 8, 1*
88 Radium **Ra** 226.03 *2, 8, 18, 32, 18, 8, 2*
89 Actinium **Ac** 227.03 *2, 8, 18, 32, 18, 9, 2*
104 Rutherfordium **Rf** 267.12 *2, 8, 18, 32, 32, 10, 2*
105 Dubnium **Db** 268.13 *2, 8, 18, 32, 32, 11, 2*
106 Seaborgium **Sg** 269.13 *2, 8, 18, 32, 32, 12, 2*
107 Bohrium **Bh** 270.13 *2, 8, 18, 32, 32, 13, 2*
108 Hassium **Hs** 269.13 *2, 8, 18, 32, 32, 14, 2*
109 Meitnerium **Mt** 277.15 *2, 8, 18, 32, 32, 15, 2*

- Alkali metals
- Alkaline-earth metals
- Transition metals
- Post-transition metals
- Metalloids
- Inner transition metals
- Nonmetals
- Halogens (also nonmetals)
- Noble gases
- Radioactive isotopes

*

58 Cerium **Ce** 140.12 *2, 8, 18, 19, 9, 2*
59 Praseodymium **Pr** 140.91 *2, 8, 18, 21, 8, 2*
60 Neodymium **Nd** 144.24 *2, 8, 18, 22, 8, 2*
61 Promethium **Pm** 144.91 *2, 8, 18, 23, 8, 2*
62 Samarium **Sm** 150.36 *2, 8, 18, 24, 8, 2*

**

90 Thorium **Th** 232.04 *2, 8, 18, 32, 18, 10, 2*
91 Protactinium **Pa** 231.04 *2, 8, 18, 32, 20, 9, 2*
92 Uranium **U** 238.03 *2, 8, 18, 32, 21, 9, 2*
93 Neptunium **Np** 237.05 *2, 8, 18, 32, 22, 9, 2*
94 Plutonium **Pu** 244.06 *2, 8, 18, 32, 24, 8, 2*

Periodic Table of the Elements

	10	11	12	13	14	15	16	17	18
									2 Helium **He** 4.00 2
				5 Boron **B** 10.81 2, 3	**6** Carbon **C** 12.01 2, 4	**7** Nitrogen **N** 14.01 2, 5	**8** Oxygen **O** 16.00 2, 6	**9** Fluorine **F** 19.00 2, 7	**10** Neon **Ne** 20.18 2, 8
				13 Aluminum **Al** 26.98 2, 8, 3	**14** Silicon **Si** 28.09 2, 8, 4	**15** Phosphorus **P** 30.97 2, 8, 5	**16** Sulfur **S** 32.06 2, 8, 6	**17** Chlorine **Cl** 35.45 2, 8, 7	**18** Argon **Ar** 39.95 2, 8, 8
	28 Nickel **Ni** 58.69 2, 8, 16, 2	**29** Copper **Cu** 63.55 2, 8, 18, 1	**30** Zinc **Zn** 65.38 2, 8, 18, 2	**31** Gallium **Ga** 69.72 2, 8, 18, 3	**32** Germanium **Ge** 72.63 2, 8, 18, 4	**33** Arsenic **As** 74.92 2, 8, 18, 5	**34** Selenium **Se** 78.97 2, 8, 18, 6	**35** Bromine **Br** 79.90 2, 8, 18, 7	**36** Krypton **Kr** 83.80 2, 8, 18, 8
	46 Palladium **Pd** 106.42 2, 8, 18, 18	**47** Silver **Ag** 107.87 2, 8, 18, 18, 1	**48** Cadmium **Cd** 112.41 2, 8, 18, 18, 2	**49** Indium **In** 114.82 2, 8, 18, 18, 3	**50** Tin **Sn** 118.71 2, 8, 18, 18, 4	**51** Antimony **Sb** 121.76 2, 8, 18, 18, 5	**52** Tellurium **Te** 127.60 2, 8, 18, 18, 6	**53** Iodine **I** 126.90 2, 8, 18, 18, 7	**54** Xenon **Xe** 131.29 2, 8, 18, 18, 8
	78 Platinum **Pt** 195.08 2, 8, 18, 32, 17, 1	**79** Gold **Au** 196.97 2, 8, 18, 32, 18, 1	**80** Mercury **Hg** 200.59 2, 8, 18, 32, 18, 2	**81** Thallium **Tl** 204.38 2, 8, 18, 32, 18, 3	**82** Lead **Pb** 207.24 2, 8, 18, 32, 18, 4	**83** Bismuth **Bi** 208.98 2, 8, 18, 32, 18, 5	**84** Polonium **Po** 208.98 2, 8, 18, 32, 18, 6	**85** Astatine **At** 209.99 2, 8, 18, 32, 18, 7	**86** Radon **Rn** 222.02 2, 8, 18, 32, 18, 8
	110 Darmstadtium **Ds** 282.17 2, 8, 18, 32, 32, 16, 2	**111** Roentgenium **Rg** 282.17 2, 8, 18, 32, 32, 17, 2	**112** Copernicium **Cn** 286.18 2, 8, 18, 32, 32, 18, 2	**113** Nihonium **Nh** 286.18 2, 8, 18, 32, 32, 18, 3	**114** Flerovium **Fl** 290.19 2, 8, 18, 32, 32, 18, 4	**115** Moscovium **Mc** 290.20 2, 8, 18, 32, 32, 18, 5	**116** Livermorium **Lv** 293.21 2, 8, 18, 32, 32, 18, 6	**117** Tennessine **Ts** 294.21 2, 8, 18, 32, 32, 18, 7	**118** Oganesson **Og** 295.22 2, 8, 18, 32, 32, 18, 8

63 Europium **Eu** 151.96 2, 8, 18, 25, 8, 2	**64** Gadolinium **Gd** 157.25 2, 8, 18, 25, 9, 2	**65** Terbium **Tb** 158.93 2, 8, 18, 27, 8, 2	**66** Dysprosium **Dy** 162.50 2, 8, 18, 28, 8, 2	**67** Holmium **Ho** 164.93 2, 8, 18, 29, 8, 2	**68** Erbium **Er** 167.26 2, 8, 18, 30, 8, 2	**69** Thulium **Tm** 168.93 2, 8, 18, 31, 8, 2	**70** Ytterbium **Yb** 173.05 2, 8, 18, 32, 8, 2	**71** Lutetium **Lu** 174.97 2, 8, 18, 32, 9, 2
95 Americium **Am** 243.06 2, 8, 18, 32, 25, 8, 2	**96** Curium **Cm** 247.07 2, 8, 18, 32, 25, 9, 2	**97** Berkelium **Bk** 247.07 2, 8, 18, 32, 27, 8, 2	**98** Californium **Cf** 251.08 2, 8, 18, 32, 28, 8, 2	**99** Einsteinium **Es** 252.08 2, 8, 18, 32, 29, 8, 2	**100** Fermium **Fm** 257.10 2, 8, 18, 32, 30, 8, 2	**101** Mendelevium **Md** 258.10 2, 8, 18, 32, 31, 8, 2	**102** Nobelium **No** 259.10 2, 8, 18, 32, 32, 8, 2	**103** Lawrencium **Lr** 262.11 2, 8, 18, 32, 32, 8, 3